WITHDRAWN

# Smoking and the Lung

**Ettore Majorana International Science Series**
Series Editor:
**Antonino Zichichi**
European Physical Society
Geneva, Switzerland

**(LIFE SCIENCES)**

*Recent volumes in the series*

A Continuation Order Plan is available for this series. A continuation order will bring delivery of
each new volume immediately upon publication. Volumes are billed only upon actual shipment.
For further information please contact the publisher.

# Smoking and the Lung

Edited by

## G. Cumming

The Midhurst Medical Research Institute
Midhurst, West Sussex, United Kingdom

and

## G. Bonsignore

University of Palermo
Palermo, Sicily, Italy

Plenum Press • New York and London

Library of Congress Cataloging in Publication Data

International School of Thoracic Medicine (7th: 1983: Erice, Sicily)
    Smoking and the lung.

    (Ettore Majorana international science series. Life sciences; 17)
    "Proceedings of the seventh course of the International School of Thoracic
Medicine, held October 9–15, 1983, in Erice, Sicily, Italy"—T.p. verso.
    Includes bibliographies and index.
    1. Lungs—Diseases—Congresses. 2. Cigarette smoke—Toxicology—Congresses.
I. Cumming, Gordon. II. Bonsignore, G. III. Title. IV. Series. [DNLM: 1. Smoking—
congresses. 2. Lung Diseases—etiology—congresses. W1 ET712M v.17 / WF 600
I6104 1983s]
RC756.I585  1983                        616.2′4                        84-18102
ISBN 0-306-41828-2

Proceedings of the Seventh Course of the International School of Thoracic
Medicine, held October 9–15, 1983, in Erice, Sicily, Italy

©1984 Plenum Press, New York
A Division of Plenum Publishing Corporation
233 Spring Street, New York, N.Y. 10013

Printed in the United States of America

# PREFACE

The role played by cigarette smoking in the causation of ill-health is arousing widespread world interest amongst physicians. The report by the Royal College of Physicians that smoking is responsible in Great Britain for 100,000 premature deaths aroused little public interest.

Thus the seventh Course in Thoracic Medicine assembled a group of experts from various parts of the world to define current problems and to establish an appropriate direction in which to proceed. These presentations and discussions form the basis of this book.

One contributor had difficulty with his three contributions and after six months delay, they were unfortunately not available. Despite this, the presentations were both illuminating and interesting, so the discussion upon them has been published though the absence of the three chapters is regrettable.

Many people have contributed to the production of this volume: our secretaries Corinne Wade, Karen Wadey and Heather Inman Beard; our translator Guiliana de Ferio; the recording technicians of the Ettore Majorana School and John Griffiths of the Reprographic Department of The Midhurst Medical Research Institute where the volume was prepared before despatch to Plenum Press.

Our thanks are due, as always, to the Ettore Majorana School of Scientific Culture and to Dr. Alberto Gabrielli for his infinite patience.

G. Cumming
G. Bonsignore

v

# CONTENTS

# STRUCTURAL ASPECTS OF CIGARETTE SMOKE-INDUCED PULMONARY DISEASE

P. K. Jeffery, D. F. Rogers, M. M. Ayers
and P. A. Shields

Department of Lung Pathology, Brompton Hospital
London

Smoking is the most important cause of preventable early death and loss of working time in the United Kingdom. Two out of five heavy smokers die before they are 65 – double the proportion of non-smokers. The most serious risk of smoking is lung cancer which kills over 38,000 people each year in the United Kingdom 80% of them men. The association of lung cancer, particularly squamous cell carcinoma, and cigarette smoking is clear cut. The death rate from the disease in heavy smokers is 30 times that in non-smokers. Other smoking-related pulmonary diseases of major significance are chronic bronchitis and emphysema. Asthma is exacerbated by smoking but is not caused by it. The pathologist faces the difficulty of only seeing the later stages in the natural history of each disease but by careful comparison of clinical material and the use of experimental animal models correlations of structure and function can be made and pathogenesis established. The structural parameters and the diseases to which they relate which will be discussed herein are: (i) mucosal sensitivity (asthma and bronchitis); (ii) cilia and mucus (bronchitis); (iii) phagocytic cells (emphysema) and (iv) squamous metaplasia/atypia (carcinoma).

## Mucosal permeability

The continuous epithelial lining of the airways and alveoli form a permeability barrier between the external (lumen) and internal (blood vessels, tissue cells and supporting stroma) environments. The various epithelial cell types are held together by three types of junction (1) terminal bars made up of "tight" and "intermediate" junctions which together form

1

attachment belts extending around the lateral surface of cells
adjacent to the airway lumen, (2) desmosomes or "spot"
junctions and (3) nexuses or "gap" junctions (fig. 1).
Normally selectively impermeable, the terminal bar prevents
excessive fluid movement across the epithelium but, after
experimental irritation by tobacco smoke (TS) or ether given to
guinea pigs, it becomes permeable to permeability markers of
40,000 daltons placed in the airway lumen (Richardson 1976;
Boucher 1980). Transport from lumen to submucosa of
potentially antigenic molecules in excess of 40,000 daltons
appears to be by a different route involving epithelial cell
transport (Richardson 1976). Changes in epithelial
permeability due to TS have also been shown to occur in man by
measurement of the half-time clearance from airway lumen
(conducting or respiratory site not yet determined) into blood
of 99m TC-DTPA as an inhaled aerosol (Minty et al 1981). In
asthma such permeability changes may be an important change
allowing allergens to gain access to mast cells, the majority
of which lie below the surface epithelium (fig. 2). En route,
allergens or irritants gain access to irritant receptors which
lie within the epithelium and close to the airway lumen
(Jeffery 1982). However, allergens need not penetrate to the
lamina propria to meet with migratory cells. To date at least
three types of intra-epithelial mononuclear migratory cell have
been distinguished on the basis of morphology and
histochemistry: (i) intra-epithelial lymphcytes (Jeffery &
Reid 1975; McDermott 1982); (ii) globular leucocytes (Kent
1966; Jeffery & Reid 1975) and (iii) intra-epithelial mast
cells. The last shares morphological features with the
subepithelial mast cell and may be capable of releasing
mediators of inflammation and affecting epithelial
permeability. In man, mast cells form up to 2% of surface
epithelial cells and a higher proportion has been found in
smokers than in non-smokers (Lamb & Lumsden 1982).

## Chronic bronchitis

Chronic bronchitis is characterised by cough associated with
the production of sputum (in the absence of cardiac disease or
localised respiratory disease) on most days for three months of
the year during at least two consecutive years. The disease
results from chronic irritation of the tracheobronchial tree by
inhaled irritants derived mainly from self pollution by tobacco
smoke or, to a lesser extent, atmospheric pollution (with
included sulphur dioxide) associated with industrialised areas.
The irritant effects are aggravated by dampness, fog or
photochemical smog. Whilst viral infection may lead to similar
changes, animal studies show clearly that chronic irritation
without infection can lead to a hypertrophy and hyperplasia of

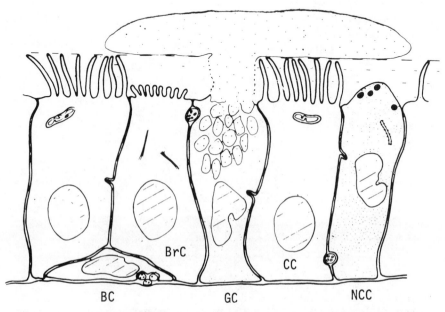

FIGURE 1:    Diagrammatic representation of surface epithelium
             illustrating ciliated (CC), goblet (GC),
             nonciliated "serous" (NCC), brush (BrC) and basal
             (BC) cells.   The cells are held together by a
             variety of adhesive structures, the terminal bar
             (arrow) separating the periciliary fluid from the
             intercellular fluid-filled spaces.

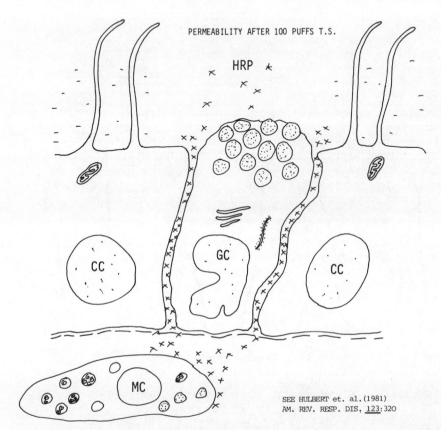

PERMEABILITY AFTER 100 PUFFS T.S.

HRP

GC

CC          CC

MC

SEE HULBERT et. al.(1981)
AM. REV. RESP. DIS, 123:320

FIGURE 2:     Diagrammatic representation of the increased
              permeability of the terminal bar after acute
              exposure of the airway mucosa to cigarette smoke.
              Permeability markers and potential allergens gain
              access to the supporting stroma in which lie the
              majority of mast cells.

mucus-secreting tissue which is the pathological hallmark of lungs obtained at autopsy from bronchitic patients (Reid 1954; Reid & Lamb 1968). Fig. 3 shows that the mucus-secreting tissue of airways consist of submucosal glands (which increase in size) and mucous and serous cells found in the lining epithelium (which increase in number).

From the functional point of view the most important change is probably the appearance of mucus-secreting cells and their increase in the most distal bronchioli where they and mucus are normally absent. Accummulation of excessive mucus due to excessive production and/or decreased mucociliary clearance leads to a tendency for colonisation of retained secretions by bacterial (especially H.influenzi and S.pneumoniae) which whilst of little importance in the prime causation of the disease, become important complicating factors. Such secondary infections, during acute exacerbations, lead to a change in sputum character, from mucoid to mucopurulent or purulent, and may also accompany mucosal inflammation, scarring and destruction of bronchiolar walls with narrowing of airways, multiple stenoses and irreversible airways obstruction.

The volume of sputum produced shows a strong correlation with smoking, frequency of chest episodes and the individual FEV1 level measured, but does not correlate with the rate of decline in FEV1 with age (Fletcher 1975). In other words whilst excessive secretion in airways causes air flow limitation, it does not of itself cause chronic progressive damage. The distinction is of value in that whilst progressive airflow limitation due to emphysematous and small airway changes is difficult to treat, treatment of the component of airflow limitation due to mucous hypersecretion is possible and likely to be of early benefit to the patient.

The bronchitic changes may be produced experimentally in specific pathogen free (SPF) laboratory animals (Lamb & Reid 1969; Jones et al 1972). When SPF rats are exposed to an atmosphere of tobacco smoke generated from 25 cigarettes per day (4 hours daily for up to 6 weeks) epithelial mucus-secreting cells are increased in number at all levels of the bronchial tree. The effect is dose related and greatest in the main extrapulmonary bronchi and intrapulmonary airways. Whilst the number of mucus-secreting cells is increased, those which contain acidic or neutral glycoprotein are affected in different ways. At each airway level there is a decrease in the number of cells containing neutral glycoprotein: in contrast there is an increase in the number of cells producing acidic glycoprotein (Fig. 4). The increase in the latter is disproportionately large so that there is an overall increase

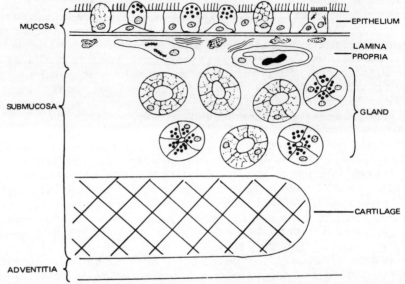

FIGURE 3:    The mucus-secreting tissue of airways consists of
             submucosal glands (wherever there is supporting
             cartilage) and secretory cells in the surface
             epithelium:  both serous and mucous cell types are
             present.

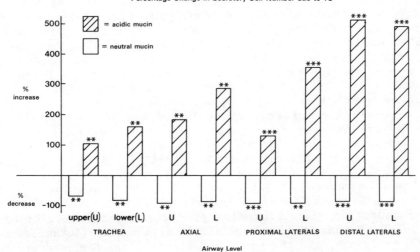

FIGURE 4: A histogram showing the percentage change due to cigarette smoke, from control values at 0, in surface secretary cell number at different airway levels of rat lung. The percentage increase of acidic secretory cells is greater than the decrease of neutral and results in an overall increase in surface secretory cell number.

in the total number of mucus-secreting cells.  The increase in
the amount of acidic secretion seen experimentally is also
characteristic of the bronchitic patient and smoker who
produces more acidic secretion rich in sulphomucin (Kollerstrom
et al 1977).  A similar increase in the number of mucus-
secreting cells, and in a shift of glycoprotein type is shown
also by electron microscopy (Jeffery & Reid 1981).

The stains used to show the presence of mucus-secreting
cells by light microscopy are alcian blue and periodic acid
Schiff, used in combination.  For a mucus-secreting cell to be
observed by light microscopy the cell must contain a reasonable
quantity of intracellular secretion in order to be counted.
Fig. 5 shows that the amount of intracellular mucin depends on
the balance of the uptake of glycoprotein precursors (and their
incorporation into the macromolecules of mucin during
synthesis) and the rate of discharge into the airway lumen of
stored mucus.  For example if the rate of discharge were to be
reduced but that of synthesis maintained, there would be an
accummulation of stored intracellular mucin.  Such an effect
would lead to an apparent increase in mucous cell number
throughout the eipthelium and this has been suggested mechanism
in some forms of mild injury (McDowell et al 1983).

Does the increase in the amount of mucus-secreting tissue
(i.e. the ´factory´), seen experimentally, correlate with
hypersecretion (i.e. increased production and output by the
factory).  We have used a modified in situ system (Turner et al
1983) to investigate the amount of airway secretion produced in
response to tobacco smoke in normal and ´bronchitic´ animals
(fig. 6).  Any secretion produced in the tracheal segment, with
its blood and nerve supply intact, is collected by a continuous
passage of physiological saline through the trachea and half-
hourly collections are made during the 5 hours of experiment.
During the experiment two acute exposures of tobacco smoke
diluted 1:3 with air are passed through the trachea at 16
ml/min for 5 minutes.  Table 1 shows the results of chemical
estimations for baseline fucose (as a marker of mucous
glycoprotein), hexose and total protein, in animals made
´bronchitic´ by inhalation of tobacco smoke for two weeks:
comparison is made with the respective levels in control
animals.  The mean baseline levels of each are raised in the
´bronchitic´ animals, fucose significantly so (P 0.01).  Only
in control animals did the first acute administration of TS
significantly increase the mean concentration of hexose above
basal levels (P 0.05).  Acute administration of air had no
statistically significant effect (Jeffery et al 1984).  The
effects of drugs in modulating the secretory response to
tobacco smoke are now being investigated.

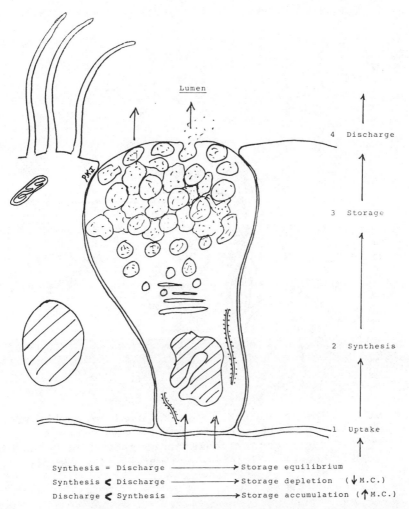

Synthesis = Discharge ──────→ Storage equilibrium
Synthesis < Discharge ──────→ Storage depletion ($\downarrow$ M.C.)
Discharge < Synthesis ──────→ Storage accumulation ($\uparrow$ M.C.)

FIGURE 5: A surface secretory (goblet) cell with intracellular secretion. The amount stores within the cell (and thus the ease with which a secretory cell can be detected by light microscopy of a paraffin section stained by stains specific for glycoprotein) depends on the balance of the rate of glycoprotein precursor uptake and synthesis and the rate of discharge from the cell.

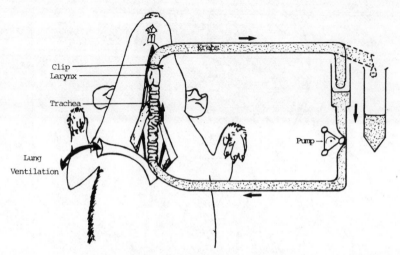

FIGURE 6:    The in situ rat system used for collection of
             tracheal secretions.  Physiological saline
             continuously passes through the trachea and half
             hourly collections are made during the five hours
             of the experiment.

TABLE 1:    Concentrations of Fucose, Hexose and Protein in Tracheal Washings.

| | Mean Basal Concentration in ug/sample ($\pm$ SEM) | | |
|---|---|---|---|
| | Control | 'Bronchitic' | P $<$ |
| fucose | 3 ($\pm$ 1) | 30 ($\pm$ 13) | 0.01 |
| hexose | 44 ($\pm$ 9) | 104 ($\pm$ 27) | NS |
| protein | 1,142 ($\pm$ 341) | 1,942 ($\pm$ 427) | NS |

The negative correlates of an increase in mucous cell number
are:

1.    Fewer ciliated cells leading to stasis of mucus.

2.    Less lysozyme allowing bacterial colonization

3.    A reduction in bronchial protease inhibitor,    and

4.    Reduced bronchial surfactant which may give rise to early
      airway closure.

    Post mortem material shows that there are fewer ciliated
cells and cilia are abnormal in patients dying of the end-stage
complications of chronic bronchitis:  it has been suggested
that tobacco smoke is the cause of such changes.  For example
Ailsby and Ghadially (1973) have shown changes in the
ultrastructure of cilia in a patient who had smoked heavily (25
cigarettes per day) for some 45 years.  However experimentally
in our S.P.F. animal studies we have never found such
structural alterations to cilia.  The cilia always appear
normal in cross section with the normal complement of tubules
and the presence of a ciliary crown (see Jeffery & Reid, 1973).
Ciliary density is normal and by scanning electron microscopy
the area covered by cilia is increased from 56% in controls to
65% in tobacco smoke exposed animals.  The microvillus length
is increased and mitochondria are increased in their maximum
length (Jeffery & Reid 1981).  it would appear from our studies
that tobacco smoke does not directly damage the structural
integrity of the cilia although it may, of course, have
functional effects (see chapter by Sleigh).  It seems,
therefore, more likely that ciliary abnormalities which are
observed at post mortem are due to secondary factors, such as
infection,  which  follow  the  primary effects of T.S.  on
respiratory secretions.  Many ciliary abnormalities have been
shown to follow infection by influenza virus, rhino virus, myxo
virus,  mycoplasm,  Bordatella sp..   Fox  et  al  (1981)  has
recently shown that a large percentage of cilia from ´normal´
airways also show structural abnormalities and that focal
changes do not necessarily reflect a generalised change in the
bronchial tree:  this underlines the need for carefully chosen
controls in studies aimed at discerning disease-associated
ciliary abnormalities.

    The diminution in the numbers of epithelial serous cells
seen experimentally (Jeffery & Reid 1981) and in their relative
numbers in hypersecretory disease is likely to result in a
reduction of serous secretion.  The secretion of serous cells

is complex and contains lysozyme and lactoferrin as well as mucous glycoprotein (Bowes & Corrin 1977; Bowes et al 1981). Both lyxozyme and lactoferrin are important anti-bacterial agents which if deficient will favour bacterial colonization of retained secretions.

Of the three anti-proteases present in lung two are serum-derived (i.e. alpha-1-antitrypsin and alpha-2-macroglobulin) and one, "low molecular weight bronchial inhibitor" (LMI) is derived from the bronchial mucosa itself. It is now thought that the last is the main anti-elastase of conducting airways (see Gadek 1980). Recent studies have localized LMI to the non-ciliated (i.e. Clara) cells of human patients and shown an inverse relationship between their number and bronchiolar diameter (Mooren et al 1983). The observation that Clara cells are reduced in number in smokers (Ebert & Terraccio 1975) suggests a reduction in available anti-protease leading to increased local proteolytic digestion.

Lastly, Clara cells have, among other things, been reported as the source of bronchiolar surfactant (Niden 1967). Surfactant is needed to maintain patent small non-cartilagenous airways especially during expiration. A loss of surfactant and its replacement by a lining of mucus would, it has been suggested, lead to early airways closure and small airway instability (Macklem et al 1970). Some or all of the above changes occur in the "bronchitic" patient who usually has a degree of emphysema also.

### Emphysema

Emphysema is defined in anatomical terms as permanent enlargement of the respiratory passages distal to the terminal bronchiolus (i.e. the last purely conductive airway). Whilst such an enlargement can be due to a failure of development or simple dilatation the discussion herein is restricted to emphysema of the destructive form. Destruction of alveolar walls can arise as a result of inhalation of noxious agents such as tobacco smoke or maybe the result of a failure (constitutional or acquired) of the alveoli to maintain an intact wall. Inhaled agents which show a close and statistically significant relationship with destructive emphysema are tobacco smoke, in the general population, and coal smoke (with included $SO_2$, $NO_2$ etc). and dust in specialised groups within the population. As might be expected, in the latter circumstances emphysema shows selective involvement of the bronchiolar portions of the lung acinus (i.e. is centrilobular or centriacinar), and often has a

preferential distribution to the upper aspects of the lung
where the ventilation/perfusion ratio is high. Niewoehner et
al (1974) and Cosio et al (1980) have described early changes
in young smokers and these include goblet cell metaplasia,
respiratory bronchiolotis, denuded epithelium, mural
inflammatory cells, peribronchiolar fibrosis, mural oedema and
smooth muscle hypertrophy. Smokers have a significant excess
of airways less than 400 um in diameter. The severity of small
airways disease is correlated with the percentage of airways
less than 400 um in diameter and with the extent of
centrilobular emphysema. In contrast, panlobular or panacinar
emphysema is more diffusely distributed throughout the acinus
and lobule with a preference for lower lobes where perfusion is
greatest. Whilst smoking seems to enhance its severity, the
latter form of emphysema shows a predilection for elderly
individuals and subjects with an inherited deficiency of the
functional serum antiprotease alpha-1-antitrypsin (Hutchinson
1973).

The protease/antiprotease inbalance hypothesis for the
pathogenesis of destructive emphysema presently receives
widespread support and is attractive in that it allows an
explanation for alveolar wall destruction by either inhaled
irritant or an intrinsic deficiency (Gadek 1982). Accordingly
if the normal balance of protease (elastase in particular) and
antiprotease is lost with a disproportionate rise in the
former, then excessive proteolysis and destruction of alveolar
walls results. Much of the support given to the inbalance
theory comes from the use of experimental animals where
instillation of proteases into the lower respiratory tract
produces structural changes to the alveolar parenchyma
resembling that of human emphysema (Lieberman 1976; Janoff et
al 1977 & Senior et al 1977). The studies clearly show that
the elastases, and particularly neutrophil elastase, are the
enzymes most relevant to alveolar wall destruction. Whilst the
extracellular connective tissue of the lung consists of
collagen and elastic fibres, proteoglycans and glycoproteins,
destruction of the elastic component is required to repoduce
the emphysematous lesion (Karlinsky & Schnider 1978). Thus
according to the inbalance theory a destructive lesion would
arise following either an overwhelming increase in elastase
burden or alternatively a decrease in functional anti-elastase.
There is evidence that irritation by tobacco smoke causes
potentially detrimental shifts of both. Tobacco smoke, for
example, has been shown to alter the morphology of resident
alveolar macrophages, figs. 7 & 8 and also their function
(Harris et al 1970; Warr & Martin 1974; Hinman et al 1980).
Either as a result of macrophage-derived chemotactic factors or
directly due to tissue damage, polymorphs (PMN) are recruited

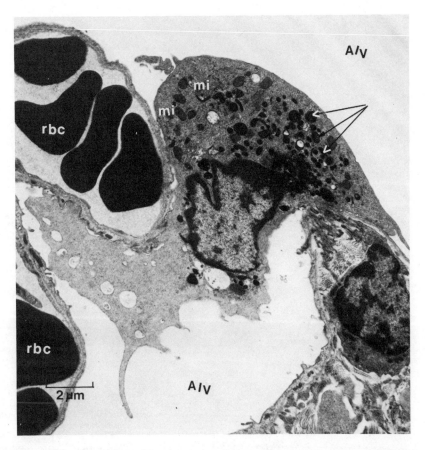

FIGURE 7: Electron micrograph of a rat alveolar macrophage (M) in alveolar space (Alv) possibly migrating through a pore of Kohn. The cell cytoplasm contains mitochondria (mi) and a large number of electron-dense lysosomal granules (arrows). Capillaries with included red blood cells (rbc). Glutaraldehyde and osmuim tetroxide: uranyl acetate and lead citrate x 7,5000.

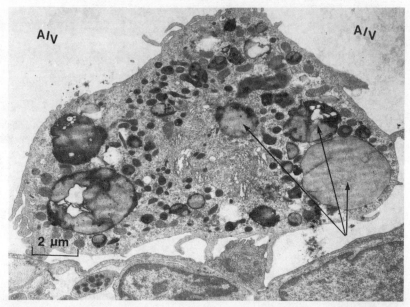

FIGURE 8:    Alveolar macrophage taken from a rat exposed to
             cigarette smoke daily for 2 weeks.  The cytoplasm
             of all such macrophages contains a number of large
             lipid-like inclusion bodies (arrows) often
             referred to as "tar bodies" x 7,500.

in large numbers to the interstitium (particularly of well perfused lower lobes) of the lung where they release their lysosomal enzymes with included elastase.

Recently, in our laboratory we (Jeffery et al 1983) have been looking at peripheral blood neutrophils (PMN) from normal and emphysematous patients and examining the effects of the water soluble gaseous fraction (WSGF) of tobacco smoke on the morphology and phagocytic ability of such cells in vitro. The most striking effect is that of intense cytoplasmic vesiculation of the cytoplasm. As expected, the WSGF of tobacco smoke significantly inhibits the number of PMN ingesting latex particles. The effects in both normal and emphysematous subjects are similar. Table 2 shows the results of an analysis of the number of lysosomal granules within cells exposed to the WSGF of tobacco smoke or to latex alone. Exposure of PMN to latex causes a 15% diminution in the number of lysosomes in both normal PMN and also those from emphysematous subjects. Exposure of these cells to latex causes a similar diminution in lysosomal granules in the normals but a two fold greater diminution in their number in patients with emphysema (P<0.001). Thus there may be inherent susceptibility of neutrophils to damage by tobacco smoke or tobacco smoke particles in the emphysematous subjects (see also Hopkin et al 1981).

**Carcinoma**

Squamous carcinoma is the commonest type of malignant tumour of both the larynx and bronchi (Coggon & Acheson 1983). At the beginning of the century it was infrequently seen but it has now become a major health problem in many parts of the world. As with other carcinomas the incidence increases with age but it is not exceptional to find that it may also be the cause of death in young men between the ages of 30 and 40. There is good statistical evidence to link the smoking habit with the incidence of bronchial carcinoma: the risk increases proportionately to the consumption of cigarettes and inversely to the length of the cigarette stub left. However, cigarette smoking is not the only factor implicated in the aetiology of lung carcinoma: atmospheric pollution plays an important role as does occupational pollution, particularly in the chromate nickel refining and, of course, asbestos industry. Squamous cell, oat cell and anaplastic carcinomas comprise over 90% or bronchial cancers. Squamous cell carcinoma is the most common histological form of bronchial cancer and is thought to arise from bronchial epithelium which has previously undergone squamous metaplasia as frequently seen in the bronchial mucosa of cigarette smokers.

TABLE 2:     Effects of Cigarette Smoke (WSGF) and/or latex on
             Neutrophil Lysosomal Number.

NUMBER ELECTRON-DENSE (LYSOSOMAL) GRANULES/CELL
       (% CHANGE FROM CONTROL VALUE)

| | N | (%Δ) | E | (%Δ) |
|---|---|---|---|---|
| Cells alone | 67± 2.0 | | 67± 1.7 | |
| Cells + WSGF | 58± 1.9* | (↓ 13) | 58± 1.7* | (↓ 13) |
| Cells + Latex | 57± 2.1* | (↓ 15) *** | 47± 1.4* | (↓ 30) |
| Cells + L + WSGF | 50± 1.7† | (↓ 25) | 54± 1.5 | (↓ 19) |

* P < 0.001 against cells alone.

† P < 0.001 against cells and latex

The induction of cancer seems to be a two-stage process: initiation may be by the direct action of the carcinogen or following its metabolic activation to a more reactive form. The reactive metabolite then combines with critical cellular macromolecules leading in some as yet undefined way to the initiation of tumour formation. Following a variable latent period, tumour development proceeds under the influence of tumour promoters which may, of themselve, not be carcinogenic. In the case of tobacco smoke which is a mixture of more than 1,200 substances (Stedman 1968), the smoke is capable of both initiating and promoting tumour development. Although the key macromolecules with which carcinogens covalently bind are not clearly known, a good correlation has often been observed between carcinogenicity and the affinity of binding to DNA. Tumour initiation requires only a single application of the subthreshold dose of a chemical carcinogen and is generally considered to be irreversible. Tumour promotion on the other hand requires multiple applications of a second non-carcinogenic chemical and is, at least in the early stages, reversible. Asbestos is a well-known promoter of human lung cancer especially mesothelioma and its effects are compounded by cigarette smoke. Table 3 shows the possible carcinogens and tumour promoters and cocarcinogens present in tobacco smoke (from Cohen 1982).

Whilst epidemiological evidence favours the close association of cigarette smoking and lung cancer and experimental studies show that both skin and lung cancer can be induced with cigarette smoke condensate, the experimental evidence for lung cancer induction by experimental inhalation of whole tobacco smoke is poor. In our experience, tobacco smoke exposure in animals, albeit for short periods of time, repeatedly demonstrates a proliferative mitotic response which occurs during the first 2 - 3 days of exposure but then falls to within control values despite continuing exposure (Ayres & Jeffery 1982; Bolduc et al 1981; Wells & Lamerton 1975). However, if the exposure period is stopped for 2 days and then recommenced there is a second wave of mitoses (Bolduc et al 1981). The dominant cell type involved in this mitotic response is the basal cell but mucous cells also divide. We have never found squamous metaplasia, carcinoma 'in situ' or invasive cancer in our experimental animals. However, recent epidemiologic evidence correlates low circulating levels of Vitamin A with an increased risk of developing lung cancer (Bjelk 1975; Wald et al 1980) and we have become interested in the possibility that vitamin A might, by inducing squamous metaplasia of the respiratory tract, so predispose the respiratory tract to the subsequent development of lung cancer.

TABLE 3:      Carcinogens, Tumour Promoters and Co-Carcinogens
              in Tobacco Smoke (From Cohen 1982).

Carcinogens

Benzo(a)pyrene, 5-methylchrysene, dibenz(a,h)anthracene, dibenzo
(a,h)pyrene, dibenzo(a,i)pyrene, dibenz(a,j)acridine,
benzo(b)fluoranthene, benzo(j)fluoranthene, benz(a)anthracene,
chrysene, methylchrysenes, methylfluoranthes,
dibenzo(c,g)carbazole, benzo(c)phenanthrene, N-nitrosonornicotine,
nitrosopiperidine, nitrosopyrrolidine, polonium-210, arsenic,
nickel, cadmium, o-toluidine, $\beta$ -naphthylamine.

Tumour Promoters

Volatile Phenols, benzo(e)pyrene.

Cocarcinogens

Catechol, 4-alkylcatechol, pyrene, fluoranthene, benzo(e)pyrene,
naphthalenes, methylfluoranthenes, 1-methylindoles, 9-
methylcarbazoles

*Most of the data obtained from activity on mouse skin

In our present studies (Shields & Jeffery 1984) specific pathogen-free rats were divided into 4 treatment groups and the synergism of vitamin A deficiency and tobacco smoke exposure (for periods of between 2 weeks and 3 months) examined. Animals were rendered vitamin A deficient by feeding on a specially prepared synthetic diet deficient in vitamin A whereas the control animals were fed on a synthetic diet supplemented with 4000 international units of vitamin A. On reaching their weight plateau the vitamin deficient animals were transferred to a maintenance dose (200 international units of vitamin A/kg diet) for the remainder of the experiment during which time a group were exposed to tobacco smoke. Fluometric analysis of their plasma retinol and liver retinol palmitate levels showed that most animals exposed to tobacco smoke but on a normal diet had significantly reduced levels of stored liver retinol palmitate ($p < 0.01$). It would be of interest to know whether smokers show a similar depletion of their liver stores. Chemical analysis confirmed the depletion of vitamin A in those animals given a vitamin A deficient diet. Examination of respiratory tract morphology revealed that cigarette smoke exposure and vitamin A deficiency act predominantly in the trachea and larynx. Vitamin A deficiency alone caused focal tracheal squamous metaplasia. Smoke exposure, in rats receiving the normal diet, produced a hyperplastic epithelium with mucous cell hyperplasia and occasional foci of stratified non-ciliated columnar cells. However, in the vitamin A deficient rats which were additionally exposed to tobacco smoke there was widespread keratinising squamous metaplasia of tracheal epithelium. Those areas which did not show squamous metaplasia showed mucous cell hyperplasia. The epithelium was significantly thickened also. Neither vitamin A deficiency nor cigarette smoke alone caused significant squamous metaplasia of the larynx but in combination they caused squamous metaplasia which covered 50-100% of the posterior wall of the infraglottal region. Thus tobacco smoke and vitamin A deficiency show striking synergism in their pathological effects on airway epithelium.

In conclusion, tobacco smoke is strongly implicated in the causation of squamous carcinoma of the lung but is likely not to act alone. Host factors including dietary factors such as vitamin A status may be important in determining the susceptibility of an individual to tumour development.

## Summary

In summary, tobacco smoke affects airway permeability and is implicated in the aetiology of chronic bronchitis, certain

**TABLE 4**    Recovery time of lung tissue components after cessation of smoking

REVERSIBILITY

| PARAMETER | TIME TO RECOVER | THERAPY |
|---|---|---|
| Permeability ( man ) | 1 week ( NC )* | ? steroids |
| Muco-ciliary clearance ( man ) | 3 months ( NC )* | ? steroids |
| Mucous cell hyperplasia ( animal ) | 3 months | anti-inflammatory drugs 'mucolytics' |
| Emphysema | none | inhibit chemotaxis anti-elastases anti-oxidants |
| Carcinoma / Carcinoma in situ | rare | surgery chemotherapy radiotherapy Vitamin A/carotene |

* incomplete recovery

forms of emphysema, and squamous carcinoma of the lung. Table 4 shows the extent to which some of these lesions are reversible and summarizes the types of therapeutic approach which might either help to reverse the lesion or stop the subsequent development or progression of the disease.

**REFERENCES**

AILSBY, R.L. & Ghadially, F.N. (1973).
    Atypical cilia in human bronchial mucosa.
    J. Pathol, 109, 75–77.

AYERS, M. & Jeffery, P.K. (1982).
    Cell division and differentiation in bronchial epithelium.
    In: Cellular Biology of the Lung (ed: G. Cumming & G.
    Bonsignore). Vol. 10, of Life Science series (ed. A.
    Zichichi). pp. 33–54. Plenum, New York.

BJELKE, E. (1975).
    Dietary Vitamin A and human lung cancer.
    Intern. J. Cancer 15, 561.

BOLDUC, P., Jones, R. & Reid, L. (1981).
    Mitotic activity of airway epithelium after short exposure
    to tobacco smoke and the effect of the anti-inflammatory
    agent phenylmethyloxidiazole.
    Brit. J. Exp. Path., 62, 461–468.

BOUCHER, R.C., Johnson, J., Inone, S. et al (1980),
    The effect of cigarette smoke on the permeability of guinea
    pig airways.
    Lab. Invest. 43, 94–100.

BOWES, D. & Corrin, B. (1977).
    Ultrastructural immunocytochemical localization of lysozyme
    in human bronchial glands.
    Thorax 32, 163–170.

BOWES, D., Clark, A.E. & Corrin, B. (1981).
    Ultra-structural localization of lactoferrin and
    glycoprotein in human bronchial glands.
    Thorax, 36, 108–115.

COGGON, D. & Acheson, E.D. (1983),
    Trends in lung cancer.
    Thorax 38, 721–723.

COHEN, G.M. (1982),
    The induction of cancer by chemicals.
    In: Cellular Biology of the Lung (ed: G. Cumming & G.
    Bonsignore), Vol. 10 of Life Sciences Series (ed: A.
    Zichichi), pp.425-432, Plenum, New York.

COSIO, M.G., Hale, K.A. & Niewoehner, D.E. (1980),
    Morphologic and morphometric effects of prolonged cigarette
    smoking on the small airways.
    Amer Rev. Resp. Dis., 122, 265-271.

EBERT, R.V. & Terracio, M.J. (1973),
    The bronchiolar epithelium in cigarette smokers:
    observations with the scanning electron microscope.
    Am. Rev. Resp. Dis., 111, 4-11.

FLETCHER, C.M. (1975),
    The natural history of chronic bronchitis.
    Community Health, 7, 70-78.

GADEK, J.E., Hunninghake, G.W., Fells, G.A., Zimmerman, R.L.
    Keogh, B.A. & Crustal, R.G. (1980),
    Evaluation of the protease-antiprotease theory of human
    destructive lung disease.
    Bull. Europ. Physiopath. Resp, 16 (suppl), 27-40.

HARRIS, J.O., Swenson, E.W. & Johnson, J.E. (1970),
    Human alveolar macrophages: comparison of phagocytic
    ability, glucose utilization and ultrastructure in smokers
    and non-smokers.
    J. Clin. Invest, 49, 2086-2096.

HINMAN, L.M., Stevens, C.A., Matthay, R.A. & Gee, J.B.L.
    (1980),
    Elastase and lysozyme activities in human alveolar
    macrophages: effects of cigarette smoking.
    Amer. Rev. Resp. Dis., 121, 263-272.

HOPKIN, J.M., Tomlinson, V.S. & Jenkins, R.M. (1981),
    Variations in response to cytotoxicity of cigarette smoke.
    Brit. Med. J., 283, 1209-1211.

HUTCHINSON, D.C.S. (1973),
    Alpha-1-antitrypsin deficiency and pulmonary emphysema: the
    role of proteolytic enzymes and their inhibitors.
    Brit. J. Dis. Chest, 67, 171-196.

JANOF, A., Sloan, B., Weinbaum, G., Daminao, V.,
    Sandhaus, R.A., Elias, J. & Kimbel, P. (1977).
    Experimental emphysema induced with purified human
    neutrophil elastase.
    Tissue localisation of the instilled protease.
    Amer. Rev. Resp. Dis., 115, 461-478.

JEFFERY, P.K. & Reid, L. (1975),
    New observations of rat airway epithelium: a quantitative
    electron microscopic study.
    J. Anat. 120, 295-320.

JEFFERY, P.K. & Reid, L. (1981),
    The effect of tobacco smoke, with or without
    phenylmethyloxadiazole (PMO), on rat bronchial epithelium:
    a light and electron microscopic study.
    J. Pathol. 133, 341-359.

JEFFERY, P.K. (1982),
    The innervation of bronchial mucosa.
    In: Cellular Biology of the Lung (ed: G. Cumming & G.
    Bonsignore), Vol. 10 of Life Science series, (ed. A.
    Zichichi), pp 1-32, Plenum, New York.

JEFFERY, P.K., Romano, H., Hutchison, D., Baum, H. (1983),
    Effect of the water soluble gaceous fraction of tobacco
    smoke on phagocytosis and morphology of human peripheral
    blood neutrophils: comparison of normal and emphysematous
    subjects.
    Proc. Brit. Thoracic Soc. Thorax 38, 229.

JEFFERY, P.K., Marriott, C., Rogers, D.F. & Turner, N.C.
    Tobacco smoke-induced hypersecretion.
    J. Physiol. (in press).

JONES, R., Bolduc, P. & Reidl L. (1972),
    Protection of rat bronchial epithelium against tobacco
    smoke.
    Brit. Med. J., 2, 142-144.

KARLINSKY, J.B. & Snider, G.L. (1978),
    State of the art.  Animal models of emphysema.
    Amer. Rev. Resp. Dis., 117, 1109-1133.

KENT, J.F. (1966),
    Distribution and fine structure of globular leukocytes in
    respiratory and digestive tracts of the laboratory rat.

KOLLERSTROM, N., Lord, P.W. & Whimster, W.F. (1977),
A difference in the composition of bronchial mucus between
smokers and non-smokers.
Thorax, 32, 155-159.

LAMB, D. & Reid, L. (1968),
Mitotic rates, goblet cell increase and histochemical
changes in mucus in rat bronchial epithelium during
exposure to sulphur dioxide.
J. Path. Bac. 96, 97-111.

LAMB, D. & Reid, L. (1969),
Goblet cell increase in rat bronchial epithelium after
exposure to cigarette and cigar tobacco smoke.

LAMB, D. & Lumsden, A. (1982),
Intra-epithelial mast cells in human airway epithelium:
evidence for smoking-induced changes in their frequency.
Thorax, 37, 334-342.

LIEBERMAN, J. (1976),
Elastase, collagenase, emphysema and alpha-1-antitrypsin
deficiency.
Chest 70, 62-67.

MACKLEM, P.T., Proctor, D.F. & Hogg, J.C. (1970),
The stability of peripheral airways.
Resp. Physiol. 8, 191-203.

McDERMOTT, M.R., Befus, A.D. & Bienenstock, J. (1982),
The structural basis for immunity in the respiratory tra t.
Intern. Rev. Exp. Pathol., 23, 47-112.

McDOWELL, E.M., Combs, J.W. & Newkirk C. (1983),
Changes in secretory cells of hamster tracheal epithelium
in response to acute sublethal injury: a quantitative
study.
Exp. Lung Res. 4, 227-243.

MINTY, A.D., Jordan, C. & Jones, J.G. (1981),
Rapid improvement in abnormal pulmonary epithelial
permeability after stopping cigarettes.
Brit. Med. J., 282, 1183-1186.

MOOREN, H.W.D.,Kramps, J.A., Franken, C., Meijer, C.J.L.M. &
Dijkman, J.A. (1983),
Localisation of a low-molecular-weight bronchial protease
inhibitor in the peripheral human lung.
Thorax 38, 180-183.

NIDEN, A.H. (1967),
    Bronchiolar and large alveolar cell in pulmonary
    phospholipid metabolism.
    Science, 158, 1323-1324.

NIEWOEHNER, D.E., Kleinerman, J. & Rice, D.B. (1974),
    Pathologic changes in the peripheral airways of young
    cigarette smokers.
    New. Eng. J. Med., 291, 755-758.

REID, L. (1954),
    Pathology of chronic bronchitis.
    Lancet 1, 275-279.

RICHARDSON, J.B., Bouchard, T. & Ferguson, C.C. (1976),
    Uptake and transport of exogenous proteins by respiratory
    epithelium.
    Lab. Invest. 35, 307-314.

SENIOR, R.M., Tegner, H., Kuhn, C., Ohlsson, K., Starcher, B.C.
    & Pierce, J.A. (1977),
    The induction of pulmonary emphysema with human leukocyte
    elastase.
    Amer. Rev. Resp. Dis., 116, 469-475.

SHIELDS, P.A. & Jeffery, P.K. (1984),
    Squamous metaplasia of respiratory tract: cigarette smomke
    anbd Vitamin A deficiency.
    J. Pathol. (in press).

STEDMAN, R.L. (1968),
    The chemical composition of tobacco and tobacco smoke.
    Chem. Rev. 68, 153-207.

TURNER, N.C., Marriott, C. & Machling, R. (1983),
    An evaluation of an in vivo model of tracheobronchial
    glycoprotein secretion.
    In: Glycoconjugates (eds. M. A. Chester et al. pp. 574-575.

WALD, N., Idle, M., Boreham, J. (1980),
    Low serum Vitamin A and subsequent risk of cancer:
    preliminary results of a prospective study.
    Lancet 2, 813-815.

WARR, G.A. & Martin, R.R. (1974),
    Chemotactic responsiveness of human alveolar macrophages:
    effects of smoking.
    Infect. & Immunity 2, 769-771.

WELLS, A.B. & Lamerton, L.F. (1975),
    Regenerative responses of rat tracheal epithelium after
    acute exposure to tobacco smoke: a quantitative study.
    J. Natl. Conc. Inst. 55, 887-891.

DISCUSSION

LECTURER: Jeffery                                    CHAIRMAN: Cumming


RICHARDSON:        You pointed out that the increase in mucous
                   cells proximally was perhaps not the most
                   important feature in chronic bronchitis, but
                   that blockage of small airways might be more
                   important.  Do you have any evidence for this
                   in the rats which you have exposed to cigarette
                   smoke?

JEFFERY:           In our exposure system the small airways are
                   little affected.  There is a non-significant
                   reduction in the number of Clara cells, yet we
                   know that the smoke reaches this site since all
                   the alveolar macrophages have ingested smoke
                   particles.  There is an extension of mucous
                   cells peripherally but this does not reach the
                   terminal bronchioles.

HEATH:             Would you like to make any comment on the
                   endocrine cells of the bronchial trees, the
                   Feyrter cells and the neuro-epithelial bodies.
                   Have you found any structural change in these
                   cells  either  histologically  or
                   ultrastructurally?

JEFFERY:           These cells are rare in the rat and we have had
                   difficulty in answering this question.  We are
                   now using a neurone-specific enolase as a
                   marker for these cells, we have exposed rats to
                   cigarette smoke with vitamin A deficiency and
                   without such deficiency but are not yet able to
                   report the results.

HIGENBOTTAM:       You mentioned the work of Jim Hogg and Gareth
                   Jones in connection with the tight junction. Do
                   you have any information about the presence of
                   tight junctions in alveolar cells?

JEFFERY:           Hogg studied tracheal permeability; the
                   permeability of Jones and his colleagues is
                   probably at the alveolar level, although they
                   are not sure of the actual site.  We have no
                   information from our studies about the tight
                   junction at the alveolar level, but it is
                   present in the epithelium at that level.  There

is a difference in the nature of the junctions in alveolar epithelium and in endothelium, the latter are more 'leaky'.

PRIDE:      Do rats show the increase in white cells which smokers manifest, and is there recruitment of alveolar macrophages and white cells into the lung periphery?

JEFFERY:      We have not looked at peripheral blood cells in our rats, but there is an increase in macrophages with peripheral recruitment of both macrophages and white cells.

FLETCHER:      I have never before seen the prevalence of chronic bronchitis shown as increasing, what is striking is the rapid decline in those under 65, and perhaps your figures included all age groups. This is in great contrast with lung cancer, and one can only suppose that it is in some way associated with general air pollution which has been decreasing. When your tracheas were perfused how did you expose them to tobacco smoke?

JEFFERY:      We interrupted the perfusate for five minutes whilst we passed a continuous stream of tobacco smoke through the trachea, diluted one to three with air.

CUMMING:      Would you like to make any comment about the decline in mortality?

JEFFERY:      I agree entirely with Professor Fletcher's comment, my slide showed crude figures for all ages and then only up to 1969.

BAHKLE:      I noted an effect of Indomethacin on one of your measured parameters. There are experiments showing co-oxidation of arachidonic acid with benzpyrene. Are there any experiments to inhibit the carcinogenicity of benzpyrene with drugs such as aspirin or indomethacin.

JEFFERY:      I do not know of any such studies but in other systems indomethacin seems to be helpful to the regression of tumours. In our system indomethacin inhibits mucous cell hyperplasia which is largely based on mitotic increase. I am unable to comment on the role of indomethacin as an inhibitor of carcinogenicity.

RAWBONE:    Have you any comments on the dose-response relationships in any of the studies?

JEFFERY:    No, no comment at all since we have only used one dose but perhaps the industry might provide information in this area.

CUMMING:    With that I will terminate the discussion of the first paper.

# THE REFLEX EFFECTS OF CIGARETTE SMOKING

P. S. Richardson

Department of Physiology

St. George's Hospital Medical School, London

Human beings smoke voluntarily whereas other animals do not. This paper will discuss the reflexes triggered by smoking, and a major part of it will concentrate on reflexes studied in animals. This may seem a perverse approach but it is so much easier to perform experiments to analyse respiratory reflexes in species other than man that these make a useful starting point when considering the effects of smoking in man. It is necessary, however, to exercise caution when extrapolating from results in animals to man, particularly because some of the experiments on the former have been performed using concentrations of cigarette smoke much higher than those that occur in the lungs of people who smoke.

A reflex arc consists of nervous receptors which detect the initial stimulus, an afferent pathway to the CNS, where interactions with other inputs can occur, and an efferent pathway (or pathways) leading to the effector organs. This paper starts by considering the afferent (sensory) nerves in the various parts of the airway which smoke may stimulate and the sort of reflexes that activity in these afferent nerves gives rise to. The description depends mainly on animal studies.

## Afferent nerves in the airways and their reflex effects

The nasal cavity contains both olfactory and trigeminal nerve endings. Stimulation of the nose with odours that stimulate the former but, because of their lack of irritant properties, are thought not to excite the latter tends to augment breathing (Widdicombe, 1964). When irritants are

33

puffed into the nose the responses include sneezing, apnoea or slowing of breathing, narrowing of the larynx and bradycardia. In most studies where the stimulus has been confined to the nose there has been no bronchoconstriction. Mechanical irritation of the nose increases the output of mucus into the lower respiratory tract but no one has tested whether chemical stimuli like smoke have the same effect (Richardson & Peatfield, 1981). Cigarette smoke in the nose typically produces a slowing of breathing or apnoea, and section of the trigeminal nerves abolishes this response (Widdicombe, 1964). No one has yet recorded from the afferent axons in the trigeminal nerve which must be at the root of these reflexes. There are reflexes from the nasopharynx in response to mechanical stimuli, and afferent nerves in the glossopharyngeal are responsible for these. Recordings from the axons of these nerves have suggested that they respond weakly if at all to chemical irritants though cigarette smoke was not tested.

There are at least 3 types of afferent nerve from the laryngeal epithelium which may be important in the reflex effects of smoking (Boushey et al, 1974):

(i)    Myelinated fibres with little or no spontaneous activity and a fairly rapid adaptation to mechanical stimuli. These endings respond to a wide range of mechanical and chemical irritations including cigarette smoke.

(ii)   Myelinated fibres with a steady tonic activity and a slow adaptation rate to mechanical stimuli. Cigarette smoke stimulates some of these and inhibits others.

(iii)  Non-myelinated fibres. These are mechanosensitive and respond to some chemical stimuli but the action of cigarette smoke on their firing has still to be tested (Boushey et al, 1974).

When undiluted cigarette smoke is puffed into the larynx of an anaesthetized cat it can cause a variety of changes in breathing including coughing, apnoea and slow deep breathing (Boushey et al. 1972). After the immediate change the result is always slow deep breathing. Narrowing of the larynx and an increase in airways resistance often accompany these changes in the pattern of breathing. When the superior and recurrent laryngeal nerves are cut cigarette smoke in the larynx no longer has these effects. Mechanical irritation of the larynx causes an increase in the rate of mucus secretion from the lower airway by a reflex (Phipps & Richardson, 1976). No one has yet tested whether chemical irritants like cigarette smoke have the same effect.

In the lower airway, that below the larynx, there are at least three types of afferent nerves (see Sant'Ambrogio, 1982 for references).

(i)     Slowly adapting receptors. These receptors are situated in the airway smooth muscle and are predominantly in the larger airways. They respond to stretching of the muscle in which they lie with a rather regular discharge. Myelinated fibres connect them to the brain via the vagus nerve. The reaction of these nerves to cigarette smoke has never been tested but as the endings lie some tens or hundreds of microns from the airway lumen the direct effects are unlikely to be great. When these nerves fire they cut short inspiration, prolong expiration and relax the bronchial tone.

(ii)    Rapidly adapting receptors. These nerves have endings in both the epithelium and smooth muscle of the airways and, like the slowly adapting receptors, are situated mainly in the trachea and larger bronchi. They adapt rapidly to mechanical stimuli such as inflation or deflation of the lungs. Their nerve fibres are myelinated. Widdicombe (1981) has subdivided the rapidly adapting receptors into "Cough" and "irritant" receptors. The former cluster in the trachea and at the points of branching of the larger bronchi and have a particularly rapid rate of adaptation. On indirect evidence their main reflex actions are believed to be coughing and airway mucus secretion (Richardson & Peatfield, 1981). Irritant receptors, on the other hand, are situated mainly in the bronchi, adapt rather less rapidly and have somewhat smaller nerve fibres. Their firing probably causes bronchoconstriction and rapid shallow breathing by a reflex, though this view has recently been challenged (see (iii) below). Of the nerve endings in the lung these are the only ones that have been tested systematically for their reaction to cigarette smoke: in the rabbit, smoke stimulated the majority but in the dog only a minority responded (Sellick & Widdicombe, 1970; Sampson & Vidruk, 1975).

(iii)   Non-myelinated nerve fibre endings. In some species the majority of afferent nerves from the lungs are non-myelinated yet their properties and reflex actions have been studied much less than those of the myelinated fibres. In the past Paintal (1969) has argued that the majority of the receptors served by these fibres were in the lung parenchyma near alveolar capillaries on the

grounds that they could be stimulated after a small
latency (insufficient to allow blood to pass from the
pulmonary to the systemic circulation) by drugs given
into the pulmonary artery and by inhaled irritants.
recently Coleridge & Coleridge (1977) have
distinguished two groups of receptors;  those more
accessible to injections given into the bronchial
circulation ("bronchial endings") and those more
accessible from the pulmonary circulation ("pulmonary
endings").  Even this classification has been
questioned on the grounds that many bronchi receive a
direct blood supply from the pulmonary circulation so a
"pulmonary ending" could be in a small bronchus
(Sant'Ambrogio, 1982). Agents which stimulate these
fibres cause apnoea (sometimes) followed by rapid
shallow breathing, laryngeal constriction,
bronchoconstriction, and increase in airway mucus
secretion, bradycardia and hypotension.  The Coleridges
have recently questioned whether many of the effects
attributed to rapidly adapting receptors could not be
explained better by stimulation of the non-myelinated
nerve fibres (Coleridge et al, 1983).  The matter is
still unresolved.  No one has yet reported whether
cigarette smoke excites these nerve endings.

## Experiments on smoking in animals

Animals can be exposed to cigarette smoke which can
either be directed to the whole or part of the respiratory
tract.  This is useful in analysing reflexes which may arise
from several different sites.

Concentrated tobacco smoke puffed into the nose tends
to cause apnoea and a rise in blood pressure.  In the larynx
undilated cigarette smoke can give rise to coughing but a more
common reaction in anaesthetized cats is slow deep breathing
(Boushey et al. 1972)  When anaesthetized animals are given
concentrated smoke to breathe into their lower respiratory
tracts through a tracheostomy tube the reactions are rapid
shallow breathing, sometimes coughing and an increase in airway
resistance (Aviado & Palacek, 1967).  When cigarette smoke is
directed through a segment of cat trachea it causes an increase
in the rate of mucus secretion and this applies even if the
smoke is diluted with 9 volumes of air (Richardson et al,
1978).

Davis et al. (1967) showed that unanaesthetized guinea pigs
that breathed puffs of cigarette smoke through the whole
respiratory tract showed a rise in airway resistance and slow

deep breathing. When the smoke was directed through a tracheostomy tube, thus by-passing the upper airway, breathing became rapid and shallow and there was little change in airway resistance. Smoke given into the upper airways alone caused a slowing and deepening of breathing, but surprisingly no rise in airway resistance. In these unanaesthetized guinea pigs, given rather unrealistically high concentrations of smoke, the upper airway reflexes seemed to dominate those from the lower respiratory tract.

## The reflex effects of cigarette smoking in man

Most studies of the effects of cigarette smoke in man have examined the acute effects of cigarette smoking on airway resistance. Early studies showed a considerable rise in airway resistance which lasted for many minutes (Nadel & Comroe, 1961). More recent studies have shown a far briefer effect with each puff, probably because the subjects smoked their cigarettes "naturally" rather than intensively. Rees et al. (1982a, b) have reported that the fall in airway conductance (increase in resistance) was maximal at 10 sec after inhalation and lasted little more than 40 sec. As might be expected, stronger cigarettes tended to cause larger changes. The effect on bronchial tone could be reduced by atropine-like drugs or salbutamol. Sodium cromoglycate did not attenuate the effect, which is surprising in view of the effectiveness of that drug against other reflexes which cause broncho-constriction (e.g. that from inhalation of sulphur dioxide, Shepperd et al. 1982).

There has been some work on the change in the pattern of breathing during and after smoking. Rees & Clark (1981) found that smoking led to deeper breathing and that the frequency of breathing tended to fall, though not significantly. They attributed these changes to upper airway reflexes as subjects who had gargled local anaesthetic no longer showed them. Filtering off the particulate phase of the smoke delayed the onset of the changes in pattern of breathing but did not prevent them, so the authors concluded that both the particulate and vapour phases contain substances that trigger reflexes.

Another effect of smoking which might have reflexes at its root is the hypertrophy of the bronchial mucous glands in many who have smoked for years. In animals, irritation of the respiratory tract causes secretion into the lower airway by reflexes. If cigarette smoke triggers similar reflexes in man could these result in hypertrophy of the bronchial glands? This is no more than speculation (Richardson & Peatfield, 1981).

## Conclusions

So far the reflex effect of smoking on man, at least those who are practiced smokers, seems to be a small and transient bronchoconstriction and a slight deepening (and perhaps slowing) of respiration. This may seem a laughably slight repayment for 25 year's work on the subject, but the results do establish that normal smoking excites afferent nerve endings in the respiratory tract of man. This may be more important than the reflexes themselves;  it is still a mystery why people should smoke.  What is the reward that smokers experience?  The answer could lie in the sensations evoked from enteroceptors by irritant chemicals in the smoke;  the respiratory equivalent of eating curry.  This hypothesis can explain, for example, why those smoking cigarettes with a low yield of tar and nicotine should inhale more than those smoking cigarettes with a higher yield (Sutton et al., 1982).

**REFERENCES**

AVIADO, D.M. & Palacek, F. (1967).
  Pulmonary effects of tobacco smoke and related substances
  II.  Comparative effect of cigarette smoke, nicotine and
  histamine in the anaesthetized cat.
  Arch. Environ. Health, 15, 194-203.

BOUSHEY, H. A., Richardson, P.S. & Widdicombe, J.G. (1972).
  Reflex effects of laryngeal irritation on the pattern of
  breathing and total lung resistance.
  J. Physiol., 224, 501-513.

BOUSHEY, H.A., Richardson, P.S., Widdicombe, J.G. &
  Wise, J.C.M. (1974).
  The response of laryngeal afferent fibres to mechanical and
  chemical stimuli.
  J. Physiol., 240, 153-175.

COLERIDGE, H.M., & Coleridge, J.C.G. (1977).
  Impulse activity in afferent vagal c-fibres with endings in
  the intrapulmonary airways of the dog.
  Respir. Physiol., 29, 143150.

COLERIDGE, H.M., Coleridge, J. C. G. & Roberts, A.M. (1983).
  Rapid shallow breathing evoked by relative stimulation of
  airway cfibres in dogs.
  J. Physiol., 340, 415433.

DAVIS, T.R.A., Battista, S.P., & Kenster, C.J. (1967).
Mechanism of respiratory effects during exposure of guinea
pigs to irritants.
Arch. Environ. Health, 15, 412419.

NADAL, J. A. & Comroe, J.H. (1961).
Acute effects of inhalation of cigarette smoke on airway
conductance.
J. Appl. Physiol. 16, 713716.

PAINTAL, A.S. (1969).
Mechanism of stimulation of type J pulmonary receptors.
J. Physiol., 203, 511532.

PHIPPS, R.J. & Richardson, P.S. (1976).
The effects of irritation at various levels of the
respiratory tract upon tracheal mucus secretion in the cat.
J. Physiol., 261, 563581.

REES, P.J., Chowienczyk, P.J., Ayres, J.G. & Clark, T.J.H.
(1982a).
Irritant effects of cigarette and cigar smoke.
Lancet, ii, 10151017.

REES, P.J., Chowienczyk, P.J. & Clark, T.J.H. (1982b).
Immediate response to cigarette smoke.
Thorax, 37, 417422.

REES, P.J. & Clark, T.J.H. (1981).
Pattern of breathing during cigarette smoking.
Clin. Sci., 61, 8590.

RICHARDSON, P.S. & Peatfield, A.C. (1981).
Reflexes concerned in the defence of the lungs.
Bull. Europ. Physiopath. Resp., 17, 9791012.

RICHARDSON, P.S., Phipps, R.J., Balfre, K. & Hall, R.J. (1978).
The role of mediators, irritants and allergens in causing
mucin secretion from the trachea.
In: Respiratory Tract Mucus,  Ciba Symposium 54 (new
series).
Ed: Porter, R., Rivers, J. & O'Connor, M.) pp. 111131,
Elsevier, Amsterdam.

SAMPSON, S.R. & Vidruk, E.H. (1975).
Properties of irritant receptors in canine lung.
Respir. Physiol., 25, 922.

SANT'AMBROGIO, G. (1982).
  Information arising from the tracheobronchial tree of
  mammals.
  Physiol., Rev. 62, 531569.

SELLICK, H. & Widdicombe, J.G. (1971).
  Stimulation of lung irritant receptors by cigarette smoke,
  carbon dust and histamine aerosol.
  J. Appl. Physiol., 31, 1519.

SUTTON, S.R., Russell, M.A.H., Iyer, R., Feeyerabend, C. &
  Saloojee, Y. (1982).
  Relationship between cigarette yields, puffing patterns and
  smoke intakes: evidence for tar compensation?
  Brit. Med. J., 285, 600603.

WIDDICOMBE, J.G. (1964),
  Respiratory reflexes.
  In: Handbook of Physiology, section 3, Respiration, Vol. 1
  Ed: Fenn, W.O. & Rahn, H.  pp. 585630. American
  Physiological Society, Washington, D.C.

WIDDICOMBE, J.G. (1981).
  Nervous receptors in the respiratory tract and lungs.
  In: Lung Biology in Health & Disease, vol. 17: Regulation of
  Breathing. T.F. Hornbein, ed: Dekker, New York. pp. 429472.

DISCUSSION

LECTURER: Richardson                           CHAIRMAN: Cumming

HIGENBOTTAM:    The prevalance of smoking amongst asthmatic
                subjects is not clearly understood but in a
                joint study with Tim Clark we found that 50% of
                asthmatics smoked. A second comment, Cockcroft
                has shown that the response to histamine of
                young smokers is less than the response of
                young non-smokers and argues that some
                constituent of tobacco smoke suppresses the
                irritant receptors, similarly in rats nicotine
                inhalations suppresses the methocholine
                response. It appears that Capsceicin will
                diminish the metaplastic changes in bronchial
                epithelium produced by exposure to cigarette
                smoke. Would you like to comment on these
                points?

RICHARDSON:     The response to cigarette smoke is very
                complicated. Nicotine first stimulates and
                then paralyses the ganglion cells in
                sympathetic and parasympathetic systems. The
                latter would be expected to produce
                bronchoconstriction though more controversially
                there may be bronchodilator fibres which
                produce bronchodilation and this may explain
                the bronchodilation seen in some smokers on
                exposure to cigarette smoke. I accept that
                there is no general agreement about the rarity
                of cigarette smokers amongst asthmatics and
                studies are contradictory. As to Capsceicin, it
                may be that action reflexes play some role but
                it is difficult to think what experiments might
                be done to test this hypothesis. Perhaps you
                can suggest some?

HIGENBOTTAM:    It is possible for humans to inhale Capsceicin
                without great hazard and this could be used
                before challenge with a bronchoconstrictor drug
                in much the same way as bronchodilators are
                sometimes used. This extract of the red pepper
                specifically affects the non-myelinated non-
                cholinergic nerve fibres, would you accept
                that?

RICHARDSON:     This has not been proved and the experiment of

recording from the nerve fibres has not been
done. In other systems non-myelinated fibres
are affected much more than the larger
myelinated fibres. This is difficult to square
with the observation that bradykinin can cause
a bronchoconstriction whilst inhalation of
Capsceicin does not, but only produces cough.
The evidence does not add up to the clear
picture as you have suggested.

DENISON:         We can suppose that evolution has endowed us
with a set of reflexes which appeared before
the industrial revolution, before the use of
tobacco, before the development of fire by man
and were intended for a life shorter than the
one we currently enjoy. I fail to understand
the survival value of rapid shallow breathing
or slow deep breathing, can you evaluate this?
You also spoke of non-myelinated receptors that
responded to blood-borne stimuli both
bronchial and pulmonary and I wonder why the
lungs have been designed to sense a blood-borne
stimulus and then to modify ventilatory
behaviour.

RICHARDSON:      The organs sensitive to blood-borne stimulants
are also sensitive to air borne stimulants,
such as chlorine and halthoane as Paintal has
shown. The immediate response is apnoea and
this may have a survival value in the presence
of chlorine gas. The same point can be made
with respect to the upper airways endings
stimulation of which makes breath holding
easier even with water and the survival value
may be to prevent drowning.

DENISON:         I understand the survival value of apnoea, it
is rapid, shallow and slow deep breathing for
which I see no survival value at the moment.

RICHARDSON:      The slow deep breathing was seen in
anaesthetised animals and may be part of a weak
reflex which is analogous to that producing
apnoea whilst upper airway stimulus would
permit breath holding if one chose to.

CUMMING:         I think one must be careful before making the
assumption that any physiological behaviour
must demonstrably have a survival value, there

is such a thing as physiological redundancy, where the development of the organism has required the presence of a certain organ in order that from it may develop a function with survival value and the vermiform appendix might fit into this category since it has no demonstrable survival value.

BAKE: You mentioned that airways constrict when cigarette smoke is inhaled yet do not constrict if inhalation is directly into the trachea. Are the receptors in the whole respiratory tract or only in the upper airways?

RICHARDSON: It has never been shown that the introduction selectively of cigarette smoke into the lower respiratory tract causes bronchoconstriction, but it does in the upper tract. However bronchoconstricting reflexes do occur in the lower airways and may be stimulated by ammonia and are abolished by vagal section so it is truly a reflex.

CUMMING: We have here an area of uncertainty, pointed out by Bjorn Bake and which is not yet clarified. If airways conductance falls on exposure to a noxious agent this might be due either to a change in the larynx directly or to laryngeal stimulation producing a reflex change in the airways and you have not yet made it clear to me which of these two mechanisms you believe to be operating.

RICHARDSON: The reason I have not made it clear to you is that it is not clear to me.

HIGENBOTTAM: Evidence about the airway response in man in the literature indicates great variation, and the methods of introducing cigarette smoke are extremely artificial. When smoking pattern is normal there appears to be no airways response.

CUMMING: Does any member of the audience have any data on the changes in airway resistance in normal subjects smoking normally?

HIGENBOTTAM: We carried out a study in which we measured airway resistance in a variety of styles of smoking and found that normal smoking did not

change airways resistance except for one
individual, in whom direct smoking, pulling the
smoke directly into the lungs caused an
increase in airways resistance.

CUMMING:            How soon after the dose was the resistance
                    measured, and by what technique?

HIGENBOTTAM:        Within one minute.

RICHARDSON:         That may not be soon enough since a recent
                    paper by Reece and Clarke showed the effect to
                    have disappeared by 40 seconds. May I return
                    to the problem of laryngeal and lower airways
                    receptors. Stirling, has shown that the
                    bronchoconstrictor response can be abolished by
                    agents which relax smooth muscle, for instance
                    salbutamol, so that it is likely that the major
                    part of the response is caused by
                    bronchoconstriction. One is unhappy at the
                    speed of response which occurs in 10 seconds
                    whereas bronchial smooth muscle is very
                    sluggish, but the pharmacological evidence
                    suggests that it is airways smooth muscle which
                    constricts.

RAWBONE:            I was interested by your point that people
                    might smoke in order to stimulate irritant
                    receptors. This is a difficult question to
                    answer since the literature indicates an
                    interaction between tri-geminal stimulation and
                    olfactory perception. The question I would ask
                    is whether any of your work on stimulation has
                    related to individual smoke components, or even
                    the separation of gas phase from particulate
                    phase.

RICHARDSON:         A little work has been attempted but no very
                    satisfactory conclusions have yet been reached.
                    As an example acrolein causes
                    bronchoconstriction as shown by Davis, Vendall
                    and Batista. Nadel and Comroe separated
                    particulate and administered the gas phase only
                    in the 1960's and concluded that the
                    particulate phase was more important. However
                    when Reeson and Clarke repeated the experiments
                    they showed that the gas phase only produced
                    changes in ventilation but more slowly. The
                    best evidence suggests that the active

principle is in both phases.

BONSIGNORE: We have investigated the effect of cigarette smoking in normal never smokers and found an increase in ventilation with a reduction in tidal volume and a fall on oxygen saturation.

RICHARDSON: It would be interesting to repeat the experiments in which the subjects gargled with local anaesthetic and see if those effects were still observed.

JEFFREY: You mentioned that in the isolated trachea stimulated with tobacco smoke that there was no increase in mucus secretion. Is that peculiar to the human airway, and did you give any positive stimulus such as ammonia to show that secretion could be elicited?

RICHARDSON: We have not tested any other species in the same way, but we have studied the cat trachea with its blood supply intact, which leads to your second question. We tested response at the end of the experiment with pilocarpine and with few exceptions all secreted.

FLETCHER: What data do you use to make the statement that asthmatics smoke less than others? Since in Price's Textbook of Medicine, Edition 1921 one of the cures for asthma was to have the feet in a mustard bath, wearing a mustard poultice and smoking a cigarette. When I was a boy at school the only smokers were asthmatics who took Potters asthma cigarettes and I wonder whether asthmatics do have an impaired responsiveness to smoke and how one would conduct an experiment to demonstrate this.

RICHARDSON: Some people have found that asthmatics smoke less frequently than predicted and others that they smoke more frequently. The problem is that asthmatic symptoms may present in patients with chronic airflow limitation from other causes such as bronchitis in which it is known that smoking incidence is large so the confusion may arise from the difficulty in defining asthma specifically.

# CAROTID BODY HYPERPLASIA

Donald Heath, Ross Jago and Paul Smith

Department of Pathology

University of Liverpool, England

The brilliant intuitative interpretation of the histology of the carotid body by de Castro in 1926 subsequently confirmed by the work of physiologists like Heymans et al (1930) and Comroe and Schmidt (1938), demonstrated conclusively that the carotid body has a chemoreceptor function. However, although half a century has passed since this discovery, the carotid body has been largely ignored by pathologists. Their interest in the organ is largely confined to its tumour, the chemodectoma, which has such a characteristic histological picture as to make it suitable for such exercises as the Final Practical Examination of the Royal College of Pathologists. Otherwise morbid anatomists have largely ignored the organ and it is exceptional to see the carotid bodies examined at necropsy even in a case of cardio-pulmonary disease. Little or nothing is written in undergraduate texts of pathology concerning the reactions of the carotid body in generalised disease and hence the ignorance is transmitted to the next generation. This has led to a peculiar imbalance in our knowledge of the carotid bodies. The physiologists have continued to carry out elegant and sophisticated studies on carotid bodies in animals but our knowledge of the condition of these organs in a wide range of diseases is meagre.

One suspects that this reluctance to study the carotid body stems from the fact that pathologists believe that time-consuming serial sections have to be cut and examined. This is quite untrue. All that are needed are a pair of pointed scissors, about five minutes, and, most important of all, the will-power to do it. Such a simple dissection will demonstrate the carotid body as a tan-coloured nodule situated within the

bifurcation of the common carotid artery. There is
considerable variation in the macroscopic appearance of the
carotid body for it may be bilobed or double. This is of some
importance for, if one finds a single carotid body, it must not
be assumed that it represents the total weight of glomic tissue
present until confirmed by further dissection.

On histological examination at low power the carotid
body is seen to consist of lobules separated by connective
tissue. At higher magnification the lobules are found to
comprise lobules about 100 um in diameter with intervening
Schwann cells and sustentacular cells. The clusters are made
of type 1, chief cells which contain neurosecretory vesicles
thought to be concerned with chemoreception. The main form of
chief cell is called the light cell on account of the
tinctorial properties of its cytoplasm. There are, however,
dark and pyknotic variants of chief cells (Smith et al 1982)
and we shall refer to them later. The cells enclosing the
chief cells within the clusters but having intimate contact
with them by cytoplasmic processes are elongated cells. Many
of these are the type 2, sustentacular cells but some are
Schwann cells and on light microscopy it is virtually
impossible to distinguish one from the other.

In view of the lack of interest of pathologists in the
carotid bodies it will come as no great surprise to learn that
the breakthrough in our understanding of the structural
reactions of these organs in generalised disease come not from
a University or Hospital Department of Pathology but from the
more exotic setting of the High Andes of Peru. Here in 1969
Arias-Stella demonstrated that the carotid bodies of Quechua
Indians living in the hypobaric hypoxia of the altiplano were
larger than those of mestizos living on the coast of Lima.
Four years later he extended these observations (Arias-Stella
and Calcarcel, 1973). We were able to show that the volume of
the carotid bodies is greater in domesticated animals living in
the hypoxic environment of high altitude compared to their sea-
level counterparts (Edwards et al., 1971a). The volume of the
left carotid body in sea-level and Andean guinea pigs,
estimated by tissue morphometry applying Simpson's rule is
shown in Fig. 1. A section of guinea pig carotid body at high
altitude is larger than one from sea level and sections stained
to demonstrate reticulin fibres show the opening out of the
cell clusters.

The enlarged carotid bodies of the guinea pig at high
altitude show characteristic ultrastructural changes (Edwards
et al., 1972). At sea level the neurosecretory vesicles
consist of a central osmiophilic core with a narrow surrounding

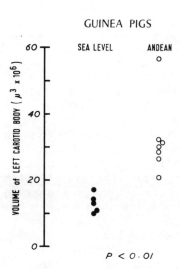

Figure 1. The volume of the left cartoid body in sea level and Andean guinea pigs.

halo and an outer thin, limiting osmiophilic membrane. At high
altitude these central cores become paler and smaller and move
to an excentric position. At the same time the surrounding
halo widens to form a microvacuole.

Studies show that prolonged stimulation by hypoxaemia
will cause the carotid bodies to enlarge when the alveolar
hypoxia is caused not only by the hypobaric hypoxia of high
altitude but by disease such as chronic bronchitis and
emphysema (Edwards et al 1971b). In one group of cases we
studied controls, patients with emphysema in whom the right
ventricle was hypertrophied, and patients with systemic
hypertension in whom the left ventricle was hypertrophied
(Edwards et al 1971b). When we examine the combined carotid
body weights in these cases we find that it is increased in
patients with emphysema complicated by right ventricular
hypertrophy in which presumably significant alveolar hypoxia
had led to pulmonary vasoconstriction (Table 1).

Study of the histopathology of the enlarged carotid
bodies yields a surprising finding for the hyperplasia is the
result of proliferation of not the chief cells containing
neurosecretory vesicles responsive to hypoxia but the elongated
cells comprising unknown proportions of sustentacular and
Schwann cells (Heath et al 1982). Such sustentacular cell
hyperplasia occurs in diseases such as the Pickwickian syndrome
and pulmonary emphysema. The proliferating elongated cells at
the peripheries of the clusters appear to compress the chief
cells. One wonders if this appearance could account for the
diminution in hypoxic drive which is found in native
highlanders, and in patients with chronic lung disease or
cyanotic congenital heart disease with significant hypoxaemia.
A study of such histological appearances allows one to draw up
criteria for the diagnosis of carotid body hyperplasia (Table
2).

It is of interest that enlargement of the carotid body
also occurs in association with a raised intravascular pressure
and its histopathological features are the same as those which
occur in association with chronic hypoxaemia. From these
histological studies one is able to list the characteristics of
carotid body hyperplasia in man (Table 3).

In a minority of subjects in which the stimuli of
raised intravascular pressure or hypoxaemia have been unusually
severe and prolonged a secondary histological change is
superimposed on the generalised hyperplasia of sustentacular
cells. This takes the form of a focal proliferation of the
dark or pyknotic variants of chief cells (Heath et al. 1983).

**TABLE 1**

Ventricular and Combined Cartoid Body Weights in Cases
of Pulmonary Emphysema and Systemic Hypertension.

| Subjects studied | No | Mean Ventricular weights (g) | | Combined weight of carotid bodies (mg) |
|---|---|---|---|---|
| | | LV | RV | |
| Free of cardio-pulmonary disease | 13 | 144 | 51 | 21.1 |
| Emphysema | 11 | 151 | 89 | 32.4 * |
| | | | | 56.2 ⁺ |
| Systemic hypertension | 5 | 262 | 73 | 51.8 |

\*   4 cases without RV hypertrophy
⁺   7 cases with RV hypertrophy

TABLE 2

**HISTOLOGICAL CRITERIA FOR THE DIAGNOSIS OF
CAROTID BODY HYPERPLASIA**

1. Combined carotid body weight          30 mg

2. Mean lobule diameter                  565 um

3. Differential cell count               47% sustentacular cells

TABLE 3

**CHARACTERISTICS OF CAROTID BODY HYPERPLASIA IN MAN**

1. Occurs in chronic hypoxaemia and systemic hypertension

2. Shows identical histopathology in the two conditions

3. Initial proliferation of sustentacular and Schwann cells
   with axons

4. Later focal proliferation of dark or pyknotic variants
   of chief cells

Sometimes some of these focal areas of dark cell proliferation show pleomorphic changes and it is conceivable that this progresses to the formation of the tumour of the carotid body, the chemodectoma.  The histology of this tumour has very characteristic histological appearances with its "zellballen", reticulin pattern, and fibrous tissue septa.  Chemodectomas are a tumour associated with high altitude in the Andes and this suggests that the prolonged stimulation of the hypobaric hypoxia may lead to the dark cell proliferation described above and thence onto chemodectoma formation.  Saldana et al (1973) mention that in the tumours from high altitude that they studied the dark cell predominated.  Inexplicably papillary carcinoma of the thyroid has been reported in association with these tumours of the carotid body (Albores-Saavedra and Duran 1968).  Chemodectomas have also been reported by Arias-Stella as occurring in cattle at high altitude (Arias-Stella and Bustos, 1976).

In the Andes it is possible to study the carotid bodies from animal species that are adapted in a Darwinian sense to hypobaric hypoxia and from others that have to acclimatize to hypoxia.  The alpaca is an adapted animal and its carotid body is small and shows a quiescent histological appearance with clusters of large chief cells with rounded nuclei separated by prominent basophilic sustentacular cells.  In contrast the carotid body of cattle which have to undergo acclimitization is enlarged and shows intense histological activity with disruption of the clusters of chief cells and proliferation of dark cells.  It is tempting to relate the hyperplasia of dark cells found in acclimatizing cattle at high altitude with the focal proliferation of dark cells seen in patients with emphysema and severe hypoxaemia.

Clearly, any understanding of the behaviour of the carotid bodies in response to disease or an adverse environment in man must first explain the initial proliferation of sustentacular cells and Schwann cells, with probably an associated proliferation of nerve axons.  It must then explain the later focal proliferation of dark or pyknotic variants of chief cells in a minority of instances.  It must account for the fact that identical histopathological changes may occur in the carotid bodies in response to systemic hypertension and hypoxaemia.

Finally, it is salutory to note how different is the histological response of glomic tissue to hypoxia in different species.  The microscopic picture is totally different between the adapted alpaca and acclimatized cattle grazing at the same altitude with presumably the same partial pressure of oxygen in

the alveolar spaces. As we have seen the response of the human
carotid body by proliferation of sustentacular and Schwann
cells and nerve axons, differs from both. Finally in our
experience the Wistar albino rat exposed to hypoxia and the
Okamoto rat spontaneously developing systemic hypertension both
exhibit enlargement of their carotid bodies. However, neither
shows any form of tissue response with proliferation of a
particular cell type. This its condition appears to be quite
distinct from that of the human carotid body. This suggests
that the rat is an unsuitable animal model for studies of the
carotid body. It would appear to be particularly unsuitable
for studies of drugs like almitrine which has a specific
stimulant effect restricted to the carotid body for murine
glomic tissue by its very nature appears to be incapable of
reacting to stimuli by tissue proliferation. This being so it
seems unsuitable to see if a new stimulant drug will give rise
to histological changes in the carotid body of the rat.

## REFERENCES

ALBORES-SAAVEDRA, J. and Duran, M.E. (1968).
    Association of thyroid carcinoma and chemodectoma.
    Am. J. Surg., 116, 887.

ARIAS-STELLA, J. (1969).
    Human carotid body at high altitudes.
    Item 150 in the sixty-ninth programme and abstracts of the
    American Association of Pathologists and Bacteriologists,
    San Francisco, California.

ARIAS-STELLA, J., Bustos, F. (1976).
    Chronic hypoxia and chemodectomas in bovines at high
    altitude.
    Arch. Path, 100, 636.

ARIAS-STELLA, J. and Valcarcel, J. (1973).
    The human carotid body at high altitudes.
    Path. Microbiol., 39, 292.

COMROE, J. H. Jnr, and Schmidt, C.F. (1938).
    The part played by reflexes from the carotid body in the
    chemical regulation of respiration in the dog.
    Am. J. Physiol., 121, 75.

de CASTRO, F. (1926).
  Sur la structure et l'innervation de la glande
  intercarotidienne (glomus caroticum) de l'homme et des
  mammiferes, et sur un nouveau, systeme d'innervation
  autonome du nerf glossopharyngie n: Etudes anatomique et
  experimentales.
  Trab. Lab Rech Biol Univ Madrid, 24, 365.

EDWARDS, C., Heath, D. and Harris, P. (1971b).
  The carotid body in emphysema and left ventricular
  hypertrophy.
  J. Path., 104, 1.

EDWARDS, C., Heath, D., Harris, P., Castillo, Y., Kruger, H.,
  Arias-Stella, J. (1971 a).
  The carotid body in animals at high altitude.
  J. Path., 104, 231.

EDWARDS, C., Heath, D. and Harris, P. (1972).
  Ultrastructure of the carotid body in high-altitude guinea-
  pigs.
  J. Path., 107, 131.

HEATH, D., Smith, P. and Jago, R. (1982).
  Hyperplasia of the carotid body.
  J. Path., 138, 115.


HEATH, D., Smith, P. and Jago, R. (1983).
  Dark cell proliferation in carotid body hyperplasia.
  J. Path.  In Press.

HEYMANS, C., Bouchaert, J. J. and Dautrebande, L. (1930).
  Sinus carotidien et reflexes respiratoires.
  11.  Influences respiratoires reflexes de l'acidose, de
  l'alcalose, de l'anhydride carbonique, de l'ion hydrogene
  et de l'anoxemie:  Sinus carotidins et echanges
  respiratoires dans les poumons et au dela des poumons.
  Arch. Internat. Pharm. Therap., 39, 400.

SALDANA, M. J., Salem, L. E. and Travezan, R. (1973).
  High altitude hypoxia and chemodectomas.
  Hum. Path., 4, 251.

SMITH, P., Jago, R. and Heath, D. (1982).
  Anatomical variation and quantitative histology of the
  normal and enlarged carotid body.
  J. Path., 137, 287.

DISCUSSION

LECTURER: Heath                                    CHAIRMAN: Cumming


CUMMING:            You have put forward a very titillating
                    hypothesis, but we need a link between the
                    changes seen at high altitude and those seen in
                    disease and this is the level of hypoxaemia.
                    Hypoxaemia at altitude is readily measureable,
                    and I wonder whether you have any measurements
                    of the hypoxaemia in the patients who showed
                    hyperplasia?

HEATH:              No, I have no measurements.

CUMMING:            So we have a rather disturbing absence of
                    connection in that area so that before drawing
                    any conclusions about the relationship between
                    hyperplasia and hypoxaemia we should require
                    some information about the latter.

HEATH:              This is an argument which is often put forward
                    to pathologists, for instance weighing the left
                    ventricle is not a measure of systemic
                    hypertension, but in the absence of any other
                    demonstrable lesion weighing the left ventricle
                    appears to me to be a much better indication of
                    an increased systemic load than one or two
                    sphygmomanometer readings.  Would you accept
                    that?

CUMMING:            I accept that these two changes are indicative
                    of a sustained increase of pressure in the
                    lesser circulation. Whether this is due to
                    hypoxaemia or to some other cause is still open
                    to doubt, and if we are strictly logical about
                    our hypothesis we should say that there is no
                    demonstrable link between right ventricular
                    hypertrophy, pulmonary artery hypertension and
                    hypoxaemia.   This is not to deny the
                    hypothesis, merely that it does not support it.

HEATH:              I would still put a lot of money on it.

CUMMING:            Prejudices are very useful.  The second point I
                    would like to raise with you concerns the
                    difference between the alpaca and beef cattle,
                    which suggest a genetic difference between

these species and there must exist a codon in one which differs from the codon in the other. This may have great relevance in the clinical situation where exposure to one section of the population produces a disastrous result whilst another part experiences no effect. This is one of the interesting factors of cigarette smoking where only a minority of smokers suffer dangerous clinical effects. Would you like to comment on the genetic aspects of the matter?

HEATH: Clearly high altitude indiginous animals have through the centuries become adapted to the environment, cattle are acclimatised.

CUMMING: Yes but what does that mean?

HEATH: I don't know.

DENISON: Have you attempted to quantify the capillarity of the carotid body in normals, those who are hypoxaemic and those with systemic hypertension, since it is conceivable that there might be carotid body hypoxaemia in the absence of systemic hypoxaemia. As a sequel to that have you thought of looking at the respiratory responsiveness in patients with systemic hypertension compared with normals?

HEATH: I have not done this myself but there is recent work in which chemoreflex activity has been studied in people with systemic hypertension and this is diminished. One might explain this by postulating an overgrowth of sustentacular cells.

DENISON: If that were so you would observe a tendency for patients with systemic hypertension to go into respiratory failure since the absence of hypoxic drive when normal carbon dioxide drive was also absent would have catastrophic results.

HEATH: The vascular system of the carotid body is intensely interesting because the vessels of the carotid body have multiple elastic laminae and nerve endings they have one histological structure which has the appearance of a baroreceptor. Thus the changes in the carotid body might be produced both by pressure or hypoxaemnia.

CUMMING:    Can you indicate where the arterial supply comes from and where the venous drainage goes to?

HEATH:      The arterial supply is variable, but usually the glomic artery (one, two or three) comes from the external carotid. They may come from the bifurcation, and rarely form the internal carotid. The appearance is unusual, consisting of multiple elastic laminae and looks like the carotid sinus.

The venous drainage comes from the upper pole of the carotid body and disappears into the connective tissue between the internal and external carotid arteries. In the literature there are numerous references to arterio-venous anastamoses and one might conceive a role for these, but close examination shows that they are not there in the human carotid body.

DENISON:    The physiological role of the carotid body is to detect $P_{O2}$ and to do this its sensors must be close to the arterial end of the capillary or the arteriole itself. This makes it very vulnerable to changes in flow, so the only way for proper detection is when there is a negligible drop in $P_{O2}$ between the artery and vein when it would be flow sensitive and perhaps this is the link with the systemic hypertension model.

BAKE:       I was intrigued by the observation that similar changes were observed in the carotid body in chronic lung disease and systemic hypertension and I wondered if there were a common denominator in cigarette smoking, which is also associated with both diseases.

HIGENBOTTAM:  Recent experiments in rabbits suggest that the effect of cigarette smoking on ventilatory rate is mediated through the carotid body. Do you have any evidence of what constituents in cigarette smoke cause the changes which you have observed in the carotid body?

HEATH:            This is why I asked earlier if tobacco smoke
                  had any effect in the bronchial tree on
                  argyrophilic endocrine cells, since the ultra
                  structure of these Feyrter cells is similar to
                  that of the carotid body.

CHRETIEN:         Have you studied carotid bodies in pulmonary
                  embolism, since with repeated embolism many
                  chemotic bodies can be found in the lung.  Also
                  have you observed chemotic transformations in
                  the lungs of smokers?

HEATH:            These small nodules of tissue occur around
                  pulmonary veins, their function is unknown and
                  they look like glomic tissue under the light
                  microscope.  Under the electron microscope,
                  instead of the cells being rounder they are
                  elongated rather like the hyperplastic cells in
                  the carotid body and are therefore unlikely to
                  be glomic.  It appears to me also that a high
                  proportion of patients who show this feature
                  have had systemic hypertension and this
                  suggests that close examination of the
                  pulmonary venules in systemic hypertension
                  might be very rewarding.

JEFFREY:          We have found para ganglia around the trachea
                  of rats, which are very similar to Type I cells
                  and I wonder whether you have looked at this
                  tissue.

HEATH:            No, I have not.

RICHARDSON:       Can I ask whether these tissue which have just
                  been discussed are innervated.

HEATH:            I don't know.

JEFFREY:          Electron microscope studies show adjacent nerve
                  fibres but no synaptic connections have been
                  found.

CHRETIEN:         I have no information  on this matter.

PRIDE:            Are there any changes in the aortic bodies, and
                  in patients with right to left shunts do they
                  have a large carotid body?

HEATH:            From your question I am glad to note that my major objective of stimulating interest in these neglected bodies has been achieved and I hope you will persuade your pathologists to look further. For years I have been trying to persuade cardiologists to look at carotid bodies in heart disease, so far without success.

At the Mayo clinic it is the policy at autopsy to examine absolutely everything, so when the question arose of the effect of cyanotic heart disease on the carotid body I telephoned Jesse Edwards for the answer, and I need not tell you that this particular examination had not been carried out.

CUMMING:   On that note with a plea for further investigation of the fascinating organ we should bring the discussion to an end and pass to the next paper by Dr. Masse.

# COCARCINOGENIC EFFECT OF TOBACCO SMOKE IN RATS

R. Masse, J. Chameaud and J. Lafuma

Commissariat a l'Energie Atomique

Fontenay Aux Roses, France

## Evaluation of the Problem

The use of an animal model for investigating the induction of lung cancer by tobacco smoke in man is open to many objections, and epidemiology has already established a connection. Society might therefore regard the use of animals in such a way as perverse and persecuting. However, the failure to resolve the problem as to why it is so difficult to induce lung carcinoma in rodents by the inhalation of tobacco smoke remains unsolved despite extensive trials and the use of ever more sophistocated smoking machines. It is however, easy to produce carcinoma in rats and hamsters by the intratracheal injection of smoke condensates.

It is likely that such tumours have some relevance to human disease, those induced by Saffioti [1] by the intratracheal injection of benzapyrene adsorbed onto particulates were of the squamous type as commonly seen in smokes. Squamous carcinomas were described in the pioneer experimental work of Kuschner and Laskin more than 25 years ago.

An experimental approach to the prediction and prevention of carcinoma would be invaluable, since the epidemiological approach measures only the failure rate in a disease which evolves very slowly from the beginning of the first exposure. We believe therefore that there should be much more work with experimental lung tumours aimed at the prevention of human lung cancer, and in this connection we refute the argument that rodents are not relevent in the production of lung cancer by the inhalation of smoke. Whilst many such well conducted studies have proved negative, others have been successful. In rats and hamsters, lung tumours are rare so that a small number of successful inductions may be significant.

In 1959 Guerin (3) observed three tumours in 68 smoking rats, in 1964, Mori et al (4) found two adenomas in 14 rats, more recently Bernfeld 1974 (5) and Dottenwill 1973 (6) induced some tumours of the upper tract in hamsters. The concept of smoke resistance of rats was definitely invalidated by the work of Dalbey et al 1980 (7).

Using a carefully selected smoking machine, depositing 0.25 mg of particulate matter from cigarette smoke in the lungs of rats, 7 cigarettes being smoked a day for up to 2.5 years, Dalbey observed ten tumours of the respiratory tract in a group of 80 smoking rats whereas 1 tumour only was observed in the 100 rat control group.

It is difficult to compare directly the "doses" smoked by rats and men, but let us suppose that oncogenesis is a cumulative process, directly related to the time integral of the quantity deposited in the lung. If r is the quantity deposited per day, per gram of lung tissue, 'x' the amount of particulate related to oncogenesis and e an effective exponential constant including all the transforming events that modify 'x', we can use the Brain Valberg (8) differential equation that describes the fate of 'x'.

$$\frac{d\,x}{d\,t} = -\,e\,X + r$$

the time integral of that equation between to and 't' is given by the authors in the form:

$$\int_0^t X\;(dt) = \frac{Xo}{\lambda_e}\left(1 - e^{-\lambda_e t}\right) + \frac{rt}{\lambda e}\left(\frac{1 + e^{\lambda_e t}}{\lambda e} - 1\right)$$

Xo referring to the quantity present a $t_o$ (or per time unit considered for deposition). Since 't' is more than 500 days and e involves probably values higher than 0.1 (9). This equation approximates to a limit which is $\frac{rt}{\lambda e}$.

If we suppose that e is similar in man and animals,

which means that there are no differences, significant for lung carcinogenesis, between handling of smoke carcinogens in man and animals, the condition for similar effects in man and in rats can be expected for:

$r_1 t_1 = r_2 t_2$, where 1 refers to man and 2 refers to rat.

If a man smokes 21 cigarettes per day during 20 years, knowing that 10 mg of particulates can be deposited in a 700 g. lung per cigarette, what time is necessary for a rat depositing 7 x 0.25 mg. per day in a 1.75 lung for a similar cumulated exposure?

$$r_1 \; t_1 = 0.3 \times 20 \; y = r_2 \; t_2 \; _{(y)} \; = \; 1 \times t_2 \; _{(y)} \cdot t_2 \neq 6 \text{yrs.}$$

It means very simply that a 20 year smoke duration in man cannot be mimicked in the rat in less than 6 years.

In other words, the very severe conditions of exposure described by Dalbey et al. are no more than the equivalent of 7 cigarettes per day in man. The tumour incidence observed by Dalbey are thus not in disagreement with what could occur in man for such a level of exposure.

Rat is no more resistant to cancer induction by tobacco smoke than is man. The main problem is that it is not feasible to deliver in rats the same cumulative dose of carcinogens during the experimental conditions, due to side effects of CO and nicotine.

In many instances lung cancer in man result from combined exposures to tobacco smoke and occupational airborne pollutants. One can expect that cumulative or synergistic effects could lower the threshold detectable level of the dose response relationship to tobacco smoke and we tried to demonstrate this in radon exposed rats.

## II – THE RADON PROBLEM

$^{222}$Radon is a natural radioactive gas which emanates slowly from rocks following the decay of natural uranium. Different human groups have been heavily exposed to radon , since the 15th Century, during the mining of copper, nickel, iron, silver, cobalt and arsenic. The ore rich country of Erzgebirge, between Germany and Czechoslovakia, later mined for

Pitchblende, became particularly infamous for a fatal lung
disease known as the "Bergkrankheit", diagnosed in 1879 by
Hartung and Hesse as lung cancer.  The incidence was thought to
be higher than 50% (10).

        Since the first hypothesis, mentioned by Uhlig in 1921,
that lung cancer was related to high concentration of $^{222}$Rn in
the Schneeberg mines, an increasing body of evidence from
uranium, fluospar and other mines, in many countries, has now
established the role of radioactive gas and allowed a
determination of the range for a risk coefficient.  Nowadays an
absolute risk coefficient of 150 - 450 cases per million and
per WL. is accepted for miners exposed to cumulative doses
averaging 100 WLM during the time of mining (11).  Most of the
miners are now smokers;  the risk seems to be higher than in
non-smokers and obviously tumours appear earlier in that group
(12).

        Due to the poor body of data available in exposed, non-
smoker groups, the life time absolute risk for non-smokers can
not be shown to be different from smokers.

        In 1971 (13) we demonstrated that lung cancer could be
induced in rats by radon , provided cumulative doses similar to
that delivered to the human lung were delivered during the same
fraction of the life time, i.e. the time equivalent for rats to
25 years spent underground by miners.  Since that time, we have
exposed thousands of rats and have derived a dose effect
relationship which fits rather well with human data.

$^{222}$Rn decays rapidly by      emission and daughter
products more or less diffusible are the main source of lung
irradiation.  Without a knowledge of the equilibrium state of
Rn with its daughters, it is impossible to extrapolate a dose
to the tissue from the activity. The unit WLM is the product of
a total potential energy of $1.3 \times 10^5$ Mev (1 WL) multiplied by
the time of working per month (170 h.) Table 1.

        We have presented the results of an experiment
conducted on a group of 1000 rats which showed a significant
excess of epidermoid carcinomas and adenocarcinomas were
induced in rats by exposure to doses as low as 40 WLM, that is
to say doses which are likely to be encountered indoor, in many
places.

        This probably suggests that the rat is more sensitive
than man to inhaled radon, and indicates also that it is a
suitable species for the study of inhaled environmental
carcinogens.

## III - CARCINOGENESIS IN THE RAT

**Materials and Methods**

The inhalation exposures to radon  and its daughters have been previously described by Chameaud (14).

A smoke box 500 l in volume, was used to expose 50 rats at a time to the tobacco carcinogens.

Cigarette smoke was produced by the simultaneous combustion of nine cigarettes (Gauloises Bleues brand). The cigarettes were placed in a cigarette holder communicating with the box. Aspiration of smoke into the chamber was ensured by means of a slight pressure differential, created in the box with a vacuum pump. A ventilation system renewed the chamber atmosphere with fresh air at the end of each session of exposure to smoke.

The smoke concentration (nine cigarettes per 500 l air) was chosen so that the animals were given 10 - 15 minute inhalation sessions delay. The rats were exposed to smoke for 1 year, 4 days a week. Rats tolerated these exposures well, and their life span was not altered. Their lungs, loaded with tars, displayed changes in broncho-alveolar structure only in the form of adenomatous metaplasia, without any malignant lesion.

Blood carbon monoxide levels in smoking animals were about 0.6% at the end of the day.

Two protocols were chosen for exposure: in the first, the effects of smoking in animals previously treated with radon was checked for three different radon  exposures as shown in Table 1. All animals were Sprague Dawley male rats, 3 months old.

In the second protocol, we studied the role of the time sequence. Since it takes a long time to expose animals to the smoke, all were exposed to radon , when only 10 months old and thus the tumour incidence in the group exposed to radon  only was not expected to be similar to the group exposed at 3 months old. (Chameaud et al, 1981. (15)).

One hundred rats were used. Fifty exposed to radon (4000 WLM) and then to tobacco smoke (300 h of exposure); 50 exposed to smoke first (300 h) and then to radon  (4000 WLM).

When animals were killed, the pulmonary circulation was

Table 1. Protocol or Exposure to Radon and Tobacco Smoke

| Experimental group | Number of animals | Radon exposure Concentration (WL) | Cumulative dose (WLM) | Duration (weeks) | Schedule | Exposure tobacco smoke (h) | Expected cancer incidence (15) |
|---|---|---|---|---|---|---|---|
| 1 | 50 | 3,000 | 4,000 | 8 | 6 h per day 4 sessions per week | 0 | 30%–40% |
| 2 | 50 | 3,000 | 4,000 | 8 | 3 h per day 4 sessions per week | 352 | 5%–10% |
|   | 28 | 3,000 | 500 | 2,5 |   | 0 |   |
| 3 | 30 | 3,000 | 500 | 2,5 | 3 h per day 4 sessions per week | 350 | 1%–2% |
|   | 28 | 300 | 100 | 4 |   | 0 |   |
| 4 | 30 | 300 | 100 | 4 |   | 350 |   |
|   | 45 | 0 | 0 | 0 | 0 | 350 | 0 |

immediately perfused in situ with physiological saline to remove blood, and the lungs macroscopically examined for lesions.

After fixation the lungs were sectioned from above downward, in a frontal plane, into two equal parts; each part was embedded in paraffin and then cut systematically in the same direction at 20 µm thickness. The 5 µm sections selected were spread and stained with haematoxylin, phloxine, alcian green, and saffron.

The lung lesions were described according to the following classification derived from the TNM (tumour-node-metastasis) classification.

| | |
|---|---|
| TO Absence of tumour | NO No lymph node involvement |
| T1 Tumour   2 mm in diameter | N1 Lymph node involvement |
| T2 Tumour 2 - 5 mm in diameter | MO No metastasis |
| T3 Tumour 5 - 10 mm in diameter | M1 Metastasis outside the |
| T4 Tumour   10 mm in diameter |    thoracic acvity |
| PO No spread to the pleura | M2 Intrapulmonary metastases |
| P1 Spread to the pleura |    or presence of several |
| |    tumours in the lungs. |
| | M3 Association of M1 and M2 |

Pleural extension was observed in 37/39 of the rats bearing tumours, whereas it was only observed in 3/11 of rats exposed only to radon and not observed in control and smoke first exposed animals.

Multiple tumours (M2) were observed in 28 of the 39 tumours bearing rat in the radon  first exposed group. They were not observed in the smoke first group.


**DISCUSSION**

In this series of experiments tobacco smoke has been shown to increase the incidence of lung cancers when exposure occurs after exposure to radon.  In rats pretreated with tobacco smoke there is a trend towards a "protection" against lung cancer:  this is not completely unexpected since exposure to smoke increases the mucus secretion and probably lowers the dose to the cells at risk.

From this observation, we conclude that tobacco smoke acts in radon  pretreated rats as a promoter.  no tumours were induced by smoke only and the effect of smoking was not important in non-induced cells.  The promoting effect was maximum in the 500 WLM group, the results are thus relevant for the prediction of possible effects in occupationally exposed miners.

It is evident from Table 2 and Table 4 that smoke promoted cancers are much bigger and more invasive. There is some radiological evidence that the doubling time for the radon alone and smoke promoted tumours are not significantly different (Chameaud et al. to be published). Thus we must admit that the major effect of tobacco smoke is to shorten the latency period, which is in agreement with the conclusion of Lundin et al. 1974 (12) for the effect of smoking in miners.

From the multifocal aspect, very typical of radon + mitogenic factor for cells induced to give later tumours. Indeed, in the animals, there may be found all the steps between metaplasia, benign and malignant tumours. With another agent, 5,6 Benzoflavone, a specific promoter of lung squamous carcinomas, an identical response was observed together with an early wave of DNA synthesis on those cells exhibiting mild squamous metaplasia.

Since this mitogenic stimulus was surprisingly specific, we consider that promotion is mainly a specific mitogenic effect in initiated cells. This theory was suggested by Barsoum and Varchawsky (16) and we consider that it is in excellent agreement with the facts observed here.

From epidemiological and experimental data, Moolgavkar and Knudson 1980 (17), derived a theory on the multi-step origin of cancer: a normal cell (1) transforms into an immediate cell (2), which can reverse, increase in number and transform again into an irreversible malignant stage (3). In this theory, there is a spontaneous rate of transformation from (1) to (2) and (2) to (3). Thus a carcinogen can act either by increasing the transformation rate, which is the most frequent investigated step, but it can act also by increasing the number of intermediate cells without interacting necessarily with the cell genome. It may be that what we observed here is relevant to this phenomenon since many foci of the multiple lesions do not share the characteristics of a malignancy.

Indeed the screening of mutagenic compounds implied in carcinogenesis is well documented but the screening of specific promoters is probably much less feasible outside the specific model: a stage 2 cell able to transform towards the stage 3. We question if this can be achieved in skin painting experiments.

Table 2. Classification and Number of Lung Cancer Induced

| T N M | Groupe 1; 4,000 WLM[a] | | Groupe 2; 500 WLM[a] | | Groupe 3; 100 WLM[a] | | Groupe 4 |
|---|---|---|---|---|---|---|---|
| | Radon only (50 rats) | Radon + smoke (50 rats) | Radon only (28 rats) | Radon + smoke (30 rats) | Radon only (28 rats) | Radon + smoke (30 rats) | smoke only (45 rats) |
| T1 | 2 | 2 | 2 | 1 | 0 | 0 | 0 |
| T2 | 4 | 4 | 0 | 3 | 0 | 1 | 0 |
| T3 | 5 | 6 | 0 | 2 | 0 | 0 | 0 |
| T4 | 6 | 22 | 0 | 2 | 0 | 0 | 0 |
| Cancer no. | 17 | 34 | 2 | 8 | 0 | 1 | 0 |
| % | 34 | 68 | 7 | 28 | 0 | 3.3 | 0 |
| P1 | 10 | 20 | 0 | 6 | 0 | 0 | 0 |
| N1 | 1 | 7 | 0 | 1 | 0 | 0 | 0 |
| M1 | 0 | 0 | 0 | 0 | 0 | 0 | 0 |
| M2 | 4 | 16 | 0 | 1 | 0 | 0 | 0 |
| M3 | 0 | 1 | 0 | 0 | 0 | 0 | 0 |

[a] Cumulative dose

Table 3.   Lung Tumor Incidence

|  | mean survival | n tumors | % |
|---|---|---|---|
| Radon alone | 789 | 11 | 22 |
| Smoke first | 713 | 8 | 16 |
| Radon first | 650 | 39 | 78 |
| Controls (n = 600) | 756 | 5 | 0.83 |

Table 4.   Histology and Size of the Tumors

| | total number of cancers observed | Squamous carcinomas (n) | $T_1 + T_2$ (n) | $T_3 + T_4$ (n) | Modes invasion (n) |
|---|---|---|---|---|---|
| Controls (n = 600) | 5 | 1 | 5 | 0 | 0 |
| Smoke first | 8 | 7 | 3 | 5 | 0 |
| Radon first | 39 | 39 | 3 | 36 | 4 |
| Radon alone | 11 | 6 | 4 | 7 | 0 |

Here, we have used the obsolete smoke chamber instead of the smoking machine, and this can be used as an argument against this animal model. We have reproduced the main features observed in smoking miners and thus we consider that the model is realistic. Moreover, the main objection against the smoke chamber is that it considerably under exposes to the tar particulates. This is true, and we have observed but little amounts of tar in the lung. As a consequence, it means that something present in the particulates or in the vapour phase, as strongly suggested by the earlier work of the Leuchtenbergers (18), is able to promote tumours at a dose which is certainly lower than that necessary for tars to initiate tumours in the lung.

We suggest that this effect exists in multi exposed people and that a screening of specific promoting agents in cigarettes is as necessary as the screening of initiating agent for the choice of the low risk cigarettes.

## REFERENCES

1. SAFFIOTTI, U., Cefis, F., Kolb, L.H., (1968),
      Cancer Res., 28, 104-124.

2. KUSCHNER, M., Laskin, S., Cristofano, E., Nelson, N., (1956),
      In: Proceedings of the third National Cancer Conference,
      485-495, Lippincott Philadelphia.

3. GUERIN, M., (1959),
      Bull. Cancer (Paris), 46, 295-309.

4. MORI, K., (1964),
      Gann., 55, 175-181.

5. BERNFELD, P., Homburger and Russfield A.B., (1954),
      J.N.C.I., 53, 1141-1157.

6. DOTTENWILL et al., (1973),
      J.N.C.I., 51, 1781-1807.

7. DALBEY, W.E., Nettesheim, P., Griesemer, R., Caton, J.E.,
      Guerin, M.R., (1980),
      In: Pulmonary toxicology of respirable particles. DOE
      Symposium Series 53, Technical Information Center US DOE
      ed., pp, 522-535.

8. BRAIN, J., Valberg, P.A., (1974),
   Arch. Environ. Health, 28, 1-11.

9. PYLEV, L.N., Roe, F.F., Turner-Warwick, J., (1969),
   Brit. J. Cancer, 23, 103-115.

10. BAIR, W.J., (1970),
    In: Inhalation Carcinogenesis. A E C Symposium Series 18,
    USAEC, Division of technical Information ed., pp, 77-101.

11. United Nations Scientific Committee on the effects of Atomic
    radiation 1977 report.
    United Nations Publication, p, 397.

12. LUNDIN, F.E., Archer, V.E., Wagoner, J.K., (1979),
    In: Energy and Health. Breslow Whitemore Eds. (Atta, Utah,
    1978). SIAM Institute for mathematics and Society, 243-
    265.

13. CHAMEAUD, J., Perraud, R., Lafuma, J., Masse, R., (1971),
    C.R. Acad. Sci. Serie D., 273, 2388-2389.

14. CHAMEAUD, J., Perraud, R., Chretien, J., Masse, R., Lafuma,
    J., (1982),
    Recent Results in Cancer Research, 82, 11-20.

15. CHAMEAUD, J., Perraud, R., Lafuma, J., Masse, R., (1982),
    In: The assessment of Radon and Daughter exposure and
    related biological effects. Proceedings of a Rome meeting,
    Clemente - Nero - Steinhausler Wrenn eds., R.D. Press,
    Utah pp, 198-204.

16. BARSOUM, J., Varshawsky, A., (1982),
    In: Gene Amplification. Schimke ed., Cold Spring Harbor
    Laboratory New York, 239-245.

17. MOOLGAVKAR, S.H., Knudson, A.G., (1981),
    J.N.C.I. 66, 1037-1052.

18. LEUCHTENBERGER, C., Leuchtenberger, R., (1970),
    In: Morphology of Experimental Respiratory Carcinogenesis.
    USA EC Symposium Series 21, Nettesheim, Hanna Deatherage
    Eds, Division of Technical Information, OAK RIDGE,
    Tenessee, pp, 329-346.

DISCUSSION

LECTURER: Masse                    CHAIRMAN: Cumming

PRODI:
Do you have any data on the location of tumour onset which might be relevant to the site of injury by the toxic agents you have used? Radon has a high diffusive coefficient with a high exposure on the upper airway whereas bronchioles and alveoli would have a higher exposure from mainstream smoke.

MASSE:
Inhaled smoke is mainly deposited in the parenchyma, but the main effect could be due to the vapour phase. We do not know whether the effect is dose related, which might explain the observation that the majority of tumours in miners occur at the tracheal bifurcation. Another explanation might be that the sensitivity of rat and man differs. The maximum dose probably occurs in the sixth generation bronchioles, and this is not where tumours are found in man, so there is probably a difference in sensitivity. The dose reaching the target cells is probably different in rats and man, and also for radon and tobacco smoke.

JEFFREY:
Your primary carcinogen was radon and tobacco smoke acted as a promotor rather than as a primary initiator, have you tried any other promoting agent such as sulphur dioxide? Regarding comparitive deposition, the animals take the dose by an entirely different method. Man by-passes the nasal mucosa whilst the animals use it exclusively. Have you looked at the nasal mucosa where tar deposition would be higher?

MASSE:
We have not used sulphur dioxide but we have used benzoflavone which induces only mitosis and squamous metaplasia. With tobacco smoke in lower dose than as an initiator it is still possible to produce tumour by promotion, which raises the question of risk in passive smoking. In animals exposed to radon and tobacco smoke we have observed tumour formation in the nasopharynx.

LEE:

I was surprised when you related risk to the product of exposure level and exposure time, since it is known epidemiologically that lung cancer risk is not linearly related to dose but to the fourth or fifth power of the duration of smoking so that a better equation would have $t^4$ or $t^5$.

MASSE:

It is not logical to suppose that the carcinogenic effect is cumulative, it is a balance between the dose rate and the total dose. It is not possible to achieve the same dose in rats and men because of the short life span of the rat.

LEE:

Two years in the rat is a larger fraction of life span. According to your formula a smoker smoking one pack a day for twenty years would have the same risk as one smoking two packs a day for ten years, whereas the first has a twenty times higher risk than the second. My second question relates to the three doses of radon with and without tobacco smoke. Were you giving both radon and tobacco smoke continuously?

MASSE:

No. Radon first and smoke after.

LEE:

Was the survival similar in the groups. Because in Dolby's experiments the tobacco exposed groups lived much longer than the non-exposed group and this may partly explain the increased incidence of tumours.

MASSE:

In Dolby's experiments there was a lot of other vascular disease and the housing and keeping of the animals was sub-optimal. In our experiments the survival was not different in the groups.

LEE:

My final question relates to the large number of squamous tumours in the radon/smoke group. It seemed to me that you showed a diminution in this group of non-squamous tumours, do you agree with this?

MASSE:

Thats right, I think it is a real finding and I think it means that the target cell is modified and perhaps the mucus cell can give rise to tumours.

CUMMING:          I suggest that you continue your discussions
                  over lunch.

# THE ACT OF SMOKING

Roger G. Rawbone

Research Division, Gallaher Limited, Belfast

## SUMMARY

Cigarette smoking is a complex activity which can be studied both in terms of detailed measurements of puffing and inhalation as well as in terms of the delivery and uptake by the smoker of a range of smoke components. Measurement of human smoking profiles have demonstrated that different products may be smoked differently and in particular that 'low tar' cigarettes are smoked more intensively than 'middle tar' cigarettes thus offsetting the reductions in tar delivery expected from standard machine smoking values. Differences in smoking patterns have been attributed to the smokers needs but evidence is presented which indicates that individual smokers smoke to a constant amount of work. This suggests that the product may have an important role with its pressure drop showing an inverse relationship to puff volume during smoking. Whatever the interaction between the smoker and the product in determining smoking behaviour however, the resulting tar intake of 'low tar' smokers remains significantly lower than that of 'middle tar' smokers.

The act of smoking a cigarette is a complex manipulation of a complex product involving the generation of a complex smoke mixture which is subsequently presented, at least in part, to the upper respiratory tract and lungs.

Any discussion of smoking should be prefaced by a consideration of the cigarette product itself as, first and formost, this is the determinant both directly, through its composition, and indirectly, through its influence on the act of smoking, of the 'dose' of smoke to which the smoker is exposed.

The cigarette is a complicated piece of design engineering and differences between the many brands available depend upon:

the filler - usually a blend of various types of tobacco and tobacco derivatives (air cured or flue cured) plus a variety of additives or tobacco substitutes.

the cigarette configuration - whether it is long and thin or short and fat.

the cigarette paper - its composition, burn rate and porosity.

the filter - where, if present, one is interested in (a) its efficiency in removing particulates and, in some cases, gas and vapour phase components and (b) the presence of associated ventilation, where small holes are 'punched' in the cork tipping paper. This latter permits air to be drawn into the smoke stream and has the effect of reducing smoke delivery per unit volume.

the cigarettes physical characteristics - its density and pressure drop.

In order to compare the variety of cigarette products on the market they are smoked in machines to standard parameters; in the United Kingdom these involve a 35cc, bell-shaped flow profile puff of two second duration every minute down to a standard butt length. The resultant smoke chemistry is then determined using standard analytic techniques with the most commonly measured components being tar (presented as either total particulate matter - TPM - or as particulate matter, water and nicotine free - PMWNF), nicotine and carbon monoxide. During the past 40 years such measurements have indicated a continuing reduction in cigarette smoke yields facilitated by incorporation of the filter into the design and more recently by the introduction of ventilation.

Human smokers do not, however, smoke in the same unvarying way as does the standard smoking machine. The intake of smoke from any brand of cigarette may span a considerable range of deliveries and not only that, but with the reduction in yields, smoking patterns of the smoker may change such as to compensate for any reduced delivery shown under the standard smoking regimen.

The study of these phenomena has involved the development of techniques in three broad areas:
(i) measurement of human smoking patterns;
(ii) measurement of the delivery of smoke to the smoker, and
(iii) measurement of the smoke uptake by the smoker.

Quantitative measurements of the smoking profile which can be made are the cigarette consumption, butt length, puffing parameters (volume, duration, number and inter-puff intervals), and inhalation. Puff parameters can be readily obtained from measurement of the pressure drop across a small resistance built into a cigarette holder. This differential pressure can be converted to flow which in turn by integration will allow derivation of the puff volume. Our current system gives a digitized flow signal together with a pressure measurement across the cigarette, relative to atmosphere, at 50 hz. Inhalation can be studied by recording external chest wall movement using one of the many available pneumograph systems or, as I have recently used, a head-out, arms-out plethysmograph. These techniques permit the shape and the relationships of the puff and inhalation/exhalation profiles to be observed when the typical two stage process of the act of smoking can be seen in the majority of smokers. In fact it might now be more appropriate to describe a three stage process for the puff and the inhalation tend to be separated by a variable time period during which a significant amount of the puff may be lost from the oral cavity. This loss of smoke, referred to as wasted smoke, is obviously important and at the same time difficult to measure; it certainly accounts for the finding of a poor correlation between measurements of smoke delivery to and smoke uptake by the smoker.

The measurement of smoke delivery has commonly been based on an analysis of the cigarette butt nicotine assuming a constant filter efficiency where this has been measured on machine smoking. This is not strictly speaking correct, although it may serve to give an approximate result, but its attraction is that it does not require the smoker to smoke in an artificial situation. In order to provide a more detailed, comprehensive and accurate analysis of smoke delivery during human smoking, machines have been developed which will exactly replicate the human smoking profiles from the flow signals obtained using the previously discussed cigarette holders. In order to measure the actual smoke uptake by the smoker, one has either to measure specific smoke components or their metabolites in a body fluid (blood, urine, saliva or expired air) or one has to calculate the figure as the difference between delivery and wasted plus exhaled smoke. It is important to emphasise that in any study of smoke delivery, or

more particularly smoke uptake, the results obtained relate
only to the smoke component measured and it is dangerous to
make unqualified extrapolations to other constituents.

Having given a broad outline of the act of smoking and
its measurement, I would like to present the results from a
series of studies designed to define further what determines
the smoking pattern for any given cigarette. One can envisage
that this might be influenced by two major factors - (i) a need
by the smoker to obtain either pharmacological or organoleptic
(sensory) satisfaction and (ii) the physical characteristics of
the product being smoked. Much early work concentrated solely
on the role of nicotine as a pharmacological agent in governing
smoking patterns and this led to the formulation of the so-
called nicotine titration hypothesis. More recently however
the over-riding influence of nicotine has been questioned and
attention has turned to the possible influence of taste
factors, either through tar or the tar:nicotine ratio. The
emphasis remains on the needs of the smoker. I would like
perhaps for the first time to turn the spotlight off the smoker
and his needs and onto the role of the product.

This work stemmed from an interest in the concept of
'work done' by the smoker while puffing and through an
examination of the inter-relationships between product pressure
drop and puff volume.

When a smoker puffs on a cigarette he expends energy
(or work) in transferring a bolus of smoke to his mouth. The
work done (dw) in displacing the gas through an infinitesimal
distance 'ds' is given by:

$$dw = F.ds$$

where F is the force applied which is in turn equal to
the pressure drop (P) multiplied by the cross-sectional area
(A).

Therefore        $dw = P.A.ds$

This may be written in terms of volumetric flowrate (Q)
at time 't' when work done during puff 'i' thus becomes

$$Wi = \int_{Di} Pi\ (t)\ Q.\ (t).dt.$$

where $\int_{Di}$ signifies integration over the duration of the
puff.

As previously discussed both the flow rate and the

pressure drop across the cigarette are provided as signals from the cigarette holder used to define the puffing parameters. Thus the work done per puff can be readily calculated. A mean work per puff has been defined by averaging the work done per puff over all puffs in the smoking session and this in turn is averaged across all subjects of a group to give the average mean work per puff on a per product basis.

The first study which we performed involved 40 smokers who simulated puffing through a series of filter rods which were constructed to be visually identical but which exhibited a range of pressure drop values (at a standard flow rate of 17.5cc/sec) from 30 to 210 mmWG at 30 mmWG intervals. The order with which the rods were presented to each smoker was randomised over a series of visits with a minimum of ten 'puffs' being taken through each rod. The cigarette holder permitting measurement of puffing parameters was used and data for subsequent analysis restricted to the last three puffs taken.

A plot of the average mean work per puff against the unlit pressure drop at 17.5cc/sec flow is presented in Fig. 1. Error bars are the 95% confidence limits for the mean values. Although a range of curves may be drawn through the confidence bands, it is hard to avoid the conclusion that when puffing on unlit rods, smokers tend to maintain a constant level of work per puff for pressure drops exceeding 90mmWG. When the pressure drop at the standard flow drops below this value the mean work per puff appears to reduce. Fig. 2 shows the relationship for puff volume against rod pressure drop when, as might be predicted from the previous finding, an inverse relationship is demonstrated with the puff volume falling as the pressure drop increases.

From these observations in a highly artificial situation we proceeded to investigate whether they had any relevance in relation to the smoking of lit cigarettes. The data analysed was from a series of three paired comparisons involving five products whose pressure drop ranged from 68 to 142 mmWG at the standard flow. Fourteen middle tar smokers took part in this series of studies. A plot of the average mean work per puff values for each product against the nominal unlit pressure drop value for each product is presented in Fig. 3. Superimposed on this is the visually fitted curve from the previous experiment which was shown in Fig. 1. It can be seen that although variation (represented by the 95% confidence limits) is appreciable, the average mean work per puff value is approximately constant over the range of pressure drop values. The line obtained from the pressure drop study on

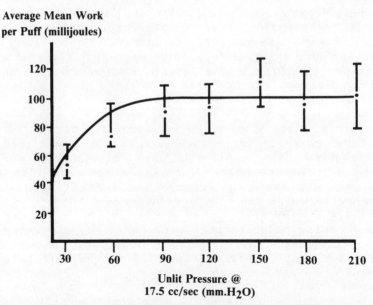

FIGURE 1:    The relationship of average mean work per puff
             during human puffing to the pressure drop measured
             at 17.5 cc/sec in a series of unlit filter rods.
             Vertical bars indicate the 95% confidence limits
             about the mean values with the superimposed curve
             restng a 'best' visual fit.

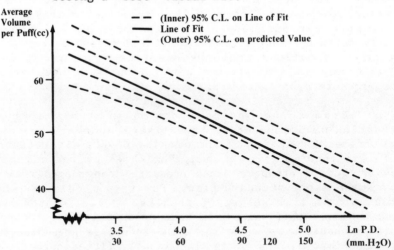

FIGURE 2:    The relationship between puff volume and pressure
             drop measured in a series of unlit filter rods.
             The mean least squares regression line is shown
             with its 95% confidence limits (inner broken
             lines) and the 95% confidence limits on predicted
             puff volume values (outer broken lines).

unlit rods apparently underestimates the average mean work per puff across all products.

When examining the responses during smoking we are interested in the pressure drop exhibited in the lit situation. This can be derived from the recorded signals as discussed earlier and presented either as average puff values or as the pressure drop at the standard flow but measured during the human profiles. In fact, for the series of products in these studies there is a close correlation between these measurements. Fig. 4 shows the lit pressure drop/puff volume values for both products in each study superimposed on the best fit curve obtained from the unlit rod study. Again the concordance between the two studies can be noted.

The data presented for lit cigarettes represents the group mean values obtained from each of the products. The lit pressure drop value over a series of smoking sessions might, however, be expected to vary from cigarette to cigarette because of variations in packing density. If this parameter from each individual cigarette smoked is therefore used as an independent variable and the range of smoking measurements plotted against it, scattergrams will result which will provide data points for intermediate pressure drop values. Such scattergrams for puff volume and the mean work per puff plotted against the lit pressure drop are presented in Figs. 5 and 6 respectively. In neither case do the correlation coefficients reach statistical significance. In an attempt further to explore the relationships the data was partitioned into 20 mmWG intervals and the mean values plotted. Whilst the correlation coefficients were improved they still failed to reach a statistical significance. The graphs show the best fit straight lines through these points which, perhaps not surprisingly, conform with those previously obtained with the slope of the volume/pressure drop relationship being similar and the mean work/ppressure drop relationship being almost flat and equal to a value of 120 millijoules.

After discussing this data one cannot but reach the conclusion that during the smoking of cigarettes there is a large variation in the results relative to the average value at each pressure drop. Although we might postulate that the mean work per puff data is fitted by a constant across the range of pressure drop values or that the puff volume is inversely related to the pressure drop, there is a lot of scatter about the mean, best fit, regression lines. This suggests the hypothesis that each smoker confines himself to a narrow band of work per puff values across a range of products of different pressure drop and that the variation in work per puff value

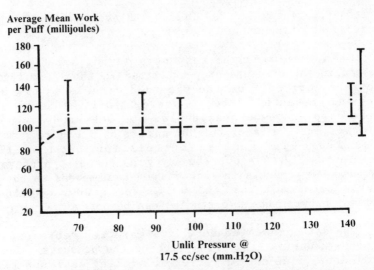

FIGURE 3:     The average mean work per puff with 95% confidence
              limits recorded during the human smoking of a
              series of different products shown in relation to
              their unlit pressure drop measured at the standard
              flow of 17.5 cc/sec.  The broken line represents
              the visually fitted curge from Fig. 1.

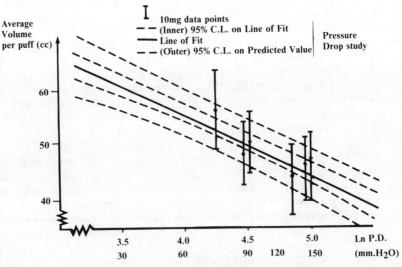

FIGURE 4:     The mean puff volume with 95% confidence limits
              recorded during the human smoking of a series of
              different products superimposed on the puff
              volume/pressure drop relationship shown in Fig. 2.

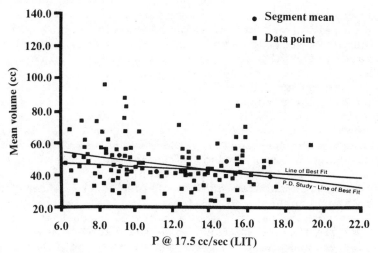

FIGURE 5:    A scattergram of mean puff volume against lit
             pressure drop during human smoking.  Solid circles
             represent the mean value on partitioning pressure
             drop at 20 mmWG intervals with the mean least
             squares regression through these points indicated.
             The regresion line from the unlit pressure
             drop/puff volume relationship shown in Fig. 2 is
             also superimposed.

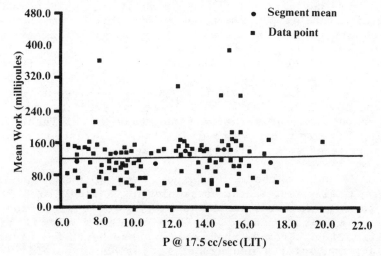

FIGURE 6:    A scattergram of mean work per puff against lit
             pressure drop during human smoking.  Solid circles
             represent mean values derived as in Fig. 5
             together with their regression line.

illustrated by these scattergrams is mainly a consequence of
person to person variation. In other words each smoker has a
defined work level with higher work done values being
associated with higher puff volumes and pressures. This
hypothesis has been tested statistically by a two way analysis
of variance on the data from the fourteen subjects, each
smoking the six products, in the current series of studies.
The results are shown in the Table. From this it is clear that
the majority of variance is between people and is highly
significant in comparison with the residual variance. On the
other hand the between product variance, which approximates in
this case to within person variation, is not significant in
comparison with the residual variance. These findings would
support the hypothesis that each individual smoker tends to
define his smoking pattern on the basis of a constant amount of
work. We are then left with a wide intersubject variability
where differences in smoking behaviour will be influenced by
the differing characteristics of individuals both in relation
to their physical make-up and in relation to their needs.

I have earlier referred to the phenomenon of
'compensation' whereby smokers of lower delivery products may
exhibit more intensive smoking patterns thereby offsetting the
reductions expected from the deliveries under standard
conditions. One might therefore ask whether the pressure drop
of low tar products is lower than that of middle tar products
so facilitating the smokers compensation. I use the terms 'low
tar' and 'middle tar' as defined by the UK Government Chemist
for products having standard tar yields of less than 10 and 17-
22 mg/cigarette respectively. An examination of pressure drop
in products within these two categories would suggest that
differences are marginal with an average figure of 135 mmWG for
middle tar and 120 mmWG for low tar.

Whatever may determine the act of smoking, and through
this the compensatory smoking of lower tar products, the final
arbiter in terms of relative reductions of tar intake from
smoking low tar relative to middle tar cigarettes has to be
some measure of smoke delivery or uptake.

We have performed a study part of which included
demographically representative samples of UK middle and low tar
smokers each returning a 24 hour butt collection from the
smoking of products equivalent to their normal brand
specification. From an analysis of these butts for nicotine
content, and using a measure of filter nicotine efficiency
defined from human smoking profiles, we have calculated the
nicotine deliveries to the smokers. An extrapolation from
nicotine to tar delivery has then been made using the product

TABLE  1:    A two way analysis of variance table of the work
             done during smoking, from a study of 14 subjects
             each smoking six products, showing the relative
             magnitudes of within and between subject
             variability.

| SOURCE | d.f. | SS | MS | F | Sig |
|---|---|---|---|---|---|
| between People | 13 | 225595 | 17353 | 16.5 | 0.001 |
| between Products | 5 | 6171 | 1234 | 1.2 | NS |
| Residual | 65 | 68236 | 1049 | | |
| Total Corrected | 83 | 300002 | | | |

tar:nicotine ratios, again defined on human smoking profiles using the human smoking duplicator.

The resultant distributions of intake from the 134 middle tar and 134 low tar smokers in this study is shown in Fig. 7. The expected wide distributions of delivery for each product can be seen but clear differences in distribution patterns are also evident with a more normal distribution of intake for the middle tar smokers and a skewed distribution for the low tar smokers. The mean tar presented to the middle tar smoker is 14.03 mg. and that to the low tar smoker 9.5 mg, a difference which is statistically significant at the 1% level.

In terms of relative intensity of smoking these average figures show a 2% oversmoking of the low tar product and an 18% undersmoking of the middle tar product, both relative to the standard smoking machine, with the consequent more intense smoking of the low tar cigarette relative to the middle tar cigarette, as reported by other workers, being demonstrated.

There was no difference between the two groups in terms of cigarette consumption and the results therefore clearly show a reduction in terms of tar delivery to the smoker of low tar cigarettes, as reported by other workers, being demonstrated.

There was no difference between the two groups in terms of cigarette consumption and the results therefore clearly show a reduction in terms of tar delivery to the smoker of low tar cigarettes, despite the more intensive smoking of these products, whatever its cause. Measurements of smoke delivery, as previously discussed, represent the maximal smoke available for uptake and differences in terms of wasted smoke and inhalation could therefore affect the quantity of tar retained by the smoker. There is, however, no evidence to suggest such differences exist and we have data from other studies which, considering measures of smoke uptake, suggest that these reductions in exposure to the low tar smoker are maintained.

I would conclude where I started – the act of smoking is a complex manipulation of a complex product – and, whilst undoubtedly the smoker may adjust his smoking behaviour to satisfy his requirements, the product may also directly influence his behaviour. The interaction between these two has now to be recognised and studied as we further try to understand the act of smoking.

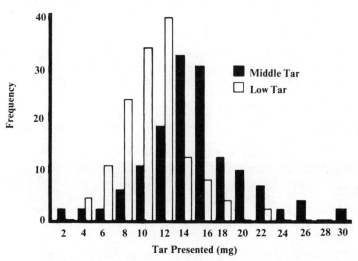

FIGURE 7:    A frequency histogram of tar intake derived from
             24 hour butt collections in 134 middle tar and 134
             low tar smokers.

**REFERENCES**

1.  RAWBONE, R. G., Coppin C. A., Guz, A.
    Carbon monoxide in alveolar air as an index of exposure
    to cigarette smoke.
    Clinical Science and Molecular Medicine (1976), 51,
    495-501.

2.  RAWBONE, R. G., Murphy, K., Tate, M.E., Kane, S.J.
    The analysis of smoking parameters: inhalation and
    absorption of tobacco smoke in studies of human smoking
    behaviour.
    In: Smoking Behaviour: physiological and psychological
    influences. (1978). Ed. R.E. Thornton, Churchill
    Livingstone, Edinburgh.

3.  RAWBONE, R. G.,
    The value of non-invasive methods for evaluating
    exposure to tobacco smoke.
    In: Smoking and Artificial Disease. (1981).
    Ed. R.M. Greenhalgh. Pitman Medical.

4.  ADAMS, L., Lee, C., Rawbone, R. G., Guz, A.
    Patterns of smoking: measurement and variability in
    asymptomatic smokers.
    Clinical Science and Molecular Medicine (1983), 65,
    383-392.

5.  ASHTON, H., Stepney, R., Thompson, J.W.
    Self-titration by cigarette smokers.
    British Medical Journal (1979), 2, 357-360.

6.  KUMAR, R., Cooke, F.C., Lader, M.H., Russell, M.A.H.
    Is nicotine important in tobacco smoking?
    Clin. Pharm. Ther., (1977), 21, 520-529.

7.  DO people smoke for nicotine?
    British Medical Journal (leader), (1977), 1041.

8.  WALD, N.J., Idle, M., Boreham, J., Bailey, A.
    The importance of tar and nicotine in determining
    cigarette smoking habits.
    J. Epid. and Community Health, (1981), 35, 23-24.

9.  SUTTON, S.R., Russell, M.A.H., Iyer, R., Feyerabend, C.,
        Saloojee, Y.
        Relationship between cigarette yields, puffing
        patterns, and smoke intake: evidence for tar
        compensation?
        British Medical Journal, (1982), 285, 600-603.

10. RUSSELL, M.A.H., Jarvis, M., Iyer, R., Feyerabend, C.
        Relation of nicotine yield of cigarettes to blood
        nicotine concentrations in smokers.
        British Medical Journal, (1980), 1, 972-976.

11. WALD, N.J., Idle, M., Boreham, J., Bailey, A.
        Inhaling habits among smokers of different types of
        cigarette.
        Thorax (1980), 35, 925-928

12. RUSSELL, M.A.H., Sutton, S.R., Iyewr, R., Feyerabend C.,
        Vesey, C.J.
        Long-term switching to low-tar, low-nicotine
        cigarettes.
        British Journal of Addiction (1982), 77, 145-158.

DISCUSSION

LECTURER: Rawbone                                    CHAIRMAN: Cumming

HIGENBOTTAM:        There is some tantalising information there.
                    When you discuss the work done drawing air
                    through the cigarette does this take into
                    account the diminution in volume as the
                    cigarette burns away? Secondly, does the puff
                    volume diminish in size as the cigarette is
                    smoked?

RAWBONE:            We have calculated the work done in all the
                    puffs and taken the mean value. Puff volume
                    does decrease as the cigarette is smoked,
                    although since resistance decreases one might
                    expect the puff volume to increase. However
                    the pressure drop in modern cigarettes during
                    smoking is much less than hitherto and in
                    ventilated cigarettes most of the pressure drop
                    is due to the filter.

HIGENBOTTAM:        You mentioned that you measured the
                    tar/nicotine ratios, do you have any
                    information on that?

RAWBONE:            I have the figures here but I have not
                    calculated the ratios, I will let you have them
                    later.

CUMMING:            Given that work is involved in determining puff
                    size what other variables influence it, and
                    what degree of importance do you attatch to
                    each?

RAWBONE:            This is a very difficult question, and one must
                    ask what causes compensation and this could be
                    a function of the product and its pressure drop
                    as I have shown, or the smokers needs, either
                    sensory or pharmacological. What we need to
                    know is the balance, and I believe that the
                    studies to illuminate this area have not yet
                    been done. The design of the experiment is
                    crucial to investigate the variability between
                    smokers and will be difficult to deliver.

WARBURTON:          Did you make any measurements of plasma
                    nicotine, salivary nicotine or urinary
                    cotinine?

RAWBONE:        We were trying to look at large populations and the meaSurements you mention are difficult to carry out in large numbers, so we did not make them, using only the butt analysis.

WARBURTON:      In the absence of such measures, and with no information on the depth of inhalation it is difficult to be sure that there was no compensation.

RAWBONE:        In a smaller population, in which we did make such measurements the results were identical and showed the lower uptake using the lower tar cigarettes compared with middle tar.

PRIDE:          The experiments you have described are reminiscent of those carried out by Moran Campbell in the 1960´s. Can you tell us what level of resistance can be detected in the mouth, are we very good at detecting small changes?

RAWBONE:        In respect of answering the question "can you tell the difference between these two devices", they appeared to be able to detect a pressure difference of 30 mm of water in the lower absolute values. Difference between 180 and 210 mm of water were however not detected, so it depends upon the absolute magnitude of the pressure drop. We observed peak pressures in excess of 200 mm during puffing.

CUMMING:        We seem to have a hypothesis of the stick and the carrot - the carrot being the gratification whilst the stick is the expenditure of work to achieve it. Can you tell us what the gratification is?

RAWBONE:        No.

CUMMING:        With that note of uncertainity I will close the discussion and move on to the next paper.

# THE QUANTITATION OF SMOKE UPTAKE

Neil M. Sinclair

Carreras Rothmans Limited
Basildon, Essex, England

## 1. INTRODUCTION

Smokers are individuals, each smoking his or her chosen product in a specific way. It follows that the delivery to the smoker's mouth, of nicotine, carbon monoxide, tar and the thousands of other chemicals in cigarette smoke will vary considerably from one person to another. For the majority of individuals delivery will be quite different, either greater or smaller, than the yields produced by analytical smoking machines which are set to give one two second puff of 35 ml every 60 seconds. Not only will smoke delivery vary in magnitude but also in quality since smoke composition may differ according to the manner in which the cigarette is smoked. For example, the tar:nicotine ratio and the carbon monoxide:nicotine ratio have been shown to depend on puff volume (1).

However, the generation of smoke, e.g. puff volume, duration etc. is probably less of a determinant for smoke uptake than the way the smoker handles the smoke once it enters the mouth. This aspect of the smoking process has been termed smoke manipulation and encompasses such factors as:

1. The position of the puff within the respiratory cycle.

2. The amount of air taken in before, during or immediately after the puff.

3. The volume of air/smoke mix inhaled into the lung and the degree of lung expansion prior to this event.

4.  The flow of air and smoke through the nose especially prior
    to inhalation;  this process offers a route for smoke in
    the mouth to be expelled without entering the lungs.

5.  The most important and possibly the most neglected aspect
    is the smoke that is taken into the mouth and immediately
    allowed to drift out or forcibly expelled prior to
    inhalation.  This is termed waste smoke.  Our own studies
    suggest that this manoeuvre alone can dispose of between 5
    and 50% of the smoke taken into the mouth.

Clearly the standard yield from the analytical smoking
machine, of tar, nicotine and carbon monoxide can tell us
little about the amounts of these substances and other smoke
components which enter the body and are retained by the smoker.
Yet it is these retained materials, which must be quantified if
we wish to relate cigarette smoking to the health and
behavioural effects attributed to this activity engaged in by
so many people worldwide.

## 2.  CONCENTRATIONS AND QUANTITIES

Cigarette smoke consists of thousands of individual
substances, the chemical and physical nature of each
determining to a large extent the site and rate of absorption
and elimination from the body.  When assessing the absorption
by the smoker, smoke components must be considered
individually.  Each component that we are able to detect in the
smoker's body provides specific information on that chemical
alone and care must be exercised when this data is evaluated in
relation to the effects of smoking.  Clearly it is not possible
to assume that the behaviour of all smoke components are
identical, each has its own chemical and physical properties
which determine uptake, metabolism and eventually the
analytical method.  Compare for example, carbon monoxide, a gas
with a strong affinity for haemoglobin, with nicotine, a water
soluble, semi-volatile liquid, which quickly enters the blood
and is then rapidly metabolised and excreted.  Each smoke
component must be considered individually, especially if a
specific behavioural or health effect is being investigated.

In practice, uptake of a smoke component is measured by
analysing a biofluid for the concentration of the component or
one of its metabolites, e.g. nicotine and cotinine in blood or
urine, carbon monoxide in blood.  However, we cannot always
extrapolate from a measured concentration to a quantity
absorbed.  Whilst in some instances the effects on the smoker
may relate directly to a concentration, either transient or

prolonged, in other cases the actual quantity of material absorbed in a defined time may be more relevent.

Table 1 lists the smoke components and metabolites which have received most attention as indicators of smoke absorption.

## 3. NICOTINE AND METABOLITES

### 3.1. Nicotine in blood and salava

Nicotine has received much attention since it is considered to be the principal pharmacologically active agent in tobacco smoke, and because it is present in sufficiently large amounts to permit detection in the biofluids of the smoker (2). Nicotine levels in blood rise rapidly in the smoker who inhales to reach a maximum within one minute of finishing the cigarette. Subsequent multiple sampling demonstrates that levels then fluctuate rapidly for a few minutes although the general trend is a rapid fall. After 15 minutes or so the decrease becomes less rapid until the pre-smoking background level is reached 1 - 2 hours later. It follows that if a smoker has a cigarette more often than every 60 - 120 minutes the background or trough levels, existing immediately before the next cigarette will steadily rise to eventually reach a plateau. A similar picture can be produced by injecting smoking doses of nicotine (3).

Plasma nicotine levels in human smokers are generally measured one or two minutes after finishing a cigarette. The increase caused by smoking is interpreted as an approximate measure of nicotine intake when comparing subjects using the same product and the same subject when smoking different products. Unfortunately, whilst this sampling time is generally that of maximum concentration in plasma it coincides with the period of maximum level fluctuation. Hence precision requires large numbers of subjects and multiple estimations per subject. The relationship between plasma nicotine levels and other measures of smoking has been examined by Russell who found "Peak blood nicotine concentrations ........bore little relation to the number of cigarettes smokes, the type of cigarette or its tar and nicotine yield" (4). More recently he concluded "the total volume of smoke puffed from a cigarette was a more important determinant of blood nicotine concentration than the nicotine or tar yield of the cigarette, its length, or the reported number of cigarettes smoked on the test day" (5). It would appear therefore that plasma nicotine

**TABLE 1:**   Smoke Components and Metabolites of Importance for
the Assessment of Smoke Component Uptake

| Smoke Component | Chemical Measured | Biofluid |
|---|---|---|
| Nicotine | Nicotine* | Blood Saliva Urine |
| | Cotinine | Blood Saliva Urine |
| | Nicotine-N-oxide | Urine |
| Carbon monoxide | Carboxyhaemoglobin | Blood |
| | Carbon monoxide | Exhaled air |
| Hydrogen Cyanide | Thiocyanate | Blood Saliva Urine |

increases may provide some comparative information on the way a particular cigarette is smoked being a function of smoke volume, which Russell measured, and presumably subsequent manipulation, which was not assessed. However, because of the short half life of nicotine in serum ($\alpha$ phase = 8-10 minutes) no reliable estimate can be made of the number of cigarettes smoked in unit time or the quantity of nicotine taken in daily from a single plasma nicotine level per se.

No absolute measure is available of the quantity of nicotine taken up by a smoker, as a result of smoking a single cigarette naturally, so it has not yet proved possible to determine if blood nicotine levels relate to total nicotine uptake.

Two practical problems exist in applying plasma nicotine analysis to a smoking study. The first is that sampling blood is inconvenient and can only be carried out by medically trained personnel. Furthermore some smokers may not agree to provide a blood sample and simply the thought of giving blood may cause many individuals to smoke in an abnormal manner. A second problem is that plasma nicotine levels are generally in the low tens of nanogrammes per millilitre and so contamination from nicotine in the environment is a major problem. This can be overcome by adopting stringent constraints during sampling and subsequent analysis. In an attempt to circumvent these problems nicotine in saliva has been studied, with special attention to its role as an indicator of plasma nicotine concentrations. Such an approach has been employed with some drugs for therapeutic monitoring (6). Whilst this method has a number of advantages, e.g. levels of nicotine are higher in saliva than plasma and saliva samples can be readily obtained, certain difficulties limit the application universally. In particular the ratio between the concentrations in saliva and plasma is highly dependent on the pH of the saliva; this cannot be readily controlled. A second confounding factor is nicotine which lodges in the mouth during smoking, which appears to be firmly absorbed, and although it gives rise to highly elevated levels in the saliva it cannot be removed by straightforward mouthwashing using either neutral or acidic solutions.

### 3.2. Cotinine in blood and saliva

Cotinine, a metabolite of nicotine, has been suggested as a useful marker for long term smoke uptake (7). Since its concentration in smoke is virtually zero that found in the plasma, saliva or urine of the smoker can only have arisen from nicotine absorbed and subsequently metabolised. The plasma half life has been reported to be 30 hours (8) although more recent work indicates much shorter times between 6 - 15 hours (9).

As with nicotine, cotinine can be detected in saliva. Two advantages over nicotine are immediately apparent. The first is that since cotinine is not present in smoke that in the mouth can only have arisen from the blood via the salivary glands. Secondly the concentration ratio between blood and saliva is essentially independent of the saliva pH. Theory suggests a ratio of 1:1. In practise Russell (10) finds ratios between 1:1 and 1.3:1 whilst in our hands ratios between 1:1 and 2:1 were obtained. Cotinine levels in both plasma and saliva can attain values as high as 400-500 ng ml$^{-1}$ so analysis is relatively straight-forward.

In summary, the potential for cotinine as a comparative indicator of nicotine retention in smokers is very good although a number of aspects need to be resolved.

(a)  The half life must be determined to provide a fuller understanding of its rate of change in response to individual smoking acts.

(b)  It must be determined if cotinine levels can be high in plasma if the smoker does not inhale. It has been shown, for example, that non-inhaling pipe smokers have high cotinine levels whilst another report demonstrates very low plasma nicotine levels for a similar group.

### 3.3. Nicotine, cotinine and nicotine-N-oxide in urine

Nicotine and two of its metabolites, cotinine and nicotine-N-oxide can be assayed in the urine of smokers. Recently Kyerematen et al (9), after intravenous administration of $^{14}$C-nicotine at 2.7 µg kg$^{-1}$ in smokers and non-smokers, have demonstrated a recovery of 72.1 and 74.3% of the nicotine injected respectively in the urine collected over the subsequent 96 hours. Earlier work by Beckett (11) has shown recovery of nicotine and cotinine to be higher when the urine is maintained at an acidic pH, when nicotine resorption in the kidney is inhibited.

If urinary levels of nicotine, cotinine and nicotine-N-oxide are to be used as a measure of nicotine retention a number of factors need to be considered.

(a) Individuals may vary in the relative proportions of nicotine which are metabolised to cotinine and nicotine-N-oxide.

(b) The decay half lives for these three substances differ, so application should be restricted to longer term studies when urine can be collected over 24 hour periods and when the smokers are stabilized, i.e. after smoking the same product for at least one week.

    This method of assessing nicotine retention differs fundamentally from either nicotine or cotinine levels in plasma. These latter approaches provide figures for concentrations achieved whereas the urine method, with urine volumes, offers a value for the quantity excreted in a defined period, e.g. 24 hours. Correlations between plasma levels and urinary levels have been demonstrated. Stepney (12) has shown a correlation between the plasma nicotine increase caused by smoking one cigarette and 24 hour urine nicotine (r = 0.66; p < 0.05) and 24 hour urine cotinine (r = 0.64; p < 0.05) levels. Likewise a good correlation has been reported between plasma cotinine and urine cotinine (7).

## 4. CARBON MONOXIDE

### 4.1. Chemistry and Physiology

    Carbon monoxide has received particular attention as a marked for the absorption of the vapour phase component of cigarette smoke. Retention is measured either as carboxyhaemoglobin in blood or as carbon monoxide in mixed exhaled or end tidal exhalate. In contrast to nicotine carbon monoxide is absorbed almost exclusively in the gas exchanging parts of the lungs; there is, for example, no carbon monoxide uptake from the last 120 ml of inspiration (13).

    Carbon monoxide has a half life of 4 - 6 hours in the sedentary human. In active subjects this can be considerably shorter whereas during sleep a half-life of about 7 hours is typical (14). During a working day half-lives of 3-4 hours are normal hence a single determination of carboxyhaemoglobin reflects smoking history and/or environmental exposure during the previous few hours. This environmental contribution to carbon monoxide body load arises from automobile exhausts, furnace gases and some forms of machinery.

## 4.2. Measurement

Carbon monoxide body load is usually assessed by measuring the concentration of carboxyhaemoglobin in blood. However, as a non-invasive alternative to blood sampling the carbon monoxide concentration in the exhaled air of the smoker may be determined. Over the complete range of values found in smokers a good correlation exists between alveolar carbon monoxide and carboxyhaemoglobin levels (15). However when the relationship is examined in more detail and related simply to the increase caused by smoking single cigarettes, it is found to vary from occasion to occasion within a subject and also from subject to subject (16). It is generally considered that methods for measuring carbon monoxide levels are not sufficiently precise to apply to situations where increases caused by smoking one cigarette are considered because of analytical and biological variability. Preference lies in assessing exposure on a daily basis by recording carbon monoxide body levels at one or more fixed times throughout the day, e.g. late afternoon, when in general, the smoker has reached a plateau.

Some workers use carbon monoxide levels as a means of estimating the absorption of both nicotine and tar (17). However, because of their differing physical and chemical properties carbon monoxide, tar and nicotine retention differ according to the way the smoke is inhaled. Stepney et al (12) have found that the intake of carbon monoxide from a single cigarette is unrelated to the intake of nicotine although the authors concluded that trough carboxyhaemoglobin concentrations might be used to give a reliable indication of the relative nicotine exposure of populations smoking cigarettes of different types. (It should be remembered that the intake of nicotine from a single cigarette, as measured by an increase in plasma nicotine, may not relate to relative nicotine exposure for a population as measured by trough plasma nicotine levels).

The use of one marker compound to assess uptake of a second is particularly undesirable when the experimental protocol calls for smokers to switch products. The act of switching may give rise to a change in smoke generation and manipulation, which would, in all likelihood, alter the uptake ratio between the marker used and the compound being estimated.

## 5. THIOCYANATE

The thiocyanate anion, a metabolite of cyanide and the end product of the body's natural chemical detoxication of cyanide, has been proposed as a "specific, simple and cheap method to assess smoking habits" (18). Thiocyanate is a vapour phase type marker, deriving principally from hydrogen cyanide and volatile nitriles in cigarette smoke. However the presence of cyanide and thiocyanate in some foods eg. cabbage, cauliflower, turnip etc. gives, in non-smokers, levels about one third that of moderate smokers. Saloojee et al (19) have compared thiocyanate and carboxyhaemoglobin as discrimators of smoking and determined that the former added little to the information obtained from the latter.

The major advantage to be gained from thiocyanate derives from its long half-life of 14 days. Hence levels are insensitive to hourly or even daily changes in smoking habits. It follows that if this marker is to be used in studies examining compensatory behaviour when changing the cigarette type being smoked subjects must remain on the new product for at least 2-3 months for thiocyanate concentrations to stabilise at a level reflecting current smoking.

As with the other marker compounds thiocyanate derives from specific chemicals in the smoke Hence body levels are a consequence of the cyanide yield of the cigarette along with the way the cigarette is smoked and the way the smoke is inhaled.

## 6. CONCLUSIONS

No single means of assessing total smoke retention in smokers exists. Indeed when considering how puffing (generation) and inhalation (manipulation) affect, in a relative manner, the uptake of smoke components it is clear no individual substance can fulcill this role.

Each method which has been employed deals with single chemicals and only provides information on the fate of that specific substance. Therefore great care must be exercised when assessing smoke exposure based on smoke marker information.

Carbon monoxide and thiocyanate reflect uptake of vapour phase components whereas nicotine (and metabolites) provide some information on the uptake of the water soluble particulate fraction. No marker yet exists for those non-volatile materials which enter the lung and are deposited there, to be cleared by mucocillary activity. Human smoking research urgently requires such a marker.

**REFERENCES**

1.  CREIGHTON, D.E. and Lewis, P.H. (1978),
    The effect of smoking pattern on smoke deliveries in
    smoking behaviour, physiological psychological
    influences.
    Ed. by R. E. Thornton, p.301, Churchill Livingstone.

2.  FEYERABEND, C. and Russell, M.A.H. (1980),
    Assay of nicotine in biological materials: Source of
    contamination and their elimination.
    J. Pharm. and Pharmacol., 32, 178-181.

3.  ROSENBERG, J., Benowitz, N.L., Jacob, P. and Wilson, K.M.
    (1980).
    Disposition kinetics and effects of intravenous
    nicotine.
    Clin. Pharmacol. Therap., 28 (4), 517-522.

4.  RUSSELL, M.A.H., Jarvis, M., Iyer, R. and Fayerabend, C.
    (1980).
    Relation of nicotine yield of cigarettes to blood
    nicotine concentrations in smokers.
    British Medical Journal, 280, 972-976.

5.  SUTTON, S.R., Russell, M.A.H., Iyer, R., Feyerabend, C. and
    Saloojee, Y. (1982),
    Relationship between cigarette yields, puffing patterns
    and smoke intake: evidence for tar compensation.
    British Medical Journal, 285, 600-603.

6.  MUCKLOW, J.C. (1982),
    The use of saliva in therapeutic drug monitoring.
    Therapeutic Drug Monitoring, 4, 229-247.

7.  HILL, P. and Marquardt, H. (1980),
    Plasma and urine changes after smoking different brands
    of cigarettes.
    Clin. Pharmacol. Therap., 27, 652-658.

8.  ZEIDENBERG, P., Jaffe, J.H., Kanzler, M., Langone, J.J.,
    Van Vunakis, H. (1977),
    Nicotine: correlation of cotinine level in blood with
    ability to stop smoking.
    Compr. Psychol., 18, 93-101.

9.  KYERENATEBM G.A., Damians, M.D., Dvorchik, B.H. and Vesell,
    E.S. (1982),
    Smoking-induced changes in nicotine disposition:
    application of a new HPLC assay for nicotine and its
    metabolites.
    Clin. Pharmacol. Therap., 32(6), 729-780.

10. FEYERBEND, C. and Russell, M.A.H., (1980),
    Rapid gas-liquid chromatographic determination of
    cotinine in biological fluids.
    The Analyst., 105, 998-1001.

11. BECKETT, A.H., Gorrod, J.W. and Jenner, P. (1971),
    The analysis of nicotine-1-N-oxide in urine in the
    presence of nicotine and cotinine and its application to
    the study of in vitro nicotine metabolism in man.
    J. Pharm. Pharmac., 23., 55S.

12. ASHTON, H., Stepney, R. and Thompson, J.W. (1981),
    Should intake of carbon monoxide be used as a guide to
    intake of other smoke constituents.
    British Medical Journal, 282, 10-13.

13. GUYATT, A.R., Holmes, M.A. and Cumming, G. (1981),
    Can carbon monoxide be absorbed from the upper
    respiratory tract in man.
    Eur. J. Respir. Disease, 62, 383-390.

14. CASTLEDEN, C.M., and Cole, P.V. (1974),
    Variations in carboxyhaemoglobin levels in smokers.
    British Medical Journal, 4, 736-738.

15. WALD, N.J., Idle, M., Boreham, J. and Bailey, A. (1981),
    Carbon monoxide in breath in relation to smoking and
    carboxyhaemoglobin levels,
    Thorax, 36, 366-369.

16. HOPKINS, B. and Sinclair, N.
    (unpublished results).

17. WALD, N.J., Idle, M., Boreham, J. and Bailey, A. (1980),
    Inhaling habits among smokers of different types of
    cigarette.
    Thorax, 35, 925-928.

18. BORGERS, D. and Junge, B. (1979),
    Thiocyanate as an Indicator of tobacco smoking.
    Prev. Med., 8, 351-357.

19.  SALOOJEE, V., Vesey, C.J., Cole, P.V. and Russell, M.A.H.
     (1982).
     Carboxyhaemoglobin and plasma thiocyanate: complementary
     indicators of smoking behaviour.
     Thorax, 37, 521-525.

DISCUSSION

LECTURER: Sinclair                          CHAIRMAN: Cumming

CUMMING:          Part of the chairman´s responsibility is to
                  identify semantic problems so that their
                  solution leads to clearer understanding. Our
                  problem concerns the word "uptake" and refers
                  to smoke which is physical description
                  involving a disperse phase within a gaseous
                  continous phase. Uptake must then be involved
                  in the measurement of that physical process,
                  to say that we will use a chemical marker and
                  assume that this represents smoke uptake is
                  dangerous and emphasises the need to be quite
                  specific in what is being measured. A second
                  comment concerns the rate of metabolism and
                  excretion of nicotine which may offer us a
                  pwerful weapon, so that administration of
                  substances which make the urine acid or
                  alkaline may permit manipulation of the average
                  serum of nicotine level. Whether this would be
                  useful or not is of course, an open question,
                  but the phenomenon may be a useful one if the
                  objective is to break the cigarette habit.

JEFFREY:          We have looked at the administration of
                  nicotine by aerosol and by intravenous
                  injection. Using the aerosol we observed no
                  goblet cell hyperplasia. By injection is very
                  high doses we likewise observed no goblet cell
                  changes in the lung. Over the last few years
                  we have found it increasingly difficult to
                  obtain a goblet cell response using modern
                  cigarettes, with their lower delivery of tar
                  and nicotine. We have therefore switched to
                  higher yields - of course our animals are not
                  compensating.

HIGENBOTTAM:      You have shown a poor correlation between
                  carbon monoxide and nicotine, what happens when
                  lung function is disturbed? Do patients with
                  emphysems alter to ratio of uptake of nicotine
                  and carbon monoxide?

SINCLAIR:         I do not think that work has been done. We are
                  trying to relate the carbon monoxide uptake to

the carboxyhaemoglobin to establish the repeatability between subjects.

WARBURTON:     You showed a poor correlation between nicotine and carboxyhaemoglobin in individuals. Do you have information on the correlation between populations? An analysis of Russells data suggests such a correlation. Would you agree that a smoker with a high level of carboxyhaemoglobin would have a larger uptake of other smoke components than someone with a lower level. We are interested in this since it is a non-invasinve way of assessing smoke uptake.

SINCLAIR:      Yes I would. The argument becomes more powerful as the size of the population increases. I cannot imagine how a smoker could retain carbon monoxide and not nicotine, but he could take up nicotine and not carbon monoxide by taking smoke only into the mouth.

CUMMING:       Running through that discussion was the implicit assumption that carbon monoxide in the blood arrives only through cigarette smoking, and whilst this is an important cause it is not the only cause and if such a cause is present in an individual it will destroy the correlation. In a population however the existance of other causes will not be dominant, so that a correlation emerges. It is important to remember that carboxyhaemoglobin is a normal constituent of human blood and the rate at which it metabolises differs between individuals. Thus we have at least two confusing factors at the outset in the use of carbon monoxide as an uptake marker.

DENISON:       The reproducibility of carbon monoxide uptake in individuals is very well known, and is $\pm$ 5%, measured over many months. In emphysems uptake is reduced, so that when they smoke a cigarette uptake will be reduced to a comparable extent. You said you failed to see how the body could handle carbon monoxide and nicotine differently, but one of the major points of your own slide was that the body did just this. How do you explain this contradiction?

SINCLAIR:      Knowing that carbon monoxide is taken up only in the lungs, but nicotine is readily absorbed from the buccal mucosa, it would require some filtration device to separate the carbon monoxide and the nicotine in order not to retain nicotine when carbon monoxide was taken up.

CUMMING:      The nature of that device is called thermal diffusion.

DENISON:      My semantic problem was how something could be manipulated by the mouth, but was defeated by the chairman manipulating blood nicotine levels.

FLETCHER:     The chairmans suggestion of maintaining a high serum level of nicotine by keeping urine alkaline has the disadvantage that the smoker would retain his dependance on nicotine whereas the objective should be to remove this. Perhaps what the smoker likes is his peaks of nicotine.

ANTHONY:      I would like to make another semantic point concerned with the word response. Would it not be better if we reserved this work for the response of the body to the smoke for instance the occurrence of a leucocytosis, reserving some other word for the components of smoke and their management.

PRIDE:        If one is interested in long term effects one requires a slow marker. You mentioned thiocyanote, is this a good long term marker?

SINCLAIR:     Yes it is, the main confounding factor being diet, which might contribute 30-40% of the total. Thiocyanote levels vary up to 40% from day to day in smokers who preserve the same smoking pattern.

CUMMING:      If gratification depends on mean serum level, which it may not; then maintaining an alkaline urine may produce the same gratification but with a lesser hazard. Whilst not an ideal solution it may have some merits.

FLETCHER:     It would save the smokers money and would be bad for the manufacturers so it would be a good idea if it worked.

RAWBONE:        Schacter has manipulated urinary pH and shown
                that cigarette consumption follows the expected
                behaviour and indeed this is one of the
                agrument in favour of the nicotine hypothesis.

SINCLAIR:       Much of the work in this area conflicts with
                that of Schacter, so the problem is not yet
                clarified.

WARBURTON:      One comment on Schacters study, they were
                unable to demonstrate any change in serum
                nicotine related to urinary pH changes.

HIGENBOTTAM:    I think there is a group of smokers who develop
                optic neuritis when they smoke strong tobacco
                and they metabolise thiocyanote very slowly, so
                that plasma levels are not so valuable.

BAHKLE:         To what extent is nicotine in the mouth
                swallowed rather than inhaled?

SINCLAIR:       We have used salivary nicotine to observe the
                difference between aerosol nicotine and
                injected nicotine, and about 30% of the
                nicotine in a cigarette can be found in the
                mouth and over 30-60 mins this is swallowed
                with the saliva.

CUMMING:        The relevant figures for bronchodilators is
                that about 90% is converted by the gastric
                route and 10% by the pulmonary route.

BAHKLE:         Then uptake by the mouth would appear to be
                faster than you are implying?

SINCLAIR:       Tissue uptake is indeed fast, I am reporting
                the speed of removal from the tissue and its
                appearance in the saliva.  Levels in the saliva
                are still high after 30 mins.

WESNES:         Blood levels from nicotine tablets have been
                measured and there was rapid uptake from the
                buccal mucosa, within 5 minutes and it remained
                elevated for up to 45 minutes.  Due to the very
                rapid absorption of nicotine from cigarettes do
                you think that area nicotine is an adequate
                indicator of the cerbral dose?  If one assumes

a peak level of 30 nanogrammes, then the average serum level would be only 0.12, assuming a 2 mgm delivery from the cigarette. Have you considered measuring the area under the curve rather than the peak height?

SINCLAIR: We have done this in one study and the correlation with peak level was very good. The amount of nicotine in the plasma is a very small proportion of that which enters the body, and the volume in which nicotine is distributed has been reported to be in excess of 100 litres. Which suggsts that there are reservoirs in the body for nicotine so that the serum level is a poor indicator of nicotine uptake.

CUMMING: This leads to the question of what is the size of the nicotine space. Keith Wesnes, you did some calculations on this, what was its size and how was it measured?

WESNES: I assumed equal distribution and came up with the value of 40 litres.

RAWBONE: I thought nicotine was selectively concentrated in certain organs such as pancreas and central nervous system.

CUMMING: This is true, so that if one finds a volume of distribution of 100 litres in a 70kg man it is clear that nicotine is concentrated in some organs. This in turn implies that the assumption of uniform distribution is necessarily incorrect and that any volume determination made on that basis is in error.

BAHKLE: What is the pH of nicotine?

SINCLAIR: 7.9 for the pyrrolidine ring and 4.3 for the pynadic ring.

CUMMING: Could you say Mick, what is the purpose of your question?

BAHKLE: The lung has a good affinity for basic compounds, but nicotine is not basic enough to be strongly retained by the lung.

CUMMING:          Having demonstrated large lacunae in our
                  knowledge of the physiology of nicotine we
                  should now close the discussion.  We do not
                  know the size of the space, the difference in
                  concentration in different tissues, our
                  measurements are rudimentary, we do not know
                  whether the congener or its metabolites are an
                  appropriate indicator, so we have plenty of
                  scope for work over the next five years.

# PHYSIOLOGICAL EFFECTS OF CHANGING CIGARETTE NICOTINE YIELD

Andrew R.Guyatt, Melanie J. McBride,
Andrew J.T. Kirkham and Gordon Cumming

The Midhurst Medical Research Institute
Midhurst, West Sussex,GU29 OBL, UK

## Introduction

Many epidemiological studies have illustrated the general hazards associated with cigarette smoking, and have also shown wide differences in the susceptibility of individuals with similar smoking histories. Much of this variability may be due to genetic or environmental factors, but another important factor concerns the quantities of the smoke constituents which reach the tissues. This is largely influenced by differences in the way that people smoke, and this paper is concerned with the problems of measuring the process and the effects of smoking two cigarettes which differ principally in their nicotine yields.

The most crucial factor in the measurement of human smoking is the comfort and cooperation of the subject. Smoking is a very labile behaviour, so every attempt should be made to provide a pleasant relaxed environment, with the measurements being as non invasive as possible. In view of this variability, it is also desirable to obtain repeated measurements on different occasions, and this limits the use of invasive techniques such as blood sampling which would discourage the volunteer from returning. There is also a serious practical constraint, arising from the need to leave the face unimpeded during smoking, which precludes conventional measurement of ventilation using a mouthpiece.

Smoking studies generate a considerable amount of data so we have planned our studies to incorporate digital processing. At present we record six different signals

113

continuously during smoking on an analogue FM tape unit, and these can then be played back, either to produce paper records to monitor the data or for off-line analysis by computer.

In this discussion we first consider the process of smoking under three headings, smoke generation, (the frequency of puffing and puff size), smoke handling, (the extent to which the smoke bolus is inhaled), and smoke uptake, (assessed by concentration changes of smoke constituents in the body). Finally we consider the effect of a change of nicotine yield on these measurements, illustrated by data from a study where nine current smokers smoked their customary brand of cigarette, or the special products with different nicotine yields.

## Smoke generation

Measurements can be made of the way in which a subject puffs a cigarette if he smokes through a special holder incorporating some flow meter such as an orifice, (Creighton, Noble and Whewell, 1978). As the subject puffs in, a pressure gradient develops across the orifice which is sensed by a transducer connected to ports on either side. The pressure signal is actually proportional to the square root of flow through the orifice, but can easy be corrected and scaled by the computer to give puff flow directly, and by integration, puff volume. The draw, (airflow), resistance of the cigarette, can be measured by comparing the flow with the simultaneous pressure gradient across the cigarette, sensed by a second transducer connected to the holder. These data are customarily expressed as the pressure across the cigarette at a standard flow rate of 17.5 ml s$^{-1}$.

The use of a holder disturbs normal smoking patterns to some extent Tobin and Sackner, (1982) have claimed that a subject puffs more when using a holder, but they only studied a small number of subjects, and their method of measuring puff volume from movements of the cheeks is not very precise. We believe that the problems are not significant; the deadspace of the holder is only about 1 ml, and with a 2.4 mm orifice the draw resistance is normally 1% or less that of the cigarette.

An example of the measurements is shown in Fig 1, which represents the smoking of one cigarette. The subject was puffing fairly intensively since he took 19 puffs over a period of about 200 seconds. The puff volume (lower symbols, scale on right, solid line least squares regression), fell significantly

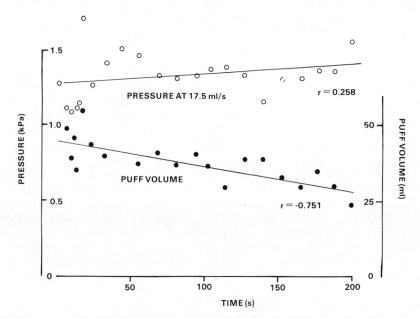

FIGURE 1.    Puff volume and draw pressure measurements made
during the smoking of a single cigarette.

by almost 50% from beginning to end.  The pressure data, (upper
symbols, scale on left), was much more variable and there was
no signicant trend with time.  This pattern was seen in the
majority of studies; the fall in puff volume has been described
before, (Higenbottam, Feyerabend and Clark, 1980), and can be
explained by a simultaneous reduction in puff duration.  These
findings appear consistent with the comment made earlier  in
this conference, (Rawbone, 1984), that "individual smokers
smoke to a constant amount of work", since the subject puffs
each time at the same rate against a constant load, but adjusts
puff volume by varying the duration. The time effect of puff
volume may also have important implications for the reporting
of smoking studies, since it be misleading to describe puff
volume changes merely by their average values.  The rate of
change of puff volume may also vary markedly between different
cigarettes.


## Smoke handling

    Whilst most smokers can give a subjective assessment  of
the   extent   to   which   they   inhale   smoke,   objective
confirmation is  difficult.  Several groups of workers have
measured the  breathing pattern  during smoking,  using some
form of indirect spirometry, (Guillerm and  Radziszewski,
1978,  Higenbottam,  Feyeraband  and  Clark,  1980,  Rawbone,
1981,  and Tobin and Sackner, 1982), or a  head-out, arms-out
body plethysmograph, (Adams, Lee,  Rawbone  and  Guz,  1983).
The latter instrument gives an accurate measurement  of volume
changes   at   the   expense   of   confining   the   subject,
limiting   his   movement   and   his   abilty   to handle the
cigarette   naturally.

    We have used the respiratory inductive plethysmograph or
RIP,  (Respitrace, Non-Invasive Monitoring Systems Inc.), which
consists  of two electrical coils held round the chest and
abdomen by a body  stocking  or  bands.   The  instrument
continuously   records   the   electrical inductance of  the
coils,   which   it   is   claimed   are   proportional   to the
cross   sectional   area   of the body at those  levels,  and
following calibration, these signals can be scaled and  summed
to  give  an  estimate  of  lung  volume changes.  Our own
experience, (Guyatt,  McBride  and  Meanock, 1983), using  a
digital  system for calibration and calculation of volume
changes, suggests  that the device is less accurate than the
body plethysmograph  but  is better tolerated by the subjects.

    Thoracic ventilation measurements do not distinguish
between   airflow   through   the  mouth and nose.   Mouth

ventilation cannot be measured directly without a mouthpiece, but can be estimated by subtracting nasal ventilation from the thoracic signal. We have developed a new non-invasive method of measuring nasal airflow for this purpose, (Guyatt, Parker and McBride, 1982). This is based on our observation that if a nasal oxygen cannula is connected to a transducer, pressure swings of about 0.1 kPa can be observed during nose breathing, since the prongs which project up the nostrils, act as pitot tubes. With suitable calibration involving digital processing, this signal can be converted into a nasal airflow record.

A preliminary impression of these studies can be gained from the paper records made before digital processing. Two examples are shown in Fig 2. The first subject only took two puffs over a minute, (top two traces). There was little interruption in the breathing pattern, as indicated by the regularity of both the rib and abdominal signals, (traces 3 and 4), and the only interruption in nasal cannula pressure occurred about the time the puff was taken. By contrast, the second subject, a female who was smoking 50 or 60 cigarettes a day, showed an extreme modification of the respiratory pattern during smoking. Breathing had become much slower and deeper than normal, and she often took a puff on each breath. Examination of the repeated studies on the nine subjects showed that each subject had his own peculiar smoking pattern, which lay somewhere in between the two extremes shown, and which was remarkably consistent from day to day.

A more detailed picture of ventilation changes was obtained, after processing the individual signals, by using computer generated graphical displays. An example, Fig 3, shows the puff flow profile, (top), and RIP estimate of total ventilation and nasal air flow (centre and bottom, inspiration upwards), for a period of 15 seconds during which a puff was taken. The puff was taken during a slow expiration, then after another inspiration, the subject held his breath for about two seconds before breathing out again. We found, (in agreement with previous workers, Guillerm and Radziszewski, 1978; Higenbottam, Feyerabend and Clark, 1980; Tobin and Sackner, 1982 and Adams, Lee, Rawbone and Guz, 1983), that almost 80% of puffs were taken during, or at the end of expiration. Occasionally a puff would be taken in during a pause in inspiration, but this was then followed by a fairly large additional inspiration. We found on average in our nine subjects, that the expiration during which the puff was taken was of normal size but took twice as long as usual, while the

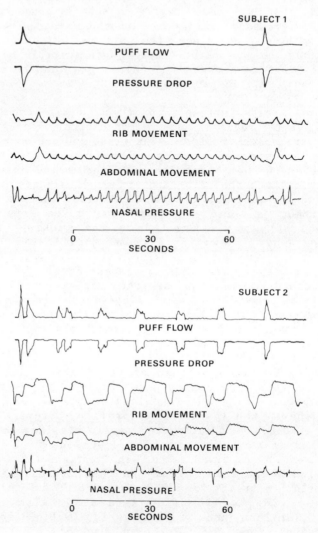

FIGURE 2.    Two paper records with raw data showing puff
             flow rates, draw pressure, rib and abdominal
             inductance and nasal cannula pressure.

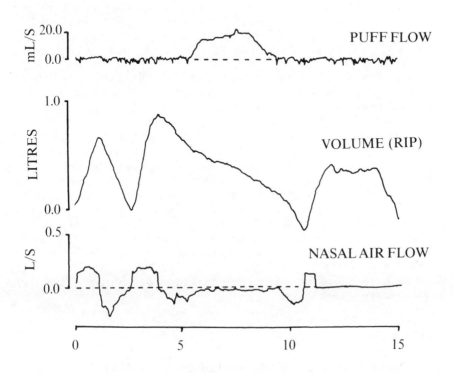

TIME (SECONDS)

FIGURE 3. Computer generated record of puff profile, with associated signals representing lung volume, recorded at the chest, and nasal air flow.

next breath was deeper but of normal duration. Occasionally subjects such as that shown in Fig 2, held their breath for several seconds after inspiration.

It is only possible to take a puff while breathing out if the buccal cavity is sealed off by closure of the oropharyngeal isthmus, expiration continuing through the nose alone. This can be seen in Fig 2, where the slow fall in lung volume occurring while the puff is being taken is matched by a flow out through the nose. In some cases, we observed that lung volume increased while the puff was being taken, and here the nasal flow was inwards. This use of the buccal cavity to generate the suction for puffing may result from the high airflow resistance of the tobacco rod, which is about 100 to 400 times that of the airways, so that it would be difficult to take a normal puff using the lungs while inspiring.

In the intervals between puffs most subjects breathed through their noses, but in the breath following the puff, nasal airflow became minimal or zero showing that the subject was breathing through his mouth alone. This behaviour would allow the bolus of smoke to be washed into the lungs from the buccal cavity, then later any residual smoke could be cleared out with a direct expiration through the mouth. This picture is incomplete however, since, in common with previous workers, we do not have any measurements relating to the waste smoking, that is the active expulsion or passive leaking of smoke from the mouth before inhalation.

**Smoke uptake**

The most important pharmacologically active constituent of tobacco smoke is nicotine. The simplest way of estimating the delivery to the mouth, is to measure the nicotine remaining in the cigarette butt after smoking and to estimate the efficency of the filter. This has been taken as an indication of the intensitity of smoking but as Schulz and Seehofer, (1978) have shown, it is very much influenced by the smoking pattern. The actual uptake of nicotine by the body has been estimated from changes in concentration in the blood. Blood sampling, as mentioned above, has the disadvantage of being unpleasant for the subject, so that we have explored the possibility of making measurements in saliva. This is very simple to perform but there remain problems of quantifying this data. Mucklow, (1982), has shown that the optimal transfer of drugs into saliva occurs when they are dissociated.

Unfortunately the pKa of nicotine is 7.9 so the final concentration is highly dependent on salivary pH, which in turn is affected by the food eaten at the previous meal.

Nicotine however is not a reliable marker of inhalation since it can be absorbed from the buccal cavity. By contrast carbon monoxide can only be taken up from the alveolar space, (Guyatt, Holmes and Cumming, 1981), so that a rise in alveolar CO concentration on smoking, (the carbon monoxide "boost") does indicate smoke inhalation. It is very difficult to quantitate this effect however since it also depends on the quantity of CO produced, (a function of the cigarette yield and manner of smoking), the amount of smoke lost before inhalation, and the body size. After inhalation, CO uptake will be affected by the breath-hold duration, and the amount washed out of the lungs by subsequent breaths. In addition there are various sources of inaccuracy in the alveolar gas method, and the boost produced by a single cigarette is close to the resolution of the technique. Therefore it is not surprising that we, in common with the previous workers mentioned above, did not find a close correlation between the measurements of CO boost and those of ventilation. This lack of agreement does emphasise the need for a reliable independent method of assessing inhalation that can be used to evaluate the ventilatory measurements.

## Effects of changes in nicotine yield

Many of the factors described above can be illustrated from data obtained from our study on the effects of smoking different types of cigarettes. We examined our nine established smokers on four separate occasions and each time asked them to smoke two identical cigarettes. The first session was used to familiarise them with the procedure. They then smoked three different types of cigarette, their usual brand, a low nicotine cigarette, LN, (nicotine, particulate and carbon monoxide yields per cigarette of 0.55, 8.1 and 10.3 mg), or a medium nicotine cigarette, MN, (yields 0.90, 7.8 and 9.8 mg respectively). The test order was randomised between subjects, and before the LN or MN sessions they were given a packet of the cigarettes to smoke in the 24 hours before the test. We used the techniques discribed above to monitor the smoking period and estimate nicotine and carbon monoxide uptake.

A summary of the some of the measurements is given in Table 1. The biggest effect on smoking behaviour of changing nicotine yield, was a significant increase in puff volume on

the LN cigarettes, associated with an increase in puff
duration. However there was considerable variation between
subjects in the size of the effect. Surprisingly there were no
significant trends in the average number of puffs taken with
the different cigarette types, though the total smoking time
was reduced with LN cigarettes. We also looked at the
ventilatory measurements, (not shown in this table), but could
find no significant differences between cigarette types in any
of the indices.

The butt nicotine and the saliva nicotine boost values,
also shown in Table 1, reflected the different nicotine
yields of the  LN and MN cigarettes, being lowest for LN.
This indicated that the delivery of nicotine to the mouth was
strongly influenced by the nicotine yield of the
cigarette smoked, but it was disappointing to find that the
individual butt nicotine and nicotine boost values were only
weakly correlated. This may indicate variations in the smoking
patterns between subjects which affect the two nicotine indices
in different ways, or be due to other factors such as the
saliva pH effect discribed above. The data on the carbon
monoxide boost showed no significant trends between cigarettes
and, as indicated above, very little relationship to changes in
the pattern of inhalation.

We also performed analyses of variance on these
measurements to   compare  the  between-subject  variability
(due  to  subject  differences), with the within-subject, (due
to differences between  cigarette types).  For most indices
there was a significant degree  of between-subject variability
suggesting  consistent  differences  between  individuals.
This agrees with earlier observations that  each subject has
his own particular  pattern of  smoking.   There  were  far
fewer  cases  of  significant  within-subject  variability,
showing that apart from a few indices such as puff volume and
the  nicotine  measurements,  changes  in these two cigarette
types had little  effect on the smoking pattern.

## Conclusions

The measurement of smoking behaviour is a complex
process.  Smokers seem to establish fixed patterns of puffing
and inhalation which are not greatly altered with the two
cigarettes.  There are still problems in estimating smoke
inhalation in the laboratory (as shown by the weak correlations
between ventilation and measurements of nicotine and carbon
monoxide uptake by the body during smoking), and more
investigations are required to produce reliable techniques.

**TABLE 1 SUMMARY OF SMOKING BEHAVIOUR, CARBON MONOXIDE AND NICOTINE MEASUREMENTS**

(Mean values for different types of cigarette, and significance levels for "t" tests).

| INDEX | OWN | LN | MN | Sig (paired "t" test) | | |
|---|---|---|---|---|---|---|
| | (1) | (2) | (3) | 1-2 | 1-3 | 2-3 |
| Number of puffs | 15.7 | 14.6 | 15.3 | - | - | - |
| Puff volume (ml) | 42.1 | 47.5 | 38.2 | * | - | - |
| Puff duration (s) | 2.14 | 2.16 | 1.89 | - | ** | ** |
| Draw pressure (kPa at 17.5 ml $s^{-1}$) | 1.41 | 1.16 | 1.26 | - | - | - |
| Inter puff interval (s) | 25.3 | 21.9 | 23.1 | - | - | - |
| Cigarette duration (s) | 352 | 285 | 316 | * | - | * |
| Initial $P_{CO}$ (ppm) | 35.6 | 33.7 | 29.0 | - | - | - |
| CO "boost" (ppm) | 4.89 | 1.89 | 3.89 | - | - | - |
| Initial saliva nicotine (ng $ml^{-1}$) | 250 | 182 | 203 | - | - | - |
| Salivary nicotine "boost" (ng $ml^{-1}$) | 1210 | 439 | 1006 | ** | - | ** |
| Butt nicotine (mg) | 1.50 | 1.23 | 1.94 | * | * | ** |

Significance levels refer to paired "t" test, 1-2 between own and LN, 1-3 own and MN, 2-3 LN and MN.

Significance Coding;- = P>0.05;* = 0.05>P>0.1; ** = P<0.01.

**REFERENCES**

ADAMS, L., Lee, C., Rawbone, R. & Guz, A. (1983) Patterns of smoking: measurement and variability in asymptomatic smokers. Clin Sci; 65, 383–392.

CREIGHTON, D.E., Noble, M.J. & Whewell, R.T. (1978) Instruments to measure, record and duplicate human smoking patterns. In "Smoking Behaviour", Ed Thornton, R.E. Churchill Livingston; Edinburgh, London, New York. pp 277–288.

GUILLERM, R. & Radziszewski, E. (1978) Analysis of smoking pattern including intake of carbon monoxide and influences of changes in cigarette design. In "Smoking Behaviour", Ed Thornton, R.E. Churchill Livingston; Edinburgh, London, New York. pp 361–370.

GUYATT, A.R., Holmes, M.A, & Cumming, G. (1981). Can carbon monoxide be absorbed from the upper respiratory tract in man? Eur J Respir Dis: 62: 383–390.

GUYATT, A.R., McBride, M.J. & Meanock, C.I. (1983) Evaluation of the respiratory inductive plethysmograph in man. Eur J Respir Dis; 64: 81–89.

GUYATT, A.R., Parker, S.P. & McBride, M.J. (1982) Measurement of human nasal ventilation using an oxygen cannula as a pitot tube. Am Rev Respir Dis; 126: 434–438.

HIGENBOTTAM, T., Feyerabend, C. & Clark, T.J.H. (1980) Cigarette smoke inhalation and the acute airway response. Thorax; 35: 246–254.

MUCKLOW, J.C. (1982) The use of saliva in therapeutic drug monitoring. Ther Drug Monit; 4: 229–247.

RAWBONE, R.G. (1981) The value of non–invasive methods for evaluating exposure to tobacco smoke. In "Smoking and arterial disease", Ed Greenhalgh, R.M. Pitman Medical, London, pp 64–73.

RAWBONE, R.G. (1984) The act of smoking. In this volume.

SCHULZ, W. & Seehofer, F. (1978) Smoking behaviour in Germany – the analysis of cigarette butts (KIPA). In "Smoking Behaviour", Ed Thornton, R.E. Churchill Livingston; Edinburgh, London, New York. pp 259–276.

TOBIN, M.J.  & Sackner, M.A.  (1982) Monitoring  smoking
    patterns of low and high tar cigarettes with inductive
    plethysmography.  Am Rev Respir Dis;  126:  258-264.

DISCUSSION

LECTURER: Guyatt                          CHAIRMAN: Cumming

RAWBONE:            I would like to comment on the use of the
                   cheeks in taking the puff. I recently examined
                   the mechanics of the puff by cine-radiography,
                   after first coating the buccal cavity with a
                   barium mixture.  Most of the puff stems from
                   a piston-like action of the tongue moving down
                   from the palate, the movement of the cheeks not
                   being very great. This study was in one
                   subject, myself, a non-smoker! A second
                   comment concerns the work done during puffing,
                   which appears  for any individual to be a
                   constant.  The puff volume resulting from
                   attempting to apply this fixed work may, in
                   practice, be limited  for certain products by
                   the size of the oral cavity. I am not
                   convinced of this but it might be worth looking
                   at.

                   You posed a question about the linearity of the
                   flow measurement from the cigarette holder –
                   when we used a laminar flow device we found a
                   linear relation to smoke up to 3 litres per
                   minute.  The error on puff volume from higher
                   flow rates was at most 4%. This has recently
                   been published in Clinical Science.

                   My question relates to the apparent exhalation
                   during the taking of the puff, does this relate
                   in any way to where the puff is taken relative
                   to F.R.C.  Could it be due to a passive
                   exhalation due to elastic recoil brought about
                   by taking the puff from above F.R.C?

GUYATT:            I am unable to say since our device has AC
                   coupling and changes in F.R.C. could not be
                   distinguished.

HIGENBOTTAM:       Your mixed expired CO results are in accordance
                   with some work by Minty and Royston who used
                   cigarettes with differing nicotine yields and
                   found the puzzling observation that the carbon
                   monoxide change was less with the lower
                   nicotine yield than with the greater.

Do you have any information on the tar/nicotine ratios, since we have argued that cigarettes with a low ratio may be more easily inhaled and the absolute value of the nicotine may not be a good guide.

GUYATT:      We have the particulate yield, water and nicotine free and the absolute nicotine yield and I might be able to let you have that information later.

HIGENBOTTAM:  Do you think it matters where the puff is taken in the breathing cycle?  The neonate takes in milk by sucking and continues to breathe quite independantly, does not a similar mechanism operate during cigarette puffing?

GUYATT:      I don't know whether it matters or not, but tried only to explain how it works rather than why, and was an attempt to put quantitative numbers to describe the process.

PRODI:       The timing of bolus intake within the respiratory cycle is important since it affects the residence time of the particles and thus dramatically affects the deposition efficiency. Similarly a long residence time in the mouth may determine the particulate content of saliva.

CUMMING:     The problem appears to be how much of the particulate is retained and what is the response of the organism to it.  Attempts to measure response have not been very successful so perhaps measurement of retention might help. Retention depends on residence time, distribution in volume, diffusivity and all the other relevant variables and is to the understanding of this process that close study of the act of smoking is directed, albeit somewhat crudely.

HIGENBOTTAM:  The reason for my blunt question was merely to elicit such a statement as you have just made. The point at which the puff is taken is not very important, save to ensure that it is ready in the mouth to be manipulated at will.

CUMMING:     We will draw the discussion to a close at that point.

# PROTEASES AND ANTIPROTEASES IN THE NORMAL HUMAN LOWER

# RESPIRATORY TRACT

R. Crystal

National Institutes of Health, Maryland, U.S.A.

## DISCUSSION

LECTURER: CRYSTAL                    CHAIRMAN: BONSIGNORE

DENISON:    You pointed out that molecule for molecule, there were many more anti elastase molecules than elastase molecules. These molecules must of course be in the right place, and some measurement of this efficiency might be their turnover. Can you tell us something about this?

CRYSTAL:    Alpha-1-Antitrypsin has a serum half life of about five days. It is difficult to measure the half life on an epithelial surface but there appears to be free access between blood and the alveolar surface. The concentration in the epithelial lining fluid is about 10% of that in the blood, which makes sense given that its molecular weight is about 50,000. Little is known about alpha 2 macroglobulin which is a larger molecule leaving the blood with difficulty but it aggregates around fibroblasts. Its turnover is unkown. However alpha-1-antitrypsin provides the bulk of the protection for the neutrophil elastase.

LAURENT:            Bronchial alveolar lavage is an important tool
                    in the investigation of disease, but looks only
                    at the lining fluid and gives no information
                    about cellular behaviour in the interstitium
                    and in particular says nothing about the
                    monocyte which contains at least two elastases.
                    Similarly alpha-2-macroglobulin is not found in
                    lavage fluid but may play an important role in
                    the interstitium.

CRYSTAL:            The best information available on alpha-2-
                    macroglobulin suggests that it is in the
                    interstitium, but only localised around the
                    fibroblasts.  Senior´s studies indicated that
                    macrophages did not produce elastase in non-
                    smokers, but did this in smokers.  This
                    elastase was a serine protease, which is
                    typical of a neutrophil elastase.  The alveolar
                    macrophage produces a metallo-enzyme, and
                    Senior showed that the macrophage has receptors
                    which recognise human neutrophil elastase,
                    probably containing galactosase.  This suggests
                    that the original study showed neutrophil
                    elastase that had been ingested by the
                    macrophage which was released in tissue
                    culture.  The non-smoker would release the
                    metallo-enzyme.  When the blood monocyte
                    differentiates into the macrophage it changes
                    its elastase type from a serine type to the
                    metallo-enzyme type.  In tissue culture this
                    occurs quickly, so the same probably occurs in
                    the interstitium and the metallo-enzyme is the
                    only one produces.  This is difficult to prove
                    since the interstitium is so difficult to
                    access.

JEFFEREY:           You have shown that the macrophage is not an
                    important producer of elastase in the normal
                    and the neutrophil is the principal cell
                    involved.   Normally the neutrophil is
                    sequestered, and is in the blood, so that we
                    tend to underestimate the protective action of
                    alpha-2-macroglobulin, which like the
                    neutrophil is sequestered in the blood.

CRYSTAL:            Alpha-2-macroglobulin is an important anti-
                    elastase in the blood, one can detect elastases
                    in the blood of non-smokers, but this is
                    combined with anti-proteases.

JEFFEREY:        Could you tell us about the partitioning of anti-elastase activity by alpha-2-macroglobulin and alpha-1-antitrypsin. What is the affinity of each and how permanent is the inhibition.

CRYSTAL:         Alpha-2-macroglobulin is like a horse shoe and when the elastase molecule enters the horse shoe it closes and the elastase is trapped. This is thought to be permanent but it is not a co-valent bond. However, the complex is taken up by the reticulo-endothelial system and disposed of. A more complex situation exists with alpha-1-antitrypsin. A methionine residue exists about 35 units from the terminus and this is the combining site for elastase - the combination constitutes a pseudo-irreversible reaction, the bond is not co-valent but is almost. A cleavage then occurs in the anti-protease leaving a moiety with a methionine terminus and a remainder of about 8000 daltons comes off. The part with the methionine terminus can be detected in various body fluids, including lavage fluid. The dissociation constant for the reaction in the reverse direction is so small that it can be considered irreversible, so that A.A.T. and proteases are thought to act in a molecule for molecule manner.

PRIDE:           Is A.A.T. involved in organs other than the lung, for instance in chronic renal infection?

CRYSTAL:         That is a very good question, and the best place to look for the answer is in patients with A.A.T. deficiency. One of the striking things is the lack of disease outside the lung, there is liver disease in children, but this may be because the A.A.T. cannot get out of the liver and there is almost no other disease. There is an association with pancreatitis, Weber Christian disease and a few other rarities. Either other parts of the body have good anti-elastases or neutrophil elastase cannot gain ready access to other parts of the body. The lung is an organ to which neutrophils can gain ready access.

PRIDE:              One example might be chronic pyelonephritis
                    where white cells are freely available.

CRYSTAL:            There is no clear evidence that the kidney
                    become involved in pyelonephritis. Clearly
                    there would be a leak of proteasy in this
                    condition, but the situation in the lung is
                    quite different as we shall see later.

# SMOKE INDUCED MODIFICATION OF ALVEOLAR MACROPHAGE ACTIVITY

M. Spadofora and R. Crystal

National Institutes of Health, Maryland

## DISCUSSION

LECTURER: Crystal and Spadofora          CHAIRMAN: Bonsignore

DENISON: My present understanding is that all oxygen which we consume is handled by enzymes which add an electron to convert it into the superoxide ion which is then used in a controlled way for many purposes. This controlled formation of superoxide is the whole basis of respiration. These oxidants have very short lives, and you now tell us that some cells such as macrophages and neutrophils are capable of producing these oxidants – are they doing this as part of their normal function or is this an expression of their disorganisation, and why is it that the body produces oxidants in this uncontrolled way.

SPADOFORA: The most important function of a phagocytic cell is to protect against bacteria and viruses. The increased production and release of toxic intermediates may be the way that the phagocytic cell protects against infection. Smoke can be considered as a non-physiological stimulus which causes the phagocytic cell to release these oxidants.

CRYSTAL:            Most of the oxygen we breathe is metabolised
                    without production of these toxic oxygen
                    radicals. Within the cell mitochondria oxygen
                    is metabolised by the cytochrome system through
                    the medium of A.T.P. directly to water.

DENISON:            May I interrupt at that point, my understanding
                    is slightly different, in that superoxide is
                    produced but only under controlled
                    circumstances and in such a way as to prevent
                    its escape. Further I understand that the
                    whole of our oxygen undergoes this process –
                    oxygen itself biochemically very inactive and
                    the first process to permit it to enter the
                    cytochrome system is to add an electron and
                    thus calls for a great deal of energy so that
                    cellular respiration can be considered to be
                    the controlled production of superoxide ions
                    where they can be put to immediate use. The
                    purpose of my question was to establish whether
                    this wild superoxide had just escaped or had
                    been deliberately thrown wide.

CRYSTAL:            The major purpose of the release by phagocytic
                    cells of oxidants is the destruction of
                    bacterial invaders. The concept of respiratory
                    burst outlined by Spadofora, was worked out by
                    studying what happened when a macrophage
                    engulfs a bacterium, in a sense it is a noble
                    purpose but some of the oxidants get free, so
                    that the phagocyte is a messy eater. Whilst
                    the system was evolved to destroy bacteria,
                    perhaps the design did not take into account
                    that we would one day smoke cigarettes or that
                    we would inhale asbestos or silica and when a
                    macrophage ingests any particulate superoxide
                    is produced, as $H_2O_2$ which can be measured. A
                    disease of childhood manifest in repeated
                    infections, results from the inability of the
                    phagocytes to produce superoxides.

CUMMING:            I would like to test this fascinating
                    hypothesis using two simple biological
                    principles. The first relates to the
                    protective function of the phagocytes. One can
                    see that the inhalation of toxic particles will
                    produce deposition on epithelial surface with
                    subsequent damage to the interstitium. To
                    prevent this the particles are ingested, but
                    despite the intervention destruction proceeds

so that the action of the phagocytes to prevent destruction is negated.  Secondly the action of ingestion is somehow imprinted in the memory of the phagocytes, which can subsequently go on to produce superoxides from the cytochrome system. My understanding is that since an enzyme is involved it must have a genetic origin, is it implied that a genetic imprint is received by the phagocyte so that it can demonstrate this unusual behaviour?

CRYSTAL:     It is a very complex question and I am not sure I can help you very much.  On a morphologic base, macrophages from smokers have a variety of inclusions including particulate.  The macrophage has a life which may extend for several years so that even on stopping smoking, the macrophage may continue in its active state which may explain why some smokers show continued deterioration after stopping smoking. The phagocyte stimulus probably remains in the macrophage for a long time.  What is not clear is whether the cells produce oxidants continually, or only intermittently. Macrophages in tissue culture also produce oxidants, so it is not a simple system and is a very difficult question to answer.  The phagocyte does have anti-oxidants in it at high levels but the oxidants and anti-oxidants are compartmentalksed anatomicaly within the cell.

FLETCHER:     One of the striking thing about emphysema is that it effects only a minority of smokers, and if the process you have described is a true representation of events then only a minority of smokers should have their macrophages activated by tobacco particulate, about 25% of heavy smokers and 10% of light smokers.  Is that the case?  Also, have you studied young smokers who have stopped smoking, and how soon do their macrophages become normal?

CRYSTAL:     Your question is a very good one, and I would imagine if we studied the production of oxidants from an equal number of macrophages we would find a distribution of amount produced, so the 25% would be those with a large production of oxidants.  Also the production of anti-oxidants would be relevant and stems from a question asked by Professor Cumming concerning the genetic basis of the development of disease and there is probably such a basis both for the amount of oxidants and the amount of anti-oxidants as well as of proteases and anti-proteases. Thus a small section of the population is at risk of the development of disease and this is a very fruitful area for future research and the tools are available to give the answers to those questions.

LAURENT: Was the viability of cells comparable between smokers and non-smokers, because if some cells died in the chromium experiment they would release many toxic products.

SPADOFORA: The viability of macrophages was equal for the two groups.

ANTHONY: What is the role of lymphocytes in this system? In the mouse mutation is associated with resistance to bleomycin-induced pulmonary fibrosis, and I wondered what part this system might play in the human.

CRYSTAL: The analogy is with the nude mouse which does not have functioning T-lymphocytes. There is varying data regarding the susceptibility of nude mice to bleomycin, so it is probably multi-factorial. Lymphocyte function can be modulated by proteases and anti-proteases as well as by oxidants and anti-oxidants. How that relates to emphysema is completely unkown. In the human system T-lymphocytes and B-lymphocytes play no role in the development in disease so far has been ascertained.

HIGENBOTTAM: In considering the role of oxidants and anti-oxidants the great danger is that we may be looking at an epiphenomenon. Tobacco smoke contains several active oxidants in the gas phase which are self generating and of short half-life. Do you get a similar effect from otherwise inert inorganic ducts?

CRYSTAL: The answer is yes. It is not an epiphenomenon but rather the common denominator of these diseases. The oxidants in the gas phase are clearly important and I will discuss that later in connection with proteases. The mechanisms of injury are similar in many diseases so we are looking for the common denominator.

# TOBACCO SMOKE AND ALVEOLAR CELL POPULATIONS

J. Chretien and C. Danel *

* Clinique de Pneumo-Phthisiologie and INSERM U. 214
Hopital Laennec 42, rue de Sevres 75007 Paris, France

The selective study of alveolar cell populations is justified by the physical characteristics of tobacco smoke and by the nature of the disorders observed in the clinic. Before developing this point several preliminary remarks concerning the deep lungs have to be made: What is the exact definition of the deep lung? If one gives an animal, for instance a rat, an aerosol with a radio-active labelled element and records the decrease of radio-activity, one obtains in semi-logarithmic co-ordinates a curve with 3 compartments: a rapid compartment which corresponds to the tracheo-bronchial clearance, an intermediate semi-slow one and a slow compartment (Fig.1). This curve is reduced to two compartments in man (Fig.1). The deep lung corresponds to this slow clearance compartment which involves a retention of particles for a long time. This is, in fact a physiological notion expressing clearance. This notion includes not only the structures for gas-exchange but also the terminal conductive air-ways. In other words there is a bronchio-alveolar unit whose cells cannot be functionally separated in terms of clearance.

The cell population which will be discussed is mainly situated in this area. This is a target area for tobacco smoke because of the conditions of clearance. This clearance depends on the physical characteristics of the particulate-phase of tobacco smoke.

## PHYSICAL AND CHEMICAL CHARACTERISTICS OF TOBACCO SMOKE

Tobacco smoke is a complex aerosol with a gas phase

137

## Lung of Rodent - Cat

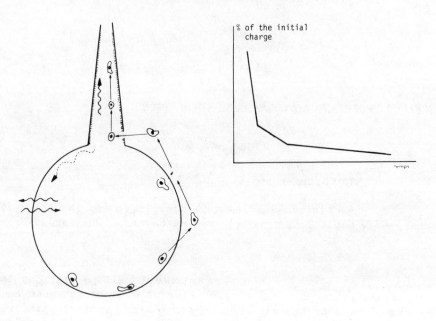

## Lung of Baboon - Man

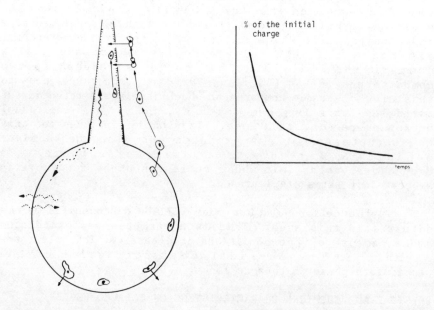

**FIGURE 1:**    Tracheobronchial clearance in different species and
in man.

and a particulate phase.    The physical clearance of these 2 phases is different.

The penetration of the particulate phase depends on several physical characteristics of particles.   The most important of which is the Diameter and more precisely the Mass-aerodynamic diameter.   For tobacco aerosol which is a multi-disperse aerosol, the size of particles is distributed between 0.1 um and 1 um. Consequently it is stable in air and the particles deposit mainly by diffusion and partly by sedimentation in the deep lung, according to the models on lung aerosol-deposition (1,2,3).

Other physical conditions may play a role such as configuration, hygrometry, ionic charge, radioactivity, but these are not as important as aerodynamic diameter.

Furthermore there is a synergy between gas phase and particulate phase.   This is important because some toxic components from the gas phase such as $SO_2$ can only penetrate into the deep lung after their adsorption on the surface of inhaled particles.

The chemical composition of the tobacco aerosol is complex, and more than 2000 components have been found (4,5). Some of them are modified by physical conditions  for example humidity.    These secondary pollutants are not always present in analysed tobacco smoke when it enters the airways.   Some others called tertiary pollutants may gain their toxicity only after having been metabolised in certain cells.   Thus, it may be difficult to evaluate with precision the nature and concentration of toxic components of tobacco smoke at the bronchio-alveolar level.

On the other hand the condition of ventilation may modify the duration of contact and the retention of toxic agents.

One can see for example using a derivation of the Landahl model that for the same aerodynamic diameter of particle in a given aerosol (0.1 m), if the time of breath-hold after inhalation increases from 1 sec. to 15 sec. nearly all the particles remain in the alveoli (Fig. 2) (6,7).

Individual factors may also modulate the toxicity of tobacco smoke on deep lung such as genetic factors or the pathological history of the individual.  For instance, in animal experiments using radioactive labelled aerosol the clearance from the deep lung is different as between germ-free rats and conventional animals (2).

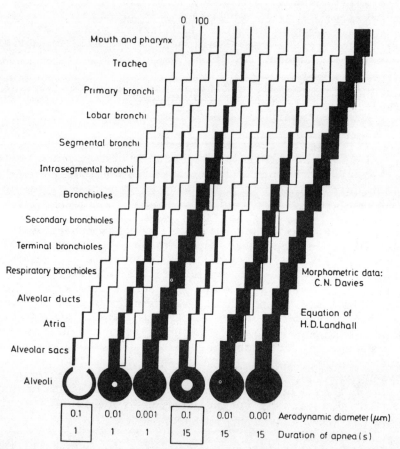

**FIGURE 2:**    Particle deposition according to aerodynamic diameter and duration of apnea.

Finally there are difficulties in the analysis of the toxicity of the tobacco smoke on the structures of the deep lung and in particular on the alveolar cells.

## METHODS OF STUDY OF CELLULAR LESIONS

A variety of techniques have allowed an approach to the cellular lesions in the bronchio-alveolar area. For in vitro studies, numerous experimental models have been used. It is uncertain that changes observed in animals would be the same in man. With all in vivo studies, the bronchoalveolar lavage (B.A.L.) is the one which allows a direct approach to the cellular changes caused by repeated inhalation of tobacco smoke in the deep lung (7). With B.A.L. one can collect and conserve the mobile cells of the deep lung. Furthermore one can study the biological environment of these cells (eg. surfactant, immunoglobulins secretion or enzymatic activities).

However, in human smokers it is difficult to have a direct observation of the behaviour of the fixed bronchoalveolar cells, such as Clara cells (C.C.) and pneumocytes type I (PI) and type II (PII) (8).

Indirect information can be obtained from the biochemical study of certain enzymes, whose secretions from specific cells may be known or supposed. Certain in vitro techniques can also be used for these indirect informations, such as cellular cultures and explants (in order to test the functional characteristics of cells) or such as the perfused lung model which leads to a better dynamic approach by sequential biochemical dosages of the toxic effects. The majority of these findings will be related or have been described in other papers, particularly in relation to protease-antiprotease balance.

## CELLULAR LESIONS – MORPHOLOGICAL AND FUNCTIONAL DATA

For direct alteration or modification of alveolar cells due to chronic inhalation of tobacco smoke, two groups among the alveolar cells have to be separated (5, 9, 10, 11).

1) The first group concerns the cells derived from blood and provisionally resident in the lung. Some of them are normally present in physiological conditions in the alveoli (such as alveolar macrophages) but some appear only in pathological conditions (lymphocytes, neutrophiles, eosinophils and basophils).

2) The second group concerns the fixed pulmonary cells, whose behaviour is less known in smokers.

## 1. MOBILE CELLS

For the provisionally resident cells derived from blood, B.A.L. helps to separate 2 groups of subjects, according to the chronic inhalation of tobacco smoke.

As shown in Fig. 3 the gross appearance of the fluid recovered by B.A.L. from asymptomatic smokers is outstandingly different from non smokers (5, 11, 12, 13): In non smokers the fluid is clear, with a serous appearance. The fluid from asymptomatic smokers is cloudy and grey-brown in colour. After centrifugation this pigment is mainly found at the level of macrophages (Fig. 4). The cellularity of the fluids recovered from asymptomatic smokers shows quantitative changes in the principal cellular groups. As shown in Table 1 the total number of cells is multiplied at least by a factor x 2. This increase is not due to the lymphocytes whose number is normal or moderately higher. In the same way the PMN leucocytes number is identical. Finally the increase of cell numbers is due to the alveolar macrophages (AM) whose number is multiplied by a factor 2 to 5 according to the series published.

The reason for this increase has been discussed. There is either a rise of blood monocytes which might migrate into the alveoli and differentiate into AM, or a local proliferation in the interstitium of monocytes under the influence of macrophage secreting factors. In vitro, exposure to oxidative factors (among other coming from nitrogen oxydes) contained in tobacco smoke, can provoke the release of these secreting factors. Another interpretation suggests a decrease of the local adherence of AM in smokers. In fact, the increase in AM observed in smokers could be due to the reduction of the mucociliary clearance of the upper respiratory tract.

In addition, chronic exposure to tobacco smoke, modifies AM morphologically and functionally (5, 11, 13, 14, 15, 16, 17, 18, 19, 20, 21, 22).

As it is known AM are not free in the alveolar lumen but enclosed within the tensio-active film, - the surfactant which contributes as environmental factor to some of AM functional properties.

Although debated, it seems that tobacco smoke reduces the rate of production of lecithins and thereby modifies tensio-activity. Its toxic agents disturbs the metabolism of AM.

**FIGURE 3:** Gross appearance of BAL fluid – on the left a smoker, on the right a non-smoker.

The exposure of tobacco smoke modifies the morphology of AM. This is observed in experimental animals as well as in cells recovered by B.A.L. in man.

On light and electron microscopy, AM in smokers have an increasing size compared to non smokers. Some of the AM are multinucleated. They may contain abnormal mitochondriae, suggesting alteration in cellular respiration. We have confirmed these alterations in animals with measurements by a Gilson oxymeter. Compared to non smokers the most striking features are the numerous dark intracytoplasmic inclusion bodies (Fig.4) seen on light or EM microscopy. (Fig. 4, 5, 6). In fact these inclusions are also found in AM from non smokers but less frequently (2% versus 65% in smokers). These inclusions correspond to the phagosomes and can be seen also in animals constantly exposed to smoke. These lipid-like vesicles may be a result of either altered cellular metabolism, or of phagocytosis. They are also due to an accumulation of particles present in cigarette smoke. Normal AM from non smokers were mixed with extracts of cigarette tars and lysed AM from smokers. After a shortwhile, these normal AM appeared similar to AM lavaged from smokers. This suggests the possibility of a transfer of pigments from one macrophage to another. These phagolysosomes contain a complex substance (23) of lipoprotein origin as shown by combined Giemsa and Soudan stains with a self-fluorescence phenomenon. This suggests that AM from smokers contain lipofuscine, which is a lipoprotein pigment, resulting from oxydative degradation. These properties are not shown in AM from non smokers. They also include polycyclic hydrocarbon and inorganic compounds. Kaolin particles have also been identified (24). These are in relation with certain components of tobacco mixture and are cytotoxic for phagocytes.

In addition to their intracytoplasmic modifications, some changes occur on the surface of AM in smokers. There is a reduction in membrame ruffles and folds, the surface is smooth, which contrasts with the highly undulated surface membrane of the non smoker AM (25). This effectively reduces the surface/volume ratio.

In addition, there is a reversible defect in the adherence of AM from smokers, on a nylon fibre column and an increase of the adherence over a glass surface. On these glass surfaces, AM might show the membrane characteristics of activated macrophages. The presence of serum factors reduces the speed of adherence of these cells to the glass surface in AM from smokers. On the other hand the supernatant of lavage

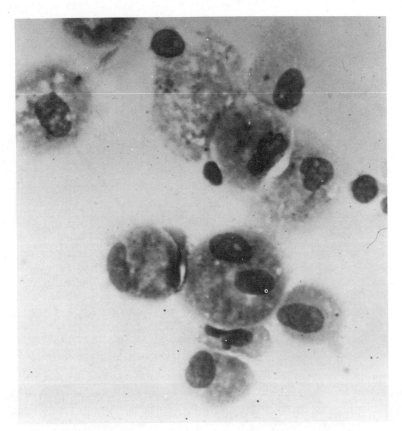

**FIGURE 4:** AM from smoker (H & E x 3000)

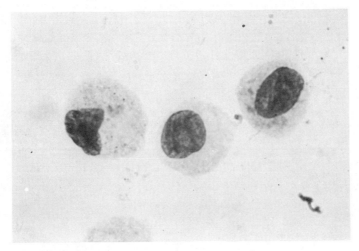

**FIGURE 5:** AM from non-smoker (H & E x 3000)

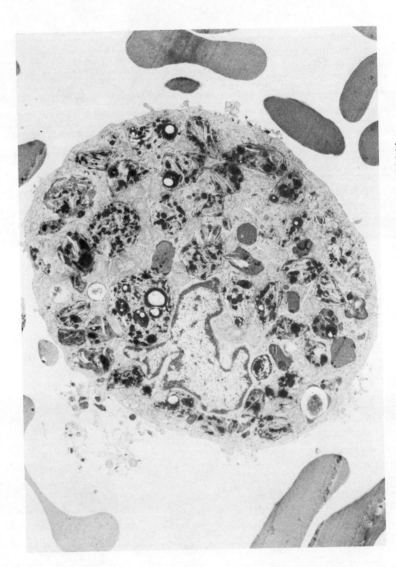

FIGURE 6:   AM from smoker (E.M. x 15000)

fluid increases the spreading of the smokers AM.

Further evidence of membrane changes is given by their enhanced agglutinability in the presence of concanavalin A after the saturation of membrane receptors by lectins. The receptor system of cyclic GMP (Guanosine monophosphatase) and cyclic AMP (Adenosine monophosphatase) regulates membrane movement. Its modification under the influence of tobacco smoke gives rise to changes of macrophage spreading to the chemotactic agents. The spontaneous mobility of smokers cells is higher than in non smokers. The mobility of non smokers macrophages in response to chemotactic factors is either identical to or higher than, in smokers.

Furthermore the in vitro action of migration inhibiting factors (MIF) on macrophages obtained by B.A.L. is neutralised. This could be due to the alteration and destruction of membrane receptors to MIF. Such alterations can be experimentally produced by the administration of oxydants ($NO_2$ or oxygen reactive radicals). Another hypothesis is that the AM membrane is in a refractory period.

All these facts concerning morphology suggest important changes in macrophage functions in smokers. These functional modifications can be regrouped under 3 headings (8, 14, 16, 20, 21, 25, 26).

(a)  Clearance functions based on pinocytosis, phagocytosis and bactericidal activity. (summarized in Table II).

(b)  Secretory functions related to enzymatic activity (summarized in Table III).

(c)  Immune and inflammatory responses (summarized in Table IV).

(a) A decrease of pinocytosis has been reported in smokers. This decrease affects the clearance of toxic substances and their antigenic presentation to lymphocytes. The decrease of phagocytosis and neutralisation of germs are more important for pulmonary defense mechanisms. Phagocytosis of microorganisms requires an initial step of opsonization by immunoglobulins and the complement system. This opsonization is followed by the adhesion of germs on receptors displayed on the cell surface of the AM (receptors for the Fc of IgG 1 and 3 and C 3b receptors). The density of Fc receptors on smokers AM is normal whereas there is a decrease of C3 b receptors. Furthermore, the smokers AM are still capable of phagocytosing microorganisms but these microorganisms stay for a longer period in the phagocytic vesicles. The number of ingested

**TABLE 1**

Cells Recovered by BAL in Smokers and Non-Smokers

| | TCN $10^6$ | Ma $10^6$ | % | Ly $10^6$ | % | PMN $10^6$ | % |
|---|---|---|---|---|---|---|---|
| Group S (n = 13) | 50.9 (34.2) | 49.3 (34.2) | 96.1 (4.2) | 1.3 (1.8) | 3.1 (4.4) | 32 (48) | 0.7 (0.8) |
| Group NS (n = 10) | 27.7 (14.4) | 25.3 (13.9) | 91.4 (5.1) | 1.9 (1.3) | 7.4 (4.2) | 30 (30) | 1.2 (0.9) |
| p | < 0.05 | < 0.05 | < C.05 | NS | < 0.05 | NS | NS |

TCN = Total cell number; Ma = Alveolar macrophages; Ly = Lymphocytes; PMN = Polymorphonuclears. Means and (standard deviations).

**TABLE 2**

A - CLEARANCE OF FOREIGN SUBSTANCES

Pinocytosis                                    ↓

Surface receptors    Fc    IgG          N

                          $C_3b$              ↓

Phagocytosis                              N or ↓

Killing of microorganisms       N or ↓

bacteria in vivo is smaller in smokers, but it is compensated for by the higher number of AM. So the total number of ingested bacteria is identical in the 2 groups.

Furthermore some primitive assertions concerning phagocytosis by AM from smokers were assessed on the basis of in vitro models. In such protocols, the conditions and details of experiments should be carefully specified.

(b) The secretory functions of AM are characterised by multiple enzymatic and biochemical activities (5,11,13,14,17,20,27,28,29,30,31). To explore these activities, protocols of experiments performed on cultured macrophages must be carefully specified. As for phagocytosis it is necessary to compare cultured AM from smokers to cultured AM from control. Both cultures must be of the same age, because the cultured macrophages show progressive metabolic abnormalities. These different enzymatic activities reported to be modified in smokers are summarized in Table III.

If the neutral proteases are normal, the concentration of acid phosphatases in lysosomes is linked to the quantity of pigment contained in AM. The secretion of lysosomal enzymes is increased by a factor x 5 in AM from smokers. Tobacco smoke exposure has also been shown to increase the synthesis of the aryl hydrocarbon-hydroxylase which is an enzyme involved in the metabolism of carcinogenic hydrocarbons. These facts were recently debated and enzymes from the cytochrome P 450 system have not been found in AM of normal smoker or non smoker subjects. The preliminary results obtained could be due to contaminations by Pn II or CC. Superoxide anions production could be increased for the elastasic activity. This also gave rise to controversial publications and one must avoid the extrapolation from animal models or from peritoneal macrophages to human AM.

Fibronectin, a protein of high molecular density which is synthetized by AM, whose importance is now well known in pulmonary structures and fibrosis, has a higher concentration in B.A.L. from smokers.

Finally, the different membrane modifications and secretory changes of AM from smokers, interact in the immune and inflammatory responses.

(c) Immune and inflammatory responses (Table IV).
Antigen presentation: After its uptake, the antigen is displayed on the macrophage surface and is closely related to HLA-DR antigens. This structure (Ag-HLA.DR) is recognized by

## TABLE 3

### B - SECRETION

| | |
|---|---|
| Lysosomial acid phosphatases | ↑ |
| Lysozyme | ↑ |
| Aryl hydrocarbure hydroxylase | ↑ |
| Superoxyde Anion | ↑ |
| Elastase | ↑ |
| Neutral proteases | N |

## TABLE 4

### C - IMMUNOLOGY

| | |
|---|---|
| . Antigen presentation | ↓ |
| . Secretion | |
| IL 1 | ? |
| Neutrophil chemotactic factor | ↑ |
| Fibronectin | N |
| Fibroblast growth factor | ? |
| Response to : | |
| Chemotactic factors | N |
| Migration inhibitory factor | ↓ |
| Macrophage activating factor | ? |

specific-antigen lymphocytes. These come into close contact with the AM prior to their activation. This phenomenon would be decreased in smokers, whereas the percentage of HLA-DR positive macrophages remains unchanged.

For interleukin 1 secretion, it is still unknown whether smoking impairs its secretions.

Neutrophil chemotactic. This factor which attracts PMN is increased. If fibronectin is increased in th B.A.L. of smoker the fibronectin production from alveolar macrophages would be unchanged in smokers in inflammation.

For fibroblast growth factor further work is needed.

Reciprocally AM are influenced by soluble mediators released by lymphocytes. In smokers, AM response to monocyte chemotactic factor seems to be normal. As has been seen above, AM from smokers is less sensitive to the MIF. To our knowledge concerning macrophages activating factor (M.A.F.) in smokers, no data are available.

The other cells of the deep lung, which are also derived from blood, have been investigated to a lesser degree in smokers (11, 13, 14, 15, 16).

If the percentage of lymphocytes in the B.A.L. from smokers is decreased, the total number is identical to non smokers. The number of T and B lymphocytes in B.A.L. (60% or T cells and 6% of B cells) is almost identical in the two groups, but there are some functional differences: some of them are due to transitory modifications of the membrane receptors. The response to mitogenic factors such as concanavalin A or phytohemaglutinin is decreased. This transitory reaction disappears 6 weeks after tobacco smoke intoxication has been stopped. Indirect information is given on B lymphocytes by the dosage of immunoglobulins in B.A.L., summarised in Table V.

For the PMN, acute exposure to tobacco smoke provokes a rapid increase of their number in animals. In human smokers and in animals which have had chronic exposure, the rate of these cells is either identical, or slightly higher than normal non smokers. Nevertheless long exposure to tobacco smoke could provoke a permanent increase of PMN.

The role of tobacco smoke on the margination and secretion of PMN in blood pulmonary circulation, is to our knowledge not seen in smokers. Alterations can be seen in PMN by electron microscope such as loss of filopodia, rounding

TABLE 5

| BIOCHEMICAL FINDINGS IN BRONCHOALVEOLAR LAVAGE OF NORMAL SMOKERS |
|---|

IMMUNOGLOBULINS

| | |
|---|---|
| IgG | N or ↑ |
| IgA | N |
| IgM | Not detectable |

COMPLEMENT                     N

Secretory                      ↓

Component

PROTEASES

| | |
|---|---|
| Elastase | Not detectable |
| Collagenase | Not detectable |

ANTIPROTEASES

| | |
|---|---|
| 2 Macroglobulin | N |
| 1 Antiprotease | N |

of cells and vesiculation of cytoplasma. They probably have a functional signification among others in phagocytic capabilities.

Apparently there is no difference in elastase activity in the supernatant of B.A.L. between smokers and non smokers, but in vitro the tobacco smoke induces a release of leucocytic elastase and other intracellular enzymes. This shows a direct toxic effect which is not physiological.

In smokers, the AM attracts the PMN with chemotactic factors. This secretion appears to be secondary to phagocytosis. However in experimental models the exposure of guinea pigs to cigarette smoke does not involve the PMN, (also in our own protocol for rats).

In summary for the immune competent cells epidemiological studies show that asymptomatic smokers have more respiratory infections than non smokers. If tobacco smoke plays an obvious role in the modifications of immune defences in the deep lung, the intermediate mechanisms are complex and far from the schematic views previously asserted.

## 2. FIXED CELLS

The effect of tobacco smoke, on fixed cells in the bronchioalveolar unit such as Pn 1 and II or CC, are much less known in man and less explored in animal experiments.

For instance, for CC, tobacco smoke effects are suspected because of its action on an enzymatic system. This system concerns the clearance of hydrophobic molecules, particularly the monooxygenases system of Cytochrome P 450 (10). Among the numerous types of cells in lung, only CC and, Pn II at a less degree, play a major role in this enzymatic system, besides the bronchial cells. Tobacco smoke may induce an increase of monooxygenase activities.

The effects of tobacco smoke on Pn II are known by indirect ways to be linked with the main functions of these cells, recalled in Table VI (8). Experimentally for surfactant, its volume is decreased in animals which have been exposed to cigarette smoke. Quantitatively nevertheless, its composition has been discussed and its rate in phospholipids does not constantly decrease.

Alveolo-capillary membranes permeability has been found to have increased in smokers but the cell junctions have not yet been studied, as they were for bronchial cells.

TABLE  6

FUNCTIONS OF TYPE II PNEUMOCYTES

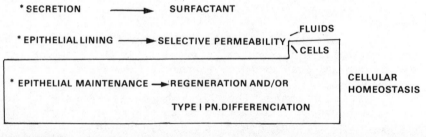

In using aerosol of technetium DTPA solution, with an aerodynamic mean diameter of 0.7 um, initially polydisperse, then after sedimentation, monodisperse. The speed of transfer through the alveolo-capillary membrane has been tested by gamma camera (32). The permeability increases in smokers compared to non smokers (9). This then rapidly returns to normal values when the subject stops smoking. The direct role of Pn I and II and the role of cell junctions in the alveolar surface and bronchiolar epithelium has been implicated in this permeability increase, which may be due to intermediate of superoxydes radicals. It probably plays a role with other factors such as vascular vaso-dilatation, histamin-secretion increase, prostaglandin liberation and surfactant modification. These factors where successively implicated in these permeability changes.

All the foregoing have shown the known direct and indirect actions of tobacco smoke on alveolar cells. Further investigation should be performed particularly concerning Pneumocytes I and II, clara cells, septal cells and endothelial cells which have never been so far performed in man.

## REFERENCES

1.    CHRETIEN, J., Metivier, H., Nolibe, D., Lafuma, J., Masse, R. (1974).
      "Methodes d'etude des passages particulaires de l'alveole a l'interstitium pulmonaire et devenir des particules".
      In: Colloques INSERM, vol. 29. Reactions broncho-pulmonaires aux polluants atmospheriques. 1 vol. INSERM. Paris. 133-162.

2.    CHRETIEN, J., Marsac, J., Huchon, G. (1980).
      Mechanismes de defense de l'arbre aerien.
      Med et Hyg., 38: 1458-1473.

3.    CHRETIEN, J. (1980)
      Penetration et devenir des aero-contaminants inhales par le poumon.
      In: Allergologie, 1: 308-323. J. Charpin Edit. Flammarion Pub. Paris.

4.     CHRETIEN, J., Hirsch, A., Thifalemont, M. (1973).
       Pathologie respiratoire du tabac. L'experimentation
       animale dans le monde.
       Objectifs et methodologie, 1: MASSON Pub. Paris.

5.     CHRETIEN, J. (1981).
       La fumee de tabac. Principaux mecanismes de toxicite.
       Bull. Europ. Physiopath. resp. 17: 135-144.

6.     CHRETIEN, J. (1982).
       "Inhalation carcinogenesis: an overview".
       Recents results in Cancer Research, 82: 1-10.

7.     BIZERTE, G., J. Chretien, C. Voisin. (1979).
       Bronchio-alveolar lavage in Man.
       1 vol. INSERM Pub. Edit. Paris.

8.     CHRETIEN, J., Jaubert, F., Basset, F. (1983).
       Type II pneumocyte injury in human respiratory diseases.
       In: The cells of the alveolar unit.
       Edit. G. Favez, A. Junod, P. Leuen Berger. 1 vol. Hans
       Huber. Pub. Bern. pp. 56-72.

9.     MASSE, R., Fritsch. P., Chretien, J. (1981).
       Renouvellement des cellules des bronches et du poumon.
       Analyse des dossiers recents.
       Rev. Fr. Mal. Resp. 9: 85-112.

10.    GAIL, D.B., Lenfant, C. (1983).
       Cells of the lung: Biology and clinical implications.
       Am. Rev. Respir. Dis. 9: 366-387.

11.    WARR, G.A. (1979).
       The biology of normal human bronchoalveolar cells.
       In: Le lavage broncho-alveolaire chez l'homme.
       1 vol. INSERM, Paris, pp. 137-158.
       G. Bizerte, J. Chretien, C. Voisin, Edit. 544 p.

12.    HUNNINGHAKE, G.W., Gadek, J.E., Kawanami, O., Ferrans,
       V.J., Crystal, R.G. (1979).
       "Inflammatory and immune processes in the human lung in
       health and disease: Evaluation by bronchoalveolar
       lavage".
       Am. J. Pathol. 97: 149-206.

13.    REYNOLDS, H.Y., Chretien, J. (1984).
       Respiratory tract fluids. Analysis of content and

contemporary use in understanding lung diseases.
In: Disease-a-Month. Ed. N.J. Cotsonas, Jr., Year Book
Medical Publishers, INC. Chicago, 30: 1-103.

14. GREEN, G., Sakab, G.J., Low, R.B., Davis, G.S. (1976-
    1977).
    Defense mechanisms of respiratory membrane.
    In: lung disease. State of the art. 245.
    1 vol. New York, Editor, Murray, J.J., Publ. American
    Lung Association 1978..

15. DANIELE, R.P., Dauber, J.H., Altose, M.D., Rowlands, D.T.,
    Gorenberg, D.J. (1977).
    "Lymphocyte studies in asymptomatic cigarette smokers. A
    comparison between lung and peripheral blood".
    Am. Rev. Respir. Dis., 116: 997-1005.

16. DANIELE, R.P., Dauber, J.H., Rossman, M.D. (1979).
    Lymphocyte populations in the bronchoalveolar air
    spaces: Recent observations in normal and asymptomatic
    smokers".
    Bull. Europ. Physiopath. Resp. 15: 26-27.

17. VENET, A., Clavel, F., Chretien, J. (1983).
    Aspects immunologiques de l'inflammation pulmonaire.
    In: Pneumopathies aigues severes de l'adulte.  J.J.
    Pocidalo et F. Vachon Edit. Librairie Arnette, Paris,
    pp. 13-34.

18. CASIO, M.G., Hale, K.A., Niewoehner, D.E. (1982).
    Morphologic and morphometric effect of prolonged
    cigarette smoking on the small airways.
    Am. Rev. Respir. Dis. 122: 265-271.

19. HILLER, F.C., Mc Cusker, K.T., Mazumder, M.K., Wilson, D.,
    Bone, R.C. (1982).
    Deposition of sidestream cigarette smoke in the human
    respiratory tract.
    Am. Rev. Respir. Dis. 125: 406-408.

20. HUBERT, G.L., Dautes, P., Zwilling, G.R., Pochan, V.E.,
    Hinds, W.C., Nicholas, H.A., Mahasan, V.K., Hayashi, M.,
    First. M.W. (1981).
    A morphologic and physiologic bioassay for quantifying
    alterations in the lung following experimental chronic
    inhalations of tobacco smoke.
    Bull. Eur. Physiopathol. Resp. 17: 270-327.

21.  ROTH, C., Arnoux, A., Huchon, G., Lacronique, J., Marsac,
     J., Chretien, J. (1981).
     Effet du tabagisme sur les cellules broncho-alveolaires
     chez l'homme.
     Bull. Eur. Physiopathol. Resp. 17: 767-773.

22.  RYLANDER, R. (1974).
     Pulmonary cell response to inhaled cigarette smoke.
     Arch. Environ. Health. 29: 4-329.

23.  CHRETIEN, J., Thieldlemont, M., Masse, R., Lebas, F.X.
     (1975).
     Action de la fumee detabac sur le macrophage alveolaire.
     Quelques donnees recentes.
     Nouv. Presse. Med. 4: 2327-2331.

24.  BRODY, A.R., Craighead, J.E. (1975).
     Cytoplasmic inclusions in pulmonary macrophages of
     cigarette smokers.
     Lab. Invest. 32: 125-132.

25.  WARR, G.A., Martin, R.R., Gentry, L.O. (1976).
     Alteration des macrophages alveolaires pulmonaires chez
     les jeunes fumeurs de cigarettes.
     Conf. Intern. de la Tuberculose. Bull. Union. Intern.
     1 Vol. 51: 569-577.

26.  CARP, H., Janoff, A. (1978).
     Possible mechanisms of emphysema in smokers.
     Am. Rev. Respir. Dis. 118: 617-621.

27.  GADEK, J.G., Fells, G.A., Crystal, R.G. (1979).
     Cigarette smoking induces functional antiprotease
     deficiency in the lower respiratory tract of human.
     Science, 206: 1315.

28.  HINMAN, L.M., Stevens, C.A., Matty, R.A., Gee, B.L.
     (1980).
     Elastase and lysozyme activities in human alveolar
     macrophages.
     Al. Rev. Respir. Dis. 121: 263-271.

29.  LAURENT, P., Janoff, D., Kagen, H.M.
     Cigarette smoke black cross linking of elastin.
     Am. Rev. Respir. Dis. (in press).

30.  LUDWIG, P.W., Hoidal, J.R. (1982).
     Alterations in leukocyte oxydation metabolism in
     cigarette smokers.
     Am. Rev. Respir. Dis. 126: 977-980.

31.  WHITE, R.R., Coggin, F.R.R. (1982).
     Effect of cigarette smoke exposure on elastase induced
     emphysema.
     Am. Rev. Respir. Dis. 125: (part 2) 214.

32.  HUCHON, G.J., Russell, J.A., Barritault, L.G., Murray,
     J.F. (1982).
     Alveolar capillary membrane permeability in chronic
     obstructive pulmonary disease and in smokers.
     Am. Rev. Respir. Dis. 125: 280 (Abstract).

DISCUSSION

LECTURER: Chretien                      CHAIRMAN: Bonsignore

JEFFREY:              According to Brain there are large numbers of
                      macrophages ascending the mucociliary
                      escalator, is there any selective inhibition of
                      clearance of macrophages which have ingested
                      smoke particles.

CHRETIEN:             Perhaps Dr. Masse could answer this question.

MASSE:                The work of Weston in 1964 indicated that large
                      numbers did indeed pass the larynx, but there
                      was no proof that these originated in the lung
                      and modern views suggest that these macrophages
                      were secreted from the epithelium of the
                      airways. We have done some work on macrophage
                      excretion following dust labelling and the
                      clearance rate is no more than 2% daily.

CHRETIEN:             We believe that the cells from broncho-alveolar
                      lavage come only from the alveoli.

JEFFREY:              You mentioned that the alveolar macrophage
                      divides once, could you review the evidence for
                      this?

CHRETIEN:             We published some work with Dr. Masse.

MASSE:                This was done by labelling the macrophages in
                      the alveolus itself with an aerosol containing
                      thymidine, after which about 2.5 to 3% of the
                      macrophages were able to incorporate the label
                      into the DNA and the curve of labelled mitosis
                      and after one division we have never observed a
                      second wa e of mitosis, so that this means that
                      macrophages probably divide only once.

HIGENBOTTAM:          I was interested in your comments about the
                      effects of cigarette smoke on surfactant
                      production, could you say how this affects the
                      quantity and quality of surfactant?

CHRETIEN:             Experimental studies on animals in which lung
                      disease was induced by external radiation were
                      followed by an investigation of the effects of
                      tobacco smoke on surfactant and this was the

work to which I referred in my talk.

LUSUARDI:       In heavy smokers we found an increase in lysosomal glycosidase activity, in particular in alpha-fructosidase, do you have any information on this?

CHRETIEN:       I´m afraid not.

LUSUARDI:       A second question. Scanning electron microscopie studies of the alveolar macrophage surface of smokers showed in some cases results similar to yours. In particular two smokers of less than 15 per day showed a macrophage surface indistinguishable from normal. Are these exceptions or is the surface of the macrophage the same in heavy smokers and non-smokers?

CHRETIEN:       I have no experience in this area.

LUSUARDI:       In addition to our s.e.m. study we made an X-ray microanalysis on groups of macrophages and in smokers who have not been exposed to inorganic dusts we found a significant correlation between the silicon/sulphur ratio and the total number of cigarettes smoked during the life of the subject. There is a suggestion that tobacco smoke contains elemental particles such as silicon and aluminium, do you have any experience of this analysis and do you think that the silicon content of the alveolar macrophage can have its origin in tobacco smoke.

CHRETIEN:       Dr. Prodi came to work with Francoise Bassett on the problems of deposition within the macrophage and he analysed tobacco samples and found a level of silicate in many samples, particularly in North Africa samples and this produced extrinsic alveolitis. 50% of the silicate was in the form of free silica capable of producing inorganic dust disease.

PROTEASES – ANTIPROTEASES IN THE LOWER RESPIRATORY TRACT OF

PATIENTS WITH ALPHA 1-ANTITRYPSIN DEFICIENCY

AND CIGARETTE SMOKERS


R. Crystal

National Institutes of Health, Maryland


DISCUSSION

LECTURER: Crystal                                    CHAIRMAN: Cumming

RICHARDSON:     Do you screen children in the United State for the Pi-Z gene?

CRYSTAL:        We do not, but I am not sure where your question is leading.  One aspect would be could we make early diagnosis and the other is can we detect alpha-one-antitrypsm deficiency in utero. The technique for the latter has been developed using short segments of DNA but has not yet been applied clinicaly.  Screening for early diagnosis is not yet done, but as therapy becomes available screening is likely to become common.

RICHARDSON:     It would be advantageous even now since it would then be possible to give strong advice not to smoke.

CRYSTAL:        I agree entirely.  For patients as a whole I give a lecture about not smoking but for those with alpha-1-antitrypsm deficiency I yell at them.  Unfortunately they begin smoking in their early teens and by the age of 35 they are devastated by the disease, before they were ever diagnosed.

163

ASSENATO:         The theory related to the pathogenesis of
                  emphysema has been confirmed indirectly in a
                  paper in the September issue of the New England
                  Journal of Medicine, jointly from the
                  University of Pittsburgh and Christchurch, New
                  Zealand. The authors describe a 14 year old
                  boy who died of haemorrhage and in whom alpha-
                  1-antitrypsin had been changed into an anti-
                  thrombin by the substitution of arginine for
                  methionine in the 358 position. There was no
                  emphysema in this case.

CRYSTAL:          I agree, that was an interesting paper in
                  which substitution of an arginine for a
                  methionine in the active site converted
                  antitrypsin into an anti-thrombin 3.
                  Unfortunately there are very few such patients
                  so that detailed study is difficult. AAT has
                  similar sequences and both probably derive from
                  an ancestral gene.

PRIDE:            Is the protein functionally active when it gets
                  into the blood?

CRYSTAL:          Yes it is. Z-protein works just as well as the
                  M protein, the problem is it cannot get to the
                  blood.

PRIDE:            Is the production in the liver normal or is
                  there a feedback which slows down production
                  when it cannot escape?

CRYSTAL:          It is not known, but the guess is that there
                  must be some feedback mechanism. It has been
                  calculated that if the Z individual produced
                  the protein at a normal rate, the liver would
                  explode, so some mechanism must slow it down.

PRIDE:            Although the Danasor effect is small, it is
                  very interesting since it allows more to get
                  out, and can that approach be developed?

CRYSTAL:          In other words does Danasor increase
                  production, of which a similar proportion gets
                  out or does it increase its secretion, or both.
                  We were worried that it would increase
                  production, overfill the hepatocyte and cause
                  more liver disease. We followed enzyme levels
                  closely and observed nothing untoward. The

ideal method of study is to take repeated liver biopsies and assay the messenger BNA level for AT using the gene probe and we hope to do this within the next few years.

GUZZARINO:      How many donors do you need to obtain a gramme of AAT?

CRYSTAL:        Many.  However it was easy to do this study because we obtained a supply of pooled plasma. The level in blood is 2g per litre and we need 4g per person, per week, or 20,000 U.S. dollars per annum which is equal to renal dialysis, so the recombinant D.N.A. technique is much more promising.

RAWBONE:        If the Z protein is as active as the M, would not the recombinant techniques be better directed to the former?

CRYSTAL:        One can do many things with genetic engineering, but only the half dozen amino acids around the active site are all that are needed to make the molecule function, so one approach is to make an eight amino acid protein and administer it by aerosol.  One might also engineer a protein with a much greater potency than the natural product.

RAWBONE:        I am not clear why people with homozygous ZZ have large numbers of neutrophils in their lungs.  Is this intrinsic to the condition or due to some external agent?

CRYSTAL:        There is evidence for two theories, one, that there is an inhibitor in serum called neutrophil chemotactic inhibitor, and this is markedly decreased in AAT deficiency.  This inhibitor prevents the chemotactic factor acting on neutrophils in the blood, but if it is deficient the neutrophils are attracted. Against that concept is that AAT has some effect either on the production or the function of the chemotactic factor.  The second hypothesis is that although there may appear to be no neutrophils in the lung, if one looks at the data they are in fact present and 1 in 200 cells are neutrophils.  Normally the AAT binds to the elastase released, but if there is deficiency of AAT there is free elastase present. From the work of Senior, the alveolar machophage has surface receptors for neutrophil elastase and when elastase binding occurs the macrophage releases the chemotactic factor. Thus small numbers of neutrophils release elastase, this binds to the macrophages, which releases the neutrophil chemotactic factor and neutrophils are attracted to the lung. This does not explain why some patients have neutrophils and some do not.

DENISON:    The only organ which appears to need AAT is the lung, is there any advantage in administering treatment to the lung alone, perhaps as an aerosol?

CRYSTAL:    There are major delivery problems, but if these could be solved, the problem of protein aggregation would occur and since these aggregates are antigenic this would be difficult. Several pharmaceutical firms are interested in this idea and such an approach may be possible in the future. There are other situations in which there is a protease imbalance, in the joints in rheumatoid arthritis, in the lungs in cystic fibrosis, and in the adult respiratory distress syndrome, in bacterial pneumonias, in gout and perhaps in acute pancreatitis. These are probably local situation in which AAT is functionally deficient and they may be candidates for AAT replacement therapy.

CUMMING:    If I may comment on the occurrence of two diseases together, my last patient with AAT deficiency also had a marked deficiency in his aortic wall from which he finally perished. One of the difficulties in getting the AAT from the blood to the alveoli is the need to traverse the barrier between the two phases and with a mass of 52,000 daltons this is very difficult. Have you thought of tackling the permeability problem?

CRYSTAL:    That is an interesting thought. We have done a study in conjunction with the liver unit looking at the permeability to PiM and PiZ and they are the same. There clearly is a permeability barrier and if this could be changed without compromising the lung it would clearly be helpful.

CUMMING:          We should thank Dr. Crystal and Dr. Spadofora for a series of fascinating presentation and I am sure you would wish to signify your appreciation.

# BREATHING OTHER PEOPLE'S SMOKE

Keith Horsfield

The Midhurst Medical Research Institute

Midhurst, West Sussex, GU29 0BL

The danger to health engendered by cigarette smoking has received a great deal of attention in the past two decades, and a wide variety of major and minor diseases has been shown to be caused by, or associated with, the habit. Many non-smokers are also exposed to cigarette smoke in the ambient air, produced by other people smoking at home, at work, on public transport, and in public places. Breathing smoke under those circumstances has been termed "passive" or "involuntary" smoking. Over the past few years there has been an upsurge of interest in the possible adverse effects on health of passive smoking, and the anti-smoking groups have attempted to make use of reports of such effects to further their cause. Although the possibility of damage to health is a legitimate cause for concern, the evidence that otherwise healthy people can be seriously adversely affected by passive smoking is not very strong. In this paper some of the health consequences of active and passive smoking are reviewed and the difficulties in studying the effects of passive smoking are examined.

## HEALTH CONSEQUENCES ATTRIBUTED TO ACTIVE AND PASSIVE SMOKING

A summary of the health consequences of active cigarette smoking is given in Table 1. Where passive smoking has been reported to have an effect, this is indicated by a plus sign.

Life expectancy is reduced in smokers, taking into account all causes of death. On average the loss of life is about 5.5 minutes per cigarette smoked. No assumptions about

169

TABLE 1.    HEALTH CONSEQUENCES OF SMOKING AND OF
BREATHING OTHER PEOPLE'S SMOKE

| Active Smokers | Passive Smokers | |
| --- | --- | --- |
|  | In health | Disease |
| Life expectancy reduced | ? | ? |
| **Cancer** | | |
| Lung | + | ? |
| Others | ? | ? |
| **Cardiovascular disease** | | |
| Ischaemic heart disease | ? | + |
| Arterial disease of legs | ? | ? |
| Aneurism | ? | ? |
| Stroke | ? | ? |
| Raynaud's diseasei | ? | + |
| **Pulmonary disease** | | |
| Bronchitis, emphysema, COLD | ? | + |
| Airflow limitation | + | + |
| Infections | ? | ? |
| Tuberculosis | ? | ? |
| Asthma    or | ? | + |
| Postoperative complications | ? | ? |
| **Others** | | |
| Delayed healing of duodenal ulcer | ? | ? |
| Mouth ulcers | ? | ? |
| Dental problems | ? | ? |
| Pregnancy and neonates | ? | ? |
| Immunological suppression | ? | ? |
| Deafness | ? | ? |
| Skin diseases | ? | ? |
| Tobacco amblyopia | ? | ? |
| Parkinson's disease | ? | ? |
| Physical fitness | + | + |
| Fires | + | + |

+ = reported association or effect
? = no reported association or effect
= exacerbated
= reduced or suppressed

causation need to be made regarding this observation. There have been no similar reports regarding passive smoking.

Cancer of the lung, lips, tongue, mouth, pharynx, larynx, oesophagus, pancreas and urinary tract are commoner in smokers, the best known and numerically the most important being cancer of the lung. This is the only cancer reported to be associated with passive smoking. Five studies have been published in which the incidence of lung cancer in non-smokers was related to their spouses' smoking habits (Garfinkel, 1981; Hirayama, 1981; Trichopoulos et al, 1981; Chan, 1982; Correa et al, 1983). Three (Correa, Hirayama and Trichopoulos) showed a statistically significant increase in those whose spouse smoked compared with those whose spouse did not smoke. One (Garfinkel) showed a non-significant increase, and one no effect (Chan). Hirayama's study of Japanese women has stimulated a great deal of interest, discussion and criticism, and has been often quoted as another reason for not smoking. A review of the study is presented in the next chapter, where the statistical problems associated with it are discussed in detail. It is sufficient to say here that the evidence suggesting a causative link between passive smoking and lung cancer has not been very convincing, but additional evidence in support of the contention is gradually being obtained.

Cardiovascular diseases together constitute numerically the most important cause of smoking related deaths. Myocardial infarction is about x3 commoner in smokers than non-smokers, although many other factors also contribute to its causation. People with ischaemic heart disease may be adversely affected by passive smoking, as described below. Arterial disease in the legs is strongly related to smoking, and aneurism, stroke and Raynaud's disease are also increased in smokers. Of interest is the example of a heavy smoker whose two successive wives suffered from Raynaud's disease. The first was relieved of her symptoms by separation from her husband, and the second by separation from the smoky environment (Bocanegra and Espinosa, 1980).

Chronic bronchitis, emphysema and chronic obstructive lung disease (COLD) are caused primarily by smoking, though other factors play a part. Patients suffering from these conditions may be affected by passive smoking, as described below. Airflow limitation, symptomless but detected by pulmonary function tests, has been described in healthy passive smokers (White and Froeb, 1980; Kauffmann et al, 1983). White and Froeb's study has been much criticised and is discussed in detail in the next chapter. Even if the small changes described were caused by passive smoking, their significance for ill health in the long term is doubtful.  Lung infections are commoner in smokers and young children whose parents smoke are twice as likely to develop acute bronchitis or pneumonia as those whose parents do not smoke.  Pulmonary tuberculosis is also commoner in smokers.  Smoking may rarely precipitate asthma, but it also has the opposite effect of suppressing it, probably by immunological suppression.  Thus asthmatics who smoke may suffer an exacerbation on stopping smoking. Occasionally an attack of wheezing may be precipitated in an asthmatic by passive smoking.  Postoperative complications are commoner in smokers.

Numerous other conditions are related to smoking. Healing of duodenal ulcers is delayed.  Gingivitis, dental caries, deafness, skin disease and tobacco amblyopia may be caused by smoking.  Mouth ulcers may be suppressed, like asthma, and exacerbate on stopping smoking.  Parkinson's disease is less common in smokers.  None of these effects have been attributed to passive smoking.  Physical fitness, as measured by cardiac and respiratory variables, may be reduced in smokers, and affected by passive smoking.

Many fires are started accidently by smokers, and the smoker and non-smoker alike may be involved in these.

### Symptoms caused by breathing other people's smoke

Many people suffer annoyance and discomfort when exposed to other people's smoke and this may be true of smokers as well as non-smokers (Table 2).  Eye irritation is the commonest symptom, accompanied by watering, excessive blinking and rubbing.  Nose and throat irritation, cough and headache may also be troublesome.  The smell of the smoke is repugnant to many, and worse still it lingers on the hair and clothes of those who have been exposed to it.  Also, a room or the inside of a car may smell for days after someone has smoked in it. Smokers, with their reduced sense of smell, may not notice this, but for non-smokers it can be most unpleasant.  Although the above do not constitute a threat to health, they may be severe enough to ruin a trip in an aeroplane or a visit to a restaurant by a non-smoker.

### Some physiological effects of breathing other people's smoke

Experimental exposure of healthy subjects to a smoky atmosphere has been shown to increase blood pressure and heart rate, and to reduce the $FEV_1$ after exercise.  Exposure to carbon monoxide (CO), in concentrations equivalent to those found in a smoky room, reduced the time of exercise to exhaustion and increased heart rate, especially in older people.

Subjects suffering from ischaemic heart disease (IHD) who were exposed to a smoky atmosphere had an increased blood pressure, heart rate and left ventricular end diastolic pressure, but no change in myocardial contractility (dp/dt and stroke index). When exposed to equivalent concentrations of CO, cardiac contractility was reduced. These effects are interpreted as showing a negative ionotropic effect of CO, and a stimulent effect of nicotine, together resulting in increased myocardial work for the same cardiac output. Exposure to smoke or CO also reduces the exercise time to onset of angina.

Exposure to CO of subjects suffering from COLD reduced their exercise time to onset of marked dyspnoea.

## CIGARETTE SMOKE

Cigarette smoke contains over 2000 compounds, many of which are potentially poisonous or carcinogenic. Smoke is produced by the burning of tobacco under two different conditions: when smouldering and during puffing. They differ in the quantity of tobacco burnt in producing the smoke, and the temperature at which it burns. Other variables affecting the composition of smoke are the type of tobacco (kind of plant, where grown, which leaf, how cured) and the manner in which it is smoked. That given off during smouldering is the sidestream smoke, and that drawn through the cigarette on puffing is the mainstream smoke. Table 3 gives some examples of how the two differ in composition, expressed as a ratio of concentration in sidestream to mainstream smoke. Of particular interest are the increased concentrations of carcinogens (eg. nitrosamine and benzo(a)pyrene) in sidestream smoke. Mainstream smoke is variably diluted by air drawn through the cigarette, and also by air through the paper, or holes in the paper and filter. Mainstream smoke may be drawn into the smoker's mouth and then exhaled. In this case at least half of the water soluble volatile compounds, and 1/5th of the water insoluble compounds and particles, are retained by the smoker. Very little CO is absorbed. Thus the exhaled smoke is altered in composition. If mainstream smoke is inhaled into the lungs, over 6/7th of the volatile and particulate compounds are retained, as well as over half of the CO. Thus the exhaled smoke is different again in composition. Other types of smoke, which are less important, are also produced (Table 4). These various kinds of smoke, of very variable composition, mix with the ambient air and are diluted to a variable degree. Both the active and the passive smoker breath the resulting smoky air,

**TABLE 2**    **SYMPTOMS IN THOSE WHO BREATH OTHER PEOPLE'S SMOKE**

Annoyance

Eye irritation, watering, blinking, rubbing

Nose and throat irritation

Cough

Headache

Smell of smoke

Smell on clothes and in room afterwards

**TABLE 3:**    **RELATIVE CONCENTRATIONS OF SOME COMPONENTS OF SIDESTREAM AND MAINSTREAM CIGARETTE SMOKE**

| Component | Ratio of concentrations sidestream/mainstream |
|---|---|
| Gas phase | |
| $CO_2$ | 8.1 |
| CO | 2.5 |
| $NH_4$ | 73 |
| Dimethylnitrosamine | 52 |
| Particulate phase | |
| Tar | 1.7 |
| Benzo(a)pyrene | 3.4 |
| Nictoine | 2.7 |
| NNK (nitrosamine) | 10 |

**TABLE 4:     TYPES OF SMOKE EMITTED BY A BURNING CIGARETTE**

**More important**

| | |
|---|---|
| Mainstream | Drawn in by the smoker during a puff |
| Sidestream | From burning cone between puffs |

**Less important**

| | |
|---|---|
| Smoulder stream | From butt end between puffs |
| Glow stream | From burning cone during puffs |
| Effusion stream | From length of paper during puffs |
| Difusion stream | From length of paper between puffs |

but the smoker, being nearer to the source of pollution, takes in much more smoke passively than does the non-smoker. This is important when considering the different degrees of exposure to which active and passive smokers are subjected. Figure 1 summaries these points.

## CIGARETTE SMOKE POLLUTION

Mainstream smoke contains up to 5 vols% of CO and sidestream smoke up to three times as much. The concentration in the ambient air varies with the number and type of tobacco products being smoked, the size of the room and its ventilation. CO mixes with the ambient air without settling, and cannot be filtered out. Fresh air replacement is therefore needed to keep the concentration down; ventilation systems which recirculate air don't help. Smoke-free air contains about 2 ppm of CO, while values up to 9 ppm have been measured at a party, 33 ppm in a conference room, 40 ppm in a submarine, and up to 110 ppm in an unventilated car. Under ordinary social conditions with ventilation, levels are usually below 10 ppm. Blood carboxyhaemoglobin levels are about 1% in a non-smoker, rising to 2.6% in experimentally produced severely polluted air.

Over 1000 $ug/m^3$ of nicotine has been measured in an unventilated enclosed space, but the concentration rapidly falls with ventilation. It settles out of the air onto clothes, hair and furnishings, so the atmospheric concentration slowly falls. The urine of passive smokers contains nicotine but at levels of less than 1% of those found in active smokers. Half of urban non-smokers have nicotine in their blood, and most have nicotine in their urine. Most urban dwellers probably have nicotine in their bodies for most of their lives, but the significance of this finding is unknown.

Benzo(a)pyrene has been found in the air of a smoky restaurant, but it may have come in part from the cooking oil. Small quantities have been found in a smoky aeroplane. Nitrosamines have been found in the air of smoky cars in concentrations which would give passive smokers as big a dose in one hour as active smokers receive from the mainstream smoke of 5 to 30 cigarettes. Just how much carcinogen is absorbed in these circumstances is not known, nor whether prolonged inhalation of low dosage carcinogen presents a danger. Exposure of passive smokers to small concentrations over long periods may have quite different effects from exposure of active smokers to high dosages for short periods. No threshold for carcinogenesis of the lung has been found in man (see Fig. 2) and because the bronchial mucosa is directly exposed to

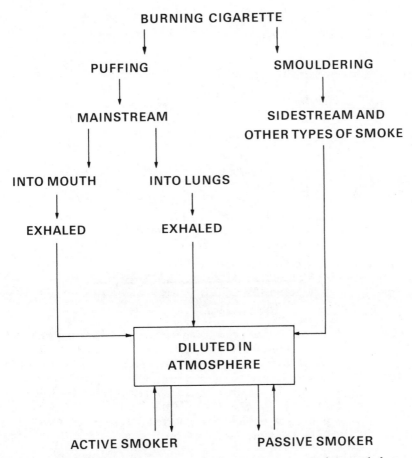

FIGURE 1:    The production of cigarette smoke and how its
composition and concentration are altered.

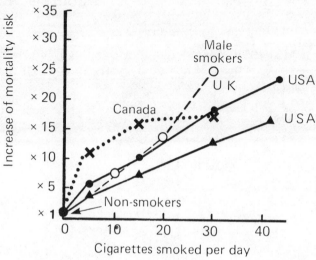

FIGURE 2:   Risk of dying from lung cancer in smokers compared
            with non-smokers, related to number of cigarettes
            smoked per day.  Results from four prospective
            studied in three countries, showing no evidence of
            a threshold for carcinogenesis.  This would be
            indicated by the plots intersecting the x axis at
            a positive value. From Royal College of Physicians
            (1977) by kind permission.

these compounds, there must be a potential for carcinogenesis. This is, however, very difficult to study because of the long latency period. If passive smoking in the U.K. had the same effect as that claimed in the Japanese study, there would be an extra 2 cases of lung cancer per 100,000 women each year. This would be difficult to detect and would require large populations for study.

Acrolein is the most important of the irritant substances found in smoky atmospheres, being present in much higher concentrations than any of the others.

Figure 3 summarises the factors which determine the level of smoke pollution in the ambient air. There is a dynamic balance between input, dilution, deposition, and ventilation, each of which is affected by several factors already discussed. When making measurements, the positions of the source and the detector, and the speed and direction of air currents, are also important variables.

Absorption by the passive smoker depends on the levels reaching him in the ambient air, and the duration of exposure. This pattern of absorption may be quite different from that of the inhaling smoker. The concentrations of CO and nicotine in the body can be measured, and therefore the absorption of these substances has been most studied. Some vapour constituents of smoke, e.g. acrolein and nitrogen oxides, remain at a constant level in an empty room. But if people go into the room, the levels fall, even if smoking is continued. This may be an indication that they are being absorbed.

## MEASUREMENT OF PASSIVE SMOKING

Attempts have been made to express passive smoking semiquantitatively in terms of exposure to smoke at home and at work. The variables include number of people smoking, number of cigarettes smoked, frequency of smoking, hours of the day exposed to smoke, and duration of exposure in years. Difficulties in interpreting such data include social customs, smoking habits of a divorced spouse, size of rooms and ventilation. Even when blood levels are measured, expression of exposure as equivalents of cigarettes smoked is obviously unsatisfactory. Because of the number of variables and difficulties already discussed, it is not surprising that there is no satisfactory way of measuring passive smoking. If this is so, then studying its long term effects must be fraught with difficulty. A whole new approach is needed, different from counting the number of cigarettes smoked.

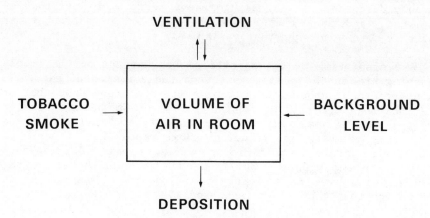

FIGURE 3:    Factors determining the level cigarette smoke
             pollution in room air.

## REDUCING EXPOSURE TO OTHER PEOPLE'S SMOKE

Any measures which reduce active smoking will also help to reduce passive exposure to smoke. These include health education, articles in the popular press, advertisements, TV films and banning the advertising of tobacco products. Clinics to help smokers to stop have had some success, and increases in taxation have been followed by reductions in sales. Teachers and parents could set better examples to children, and direct selling of cigarettes to children should be more effectively prevented. Doctors should take a lead in persuading their patients to stop smoking. Such measures have had some success, as only one third of adults in the U.S.A. and U.K. are now smokers.

Forbidding smoking in enclosed public places would help to lessen the exposure of non-smokers to smoke. These include theatres, restaurants, hospitals, aeroplanes, buses and trains. If a total ban cannot be achieved, then at least there should be no-smoking areas. Better would be a change in attitude towards regarding non-smoking as the norm, with small areas put aside for smokers and labelled as such. The place of work is especially important, because of the potential duration of exposure to smoke. Where banning smoking appears difficult, adequate ventilation is most important. People should have the right to work in a smoke-free atmosphere.

Some of these measures can be carried out without legislation, e.g. non-smoking areas in restaurants. Others may require legislation, e.g. no smoking on public transport or in public buildings. Enforecment is always difficult, and requires the co-operation of the smoker.

## RIGHTS OF SMOKERS AND NON-SMOKERS

I believe that non-smokers, who now outnumber smokers by 2:1 in the U.S.A. and U.K., should have the right to work, travel, entertainment and hospitalisation in smoke-free air. Smokers may smoke in private, and if they must smoke in public special zones should be provided for them. Non-smoking should be considered the norm.

## REFERENCES

BOCANEGRA, T.S. and Espinosa, L.R. (1980).
Letter, The New England Journal of Medicine, 303, 24.

CHAN, W.C. (1982).
    Figures from Hong Kong.
    Munchner Medizinische Wochenschrift, 124, 16.

CORREA, P., Pickle, L.W., Fontham, E., Lin, Y. and Haenszel
(1983).
    Passive smoking and lung cancer.
    The Lancet, II, 595-597.

GARFINKEL, L. (1981).
    Time trends in lung cancer mortality amoung non-smokers, and
    a note on passive smoking.
    Journal of the National Cancer Institute, 66, 1061-1066.

HIRAYAMA, T. (1981).
    Non-smoking wives of heavy smokers have a high risk of lung
    cancer: a study from Japan.
    British Medical Journal, 282, 183-185.

KAUFFMANN, F., Tessier, J-F. and Oriol, P. (1983).
    Adult passive smoking in the home environment: a risk factor
    for chronic airflow limitation.
    American Journal of Epidemiology, 117, 269-280.

ROYAL College of Physicians (1977).
    Smoking or Health.
    Pitman Medical, Tunbridge Wells, p. 54.

TRICHOPOULOS, D., Kalandid, A., Sparros, L. and MacMahon, B.
    (1981).
    Lung cancer and passive smoking.
    International Journal of Cancer, 27, 1-4.

WHITE, J. R. and Froeb, H. (1980),
    Small-airways dysfunction in non-smokers chronically exposed
    to tobacco smoke.
    New England Journal of Medicine, 302, 720-723.

**SOURCES OF INFORMATION**

ROYAL College of Physicians (1983).
    Health or Smoking: a Follow-up Report, 1983.
    Pitman Medical, Tunbridge Wells.

SHEPHARD, R.J. (1982).
    The Risks of Passive Smoking.
    Croom Hehn, London.

U.S. DEPARTMENT of Health, Education and Welfare (1979).
    Smoking and Health.   Public Health Service.
    DHEW publication No. (PHS) 79-50066.   pp.11-1 to 11-41.

DISCUSSION

LECTURER: Horsfield                                    CHAIRMAN: Cumming

BAKE:              I would reinforce the view that we lack
                   information about passive smoking. There are
                   three prinicipal sources, in childhood when one
                   or both parents smoke, in adult life at home,
                   or in the place of work. Most studies have
                   related to only one of these, none to all
                   three. A study published in Science measured
                   passive smoking in terms of particulate dose
                   and the most extreme circumstance was in a
                   badly ventilated night club, and living with a
                   chain smoker; his consumption was equivalent to
                   twenty cigarettes a day, so in passive smoking
                   the dose can be quite high.

CHRETIEN:          I understand the difficulty in making a
                   measurement of passive smoking, but do you have
                   a clear definition of the term "passive
                   smoker"?

HORSFIELD:         The inhalation of cigarette smoke when not
                   smoking a cigarette.

CHRETIEN:           Then we are all passive smokers.

HORSFIELD:         To a degree, yes. That is why most of us have
                   nicotine in our bodies for most of our lives.

GUYATT:            Is there any possibility of nicotine absorption
                   by any route other than the lung, such as the
                   skin?

HORSFIELD:         Not that I know of.

CUMMING:           In common with other alkaloids, if nicotine is
                   placed upon the skin some is absorbed.

ASSENATO:          Your tables only showed a yes or no answer. Do
                   some of your negative results indicate that you
                   do not know, and should not your evidence rely
                   only on true positive results? The population
                   structure exposed to smoking is very
                   heterogeneous and this should be taken account
                   of in the analysis. As has been mentioned
                   before it is difficult to have a working

definition of passive smoking and this couple
with population heterogeneity makes
correlations difficult to derive.

HORSFIELD:      The negative sign on the slides indicated no
                information available, where the results were
                actively negative, as in the case of lung
                cancer, I pointed it out.  I agree with your
                other comments.

CUMMING:        This makes the point that it is a very
                difficult area.

# PASSIVE SMOKING

P. N. Lee

Independent Consultant in Statistics & Epidemiology

25 Cedar Road, Sutton, Surrey

## INTRODUCTION

Passive smoking can be defined as the inhalation of tobacco smoke other than by puffing on a cigarette, cigar or pipe. Study of it is relatively new with few references in the literature before 1970. In this review we start by considering briefly a number of types of accusation that had been levelled against passive smoking up until 1979. We then consider dosimetric aspects, understanding of which is fundamental to sensible evaluation of the epidemiological evidence. Finally we look in some detail at the recent suggestions that passive smoking might be a more serious health hazard than hitherto considered likely.

## EARLY CLAIMS

### Annoyance and irritation

It is common experience that passive smoke exposure, especially under conditions of poor ventilation, can be annoying and irritating. Speer [1] interviewed 250 non-allergic patients about their reaction to cigarette smoke and found that 69% reported eye irritation, 32% headache, 29% nasal symptoms and 25% cough. Weber [2] found that the frequency of reported eye, nose and throat irritation increased with increasing smoke concentrations of smoke in a sealed chamber. They suggested acrolein was the major offending substance, but, a subsequent study by Hugod, Hawkins and Astrup [3], showed that although a gas-phase polluted atmosphere was as annoying as one polluted with whole sidestream smoke, air pollution with acrolein at three times the concentration present in sidestream smoke caused considerably less discomfort.

## Allergy

The 1979 U.S. Surgeon-General's Report (4) devoted a chapter to the subject of allergy and tobacco smoke. It concluded that the existence of such an allergy was not clearly established but that those with a history of allergies to other substances, especially those with rhinitis or asthma, were more likely to report irritating effects of tobacco smoke. Whether this was a psychological, rather than a physiological, response is open to question.

More recently Sudan (5) has reported that five of his family suffered from an atopic dermatitis which could be brought on by a single puff of tobacco smoke and which was cured by hyposensitization with extracts of tobacco leaves. Whether this is of only anecdotal interest is not clear.

## Respiratory effects in children

In 1974 Colley (6) published an important paper showing that in children in the first year of life, but not in the second to fifth year, prevalence of cough was significantly higher in children of parents who smoked, the excess being still significant if the analysis was restricted to those parents who did not have phlegm. These findings have been confirmed by further studies by Leeder (7) and by Fergusson (8). Tager (9) and Hasselblad (10) have also reported a small association between parental smoking and measures of pulmonary function in children.

The interpretation of these associations is fraught with difficulties. Childhood respiratory disease is known to be strongly social class related and removal of its confounding effects is notoriously difficult. Other factors not generally taken account of in these studies, but of possible importance as confounders, are gas cooking in the home (Melia (11)), parental neglect (4), nutrition – money spent on cigarettes may reduce money spent on food – and cross-infection. Not only do smokers themselves have more respiratory infections than non-smokers, thus increasing the chance of cross-infection in their children, but as smokers are more sociable their children are liable to have more opportunity to come into contact with infection from outside sources. It also seems very difficult to disentangle possible effects of maternal smoking in pregnancy and of true passive smoke exposure from the mother during childhood. Certainly authors tend to assume one or other is the cause of the slight deficit in stature reported in children of smoking mothers often without considering the alternative.

## Psychomotor effects

There has been some concern that relatively low levels of carbon monoxide (CO) may have an effect on psychomotor functions, especially in relation to driving a car. Summarizing the literature, which shows a considerable discrepancy in the level at which blood carboxyhaemoglobin (COHb) may affect vigilance, the U.S. Surgeon-General (4) concluded that effects seen at levels of COHb found in passive smoking conditions are measurable only at the threshold of stimuli perception and that effects of CO on driving performance and interactive effects of CO and alcohol are only found at higher COHb levels. A well executed study by Guillerm (12) in which subjects drove a specially equipped car for 5 hours during the night, exposed either to air or CO sufficient to produce blood levels of 7 or 11% COHb, found no effect of even the higher level on driving precision or visual reaction time. This level exceeds that achieved by all passive and indeed most active smokers.

## Exercise tolerance

Aronow[13] examined the effect of passive smoke exposure on 10 patients (two smokers, eight non-smokers) with angina pectoria. Mean time of exercise until onset of angina in control conditions (COHb level 1.3%) was reduced by 22% after exposure to passive smoke in a ventilated room (COHb level 1.8%) and by 38% after exposure in an unventilated room (COHb level 2.3%). He attributed this to the possible absorption of nicotine, though he did not measure blood levels. It seems unlikely that these very low levels of nicotine absorption could be responsible for this physiological effect and the alleged response may be due to stress following anxiety or aggravation induced by the smoke-filled room. An alternative explanation is indicated by a recent report [14] that Aronow has removed himself from future research in the face of evidence that he submitted false data to the U.S. Food and Drug Administration. In the light of this, the association reported by Aronow must be suspect without independent confirmation.

## Summary of evidence to 1979

From the evidence described above, it seemed clear

that, while smoking was a source of annoyance to some, although not perhaps very annoying for many, the grounds for believing it to be a health hazard were somewhat thin. Where adverse effects were claimed, they did not apply to the normal healthy adult non-smoker and/or were not backed by particularly solid evidence. This view was illustrated by some statements made at the time:

"Healthy non-smokers exposed to cigarette smoke have little or no physiological response to the smoke, and what response does occur may be due to psychological factors"

(U.S. Surgeon-General's Report in 1979 (4))

"For the moment most – but not all – of the pressure for people (including many smokers) to have the right to breathe smoke-free air must be based on aesthetic considerations rather than on known serious risks to health."

(British Medical Journal Leading Article in 1978 (15))

## DOSIMETRY

### General

Without some idea of the relative amounts of smoke constituents to which passive smokers and smokers are exposed, it is impossible to judge adquately evidence of the alleged health effects of passive smoking. It is not always easy to gain an accurate idea of this from some of the totally misleading statements that are made. One classic example is the paper by Repace and Lowrey (16) who, using a theoretical model combined with measurements of cigarette smoke particulate matter (PM) in various different environments, estimated that a non-smoking office worker exposed to moderate passive smoke inhaled the equivalent, in PM terms of 5 cigarettes a day while a very heavily exposed non-smoking musician working in a night-club with a chain smoker for a roommate inhales the equivalent of 27 cigarettes a day. However, as noted by Bock (17), Repace and Lowrey had used a base for comparison a very low yielding cigarette with a PM yield of 0.55 mg/cig, whereas it would have been correct to use an average figure of 16 mg/cig. When the passive smoke exposures were recalculated on this basis, they reduced from 5 or 27 cigarettes a day to 1/6 or 1 cigarette a day.

More ridiculous still was the claim by Lane in the national press in the UK (Daily Mail and Daily Telegraph both of 2 June 1981) that "there is now medical evidence to show that the smoke breathed in by non-smokers is 18 times higher in tar and 12 times higher in nicotine than the smoke breathed by smokers...". The source of this claim undoubtedly came from a table published by the Laboratory of the Government Chemist (18), which showed that the ratio of sidestream to mainstream yield was 18 for tar and 12 for nicotine when a very low tar cigarette was smoked under machine conditions. Not only had the cigarette used as a basis for comparison a tar level some 10 to 15 times less than that normally smoked, but the fundamental error of confusing sidestream yields and ambient concentrations had also been made. The concentration of sidestream smoke is measured as it leaves the burning cone of tobacco between puffs, whereas what is relevant to the passive smoker is the concentration of smoke as it reaches him after dilution by room air. Ambient concentrations vary drastically depending on the degree of room ventilation, but even under conditions of poor ventilation, will be very considerably less than sidestream concentrations, which a non-smoker would only receive if he were to keep his nose right on top of the cigarette.

A number of workers have made measurements of concentration of smoke constituents in ambient air and body fluids. Hugod, Hawkins and Astrup (3) measured air concentrations of a number of constituents in a closed, unventilated room in which ten volunteers were exposed to quite severe passive smoke exposure conditions in which the air CO concentration was kept at 20 ppm. Comparing the estimated inhaled amounts of each constituent with those inhaled by a smoker, they calculated cigarette equivalent times (CET) in hours for seven different constituents (Table 1). These estimates of the time taken to inhale the equivalent of one cigarette vary widely according to the particular constituent.

**Nicotine**

The figure of 50 hr for nicotine in Table 1, indicating a fairly negligible intake of nicotine from passive smoking, is consistent with a study by Russell and Feyerabend (19), who found that non-smokers exposed experimentally in an almost intolerably smoky room had urinary nicotine levels 15 times lower than average smokers and also with a later study by Feyerabend, Higenbottam and Russell (20), who found that non-smokers exposed to smokers during their normal morning's work had nicotine levels of the order of 100 times lower than those found in smokers, although there was large inter-individual variation.

Table 1.   Comparison of Uptake of Smoke Constituents in Smokers and Passive Smokers*

| Smoke constituent | Mainstream yield inhaled by smoker (mg/cigarette) | Inhaled amount in passive smoking conditions (mg/hr)† | Cigarette equivalents/hr | Cigarette equivalent time (hr) |
|---|---|---|---|---|
| NO | 0.30 | 0.182 | 0.61 | 1.6 |
| CO | 18.40 | 9.160 | 0.50 | 2.0 |
| Aldehyde | 0.81 | 0.214 | 0.26 | 3.8 |
| Acrolein | 0.09 | 0.013 | 0.14 | 7.1 |
| TPM | 25.30 | 2.300 | 0.09 | 11.1 |
| Nicotine | 2.10 | 0.041 | 0.02 | 50.0 |
| Cyanide | 0.25 | 0.005 | 0.02 | 50.0 |

TPM = Total particulate matter

* Data from Hugod, Hawkins & Astrup (1978).

† Volunteers were exposed in a closed, unventilated room to quite severe passive smoke conditions in which the air CO concentration was kept at 20 ppm over a 3-hr period.

## Total particulate matter (TPM)

Hugod (3) and his co-workers concluded that for TPM, the constituent usually considered to be related to the excess of lung cancer risk in smokers, the CET value is "so high that the passive smoker will never inhale more than what equals 1/2 to 1 cigarette per day" - a finding consistent with the amended conclusion of Repace and Lowrey's study (16) by Bock (17).

## Carbon Monoxide

The conclusions from Table 1 for CO are similar to those of Russell, Cole & Brown (21) who, under more extreme conditions involving twice the exposure level for CO, found half the CET value (i.e. 1 hour). Even despite this relatively low CET value, it is most unlikely that passive smokers will achieve blood COHb levels as high as 3%, claimed to decrease the threshold for intermittent claudication and angina pectoria in patients with obliterating arterial disease (22,23).

## N-Nitrosodimethylamine

N-Nitrosodimethylamine (NDMA) merits mention in the context of passive smoking because of its unusually high ratio of sidestream to mainstream smoke deliveries and of its known biological activity.

There is some conflict in the literature about levels of NDMA in the atmosphere. Brunnemann (24) reported concentrations of 0.24 ng/litre in a bar and calculated that a non-smoker in this situation would inhale in 1 hour an amount of NDMA equivalent to that inhaled by a person actively smoking 17-35 filter cigarettes. However, Stehlik (25) found in controlled experiments under extreme conditions that NDMA levels above 0.07 ng/litre were associated with a highly irritating atmosphere which was scarcely tolerable to those present, while in smoke filled rooms under natural conditions NDMA levels ranged from 0.02 to 0.05 ng/litre. The significance of these low levels of NDMA is not clear, no epidemiological data existing to link human respiratory cancers to volatile nitrosamines.

## Other considerations for epidemiological interpretation

The above brief review of dosimetric evidence suggests that, however assessed, exposure of non-smokers in average circumstances from passive smoke will be substantially less than exposure due to smoking cigarettes actively. The difference in exposure will be increased when one remembers

that, in smokers, passive exposure to their own smoke doubtless
exceeds the passive exposure they pass on to non-smokers.

Before assessing the epidemiological evidence, it is worth
realising at the outset that no data exist directly linking
level of intake or uptake of smoke constituents in non-smokers
with risk of disease. In all the studies to be cited relating
passive smoking to effects on the small airways or to lung
cancer, smoking by the spouse has been used as the main index
of passive smoke exposure. The weakness of this as an
indicator of total passive smoke exposure is underlined by a
recent paper by Friedman (26) which collected information on
35.000 non-smoking subjects which included spouse smoking
status and hours per week passive smoke exposure at home, in a
small place other than home (airplane, office, car, ect.) and
in a large indoor area (restaurant, hotel lobby, lecture hall,
etc). The results for both sexes were similar, results for
female subjects being presented in Table 2. It can be seen
that over 40% of women with non-smoking spouses reported some
passive smoke exposure while as many as 47% of women married to
smokers reported no (actually less than one hour per week)
passive smoke exposure at home and as many as 35% of women
married to smokers reported no total passive smoke exposure.
The mean weekly exposure outside the home was greater in those
married to smokers (7.8 hours) than in those married to non-
smokers (4.6 hours), but the difference in terms of cigarette
equivalents would appear very small.

**NEWER EVIDENCE**

**Effects on the small airways**

In the last 3 or 4 years, new evidence has caused a
considerable amount of re-thought on the passive smoking issue.
The first such evidence came from a study by White and Froeb
(27) of the relationship between various pulmonary function
indices and passive smoking. 3002 men and women who had been
physiologically evaluated during a "physical fitness profile"
course who were without a history of relevant cardiorespiratory
disease, occupational exposure to dust or fumes or severe
exposure to pollution at home or at work were divided into 6
groups according to their exposure to tobacco smoke. No
significant difference was found between non-smokers exposed to
a smoky environment for more than 20 years (group 2) and non-
smokers never so exposed (group 1) as regards forced vital
capacity (FVC) and forced expiratory volume in 1 second ($FEV_1$);
but non-smokers exposed to passive smoke had statistically
significant reductions in forced mid-expiratory flow (FEF 25 to

Table 2.  Hours Passive Smoke Exposure in Women According to the
          Smoking Habits of Their Husbands*

| Source of Exposure | Spouse | % with no[+] passive exposure | Mean hours/week passive exposure |
|---|---|---|---|
| At home | Smoker | 47.4 | 12.7 |
| | Non-smoker | 91.9 | 1.0 |
| Total | Smoker | 35.4 | 20.5 |
| | Non-smoker | 59.4 | 5.6 |

* Data from Friedman, Pettiti and Bawol[26].

+ None implies less than 1 hour per week.

75%) and in forced end-expiratory flow (FEF 75 to 85%; Table 3).

There were two odd features about White and Froeb's results that are difficult to explain. The first relates to the fact that the reductions in FEF seen in group 2 were generally very similar to those seen in group 4. Why should a relatively large difference (group 2 vs group 1) in airways dysfunction be seen as a result of an apparently relative small difference in exposure to smoke constituents when only a relatively small difference (group 4 vs group 2) is seen in response to what was in all probability a much larger difference in exposure?

The second odd feature relates to the classification of all the subjects into one of the six groups. Close reading of the paper shows that there are a very considerable number of people who do not fit into any of the classifications. Among people who appear to have vanished are:

(i)    any non-smoker living in a house where smoking was permitted.

(ii)   anyone who changed smoking habits in the last 20 years.

(iii)  any cigarette smokers not permitted to smoke at work.

(iv)   any inhaling pipe and cigar smokers.

Without an adequate explanation of this anomaly, one can have little confidence in these findings.

Kauffmann, Tessier and Oriol (28) based on a survey of more than 7800 adult residents of seven cities throughout France compared the spirometric measurements of two groups of non-smokers: those with and without exposure to passive smoking in the home. They claimed a significant reduction in FEF 25 to 75% in passive smokers among women aged 40 or over, a difference not explained by adjustment for social class, educational level, air pollution or family size. Detailed inspection of their results (Table 4) puts their conclusion in a somewhat different light. One sees that the difference between true non-smokers and passive smokers is inconsistent over age. Thus passive smokers are worse off at ages 30-34, 40-44, 45-49 and 50-54 and better off at ages 25-29, 35-39 and 55-59 as regards FEF. While when the data are restricted to those aged 40 or over, a "significant" difference may emerge it is also true that, of all the 5 year age points that could have been chosen, this cut-off is the one that accentuates the

Table 3. Vital Capacities and Expiratory Flow Rates in Smokers and Non-Smokers*

| Sex | Group number | Smoking habits† | Percentage of predicted | | | |
|---|---|---|---|---|---|---|
| | | | FVC | $FEV_1$ | FEF 25-75% | FEF 75-85% |
| Male | 1 | Non-smokers, no smoky environment | 102 | 103 | 104 | 120 |
| | 2 | Non-smokers, smoky environment | 99 | 98 | 91 | 95 |
| | 3 | Smokers not inhaling | 96 | 99 | 92 | 87 |
| | 4 | Smokers: 1-10 cigarettes/day | 95 | 97 | 89 | 77 |
| | 5 | Smokers:11-39 cigarettes/day | 84 | 86 | 76 | 68 |
| | 6 | Smokers: > 40 cigarettes/day | 82 | 77 | 72 | 60 |
| Female | 1 | Non-smokers, no smoky environment | 102 | 104 | 108 | 112 |
| | 2 | Non-smokers, smoky environment | 98 | 99 | 93 | 85 |
| | 3 | Smokers not inhaling | 97 | 99 | 92 | 85 |
| | 4 | Smokers: 1-10 cigarettes/day | 96 | 98 | 89 | 83 |
| | 5 | Smokers:11-39 cigarettes/day | 85 | 85 | 78 | 69 |
| | 6 | Smokers: > 40 cigarettes/day | 78 | 80 | 72 | 62 |

* Data from White & Froeb [27]

† Exposure to a smoky environment or consumption of cigarettes was for more than 20yr. Group 3 includes pipe, cigar or cigarette smokers who did not inhale. Groups 4, 5 and 6 were all inhaling cigarette smokers.

Table 4.  FEF 25 to 75% According to Passive Smoking and Age Among 1985 Women*

| Age | FEF (number of subjects) | | Difference in FEF |
|-----|-------------------|-----------------|-------------------|
|     | True non-smokers  | Passive smokers |                   |
| 25 - 29 | 3.38 (154) | 3.46 (219) | + 0.08 |
| 30 - 34 | 3.40 (166) | 3.18 (224) | - 0.22 |
| 35 - 39 | 3.00 (179) | 3.12 (214) | + 0.12 |
| 40 - 44 | 3.08 (136) | 2.74 (173) | - 0.34 |
| 45 - 49 | 2.86 ( 76) | 2.64 (123) | - 0.22 |
| 50 - 54 | 2.62 ( 86) | 2.38 (149) | - 0.24 |
| 55 - 59 | 2.22 ( 30) | 2.60 ( 57) | + 0.38 |

* Data adapted from graph by Kauffmann, Tessier and Oriol[28].

difference most markedly and probably the only one which would
have given "significant" p values.

In another study Comstock (29) related $FEV_1$ and FVC in
never smokers to the number of other cigarette smokers in the
household and to the use of gas or electricity for cooking. In
almost 400 men, impaired ventilatory function was significantly
associated with use of gas for cooking, but was not related to
passive smoking. The number of women was too small for
reliable analysis. No evidence was given relating smoking
habits to use of gas or electricity, but the possibility of
effects of gas cooking confounding effects of passive smoking
should be borne in mind.

**Lung Cancer**

Whilst the findings of White & Froeb (27) and of
Kauffmann et al (28) relate to an index (FEF) which is
contentious and certainly not an accepted reliable indicator of
an increased health risk, two studies published in January
1981, by Hirayama (30) and by Trichopoulos et al (31) caused
much more attention, as both claimed that non-smoking wives of
smokers had a significantly greater risk of lung cancer than
non-smoking wives of non-smokers. Subsequent papers by
Garfinkel (32), Chan (33), Correa (34) and Knoth (35), have all
presented data relating lung cancer risk to spouse smoking
habits, while Hirayama (36) has presented additional data from
his Japanese study.

**Japanese Study**

Of the 7 studies, 2 are based on large prospective
studies, the epidemiological technique normally thought most
likely to give reliable results. That by Hirayama followed up
91,540 japanese non-smoking married women aged 40 years or over
in 1965 for 14 years. The latest findings from his study are
summarized in Table 5. The results showed a quite highly
significant trend in the risk of lung cancer with increasing
smoking by the husband, with wives of heavy smokers having
almost twice the risk of wives of non-smokers. The findings
for women extend slightly the results given in his original
paper (30) and also show results for men, not given before,
which show a similar trend, though based on smaller numbers and
of marginal statistical significance. In similar analyses he
found no association between husband's smoking and risk of
cancer of other sites and a small positive association between
husband's smoking and risk of ischaemic heart disease (relative
risk of 1.31 in non-smokers married to heavy smokers compared
with non-smokers married to non-smokers), chronic bronchitis

Table 5.   Active and Passive Smoking and Lung Cancer Mortality in Japan*

| Sex | Smoking habit | | Number of subjects | Number of deaths | Relative risk[+] (90% limits) |
|---|---|---|---|---|---|
| | Own | Spouse | | | |
| Male | None | None | 19279 | 57 | 1 |
| | None | Smoker | 1010 | 7 | 2.25 (1.19 – 4.22) |
| | 1-19/day | | 59077 | 611 | 3.89 (3.15 – 4.80) |
| | 20+ /day | | 35454 | 557 | 6.61 (5.45 – 8.03) |
| Female | None | None | 21895 | 37 | 1 |
| | None | 1-19/day | 44184 | 99 | 1.41 (1.00 – 1.99) |
| | None | 20+ /day | 25461 | 64 | 1.79 (1.23 – 2.62) |
| | 1-19/day | | 16080 | 110 | 3.67 (2.73 – 4.93) |
| | 20+ /day | | 1286 | 11 | 5.26 (3.18 – 8.70) |

* Data from Hirayama [36]

+ Standardised for age and occupation.

and emphysema (corresponding relative risk 1.60) and suicide (1.92).

Hirayama's original findings for lung cancer in women were subject to intense scrutiny in correspondence in the British Medical Journal and Munchner Medizinische Wochenschrit. Partly due to inadequate presentation of results in the initial paper (30), some confusion arose and it was suggested that the original claim of statistical significance ($X^2$ for trend = 10.88, p = 0.001) may have been due to an error in calculations. In the end, it seems probable that the association found was a statistically significant one, though of course a significant association need not imply a significant casual effect of passive smoke exposure.

In interpreting the Japanese results a number of points should be borne in mind:

(a)     Smoking habits were determined only at the beginning of the period and are likely to have changed, in view of the fact that annual cigarette comsumption per adult in Japan rose by some 50% over the period of the study.

(b)     The great majority of the lung cancers seen, 17 out of 23 in a sample, were adenocarcinomas, a type of lung cancer generally believed to be much more weakly related to smoking than squamous cell carcinoma, the commonest lung cancer in the West.

(c)     Evidence of trends in lung cancer rates in Japan suggest that there may be some other important cause of lung cancer which was not studied. Over the period 1950–1980 age-standardised risk of lung cancer in Japan rose by just over 9-fold for men and 8-fold for women. From Hirayama's own data, the relative risk of the total population to that of non-smokers married to non-smokers is only 4-fold for men and just under 2-fold for women. The conflict between these estimates and the rises from 1950-1980 not only in the sexes individually, but also relatively, requires explanation.

(d)     The index of passive exposure used is not likely to be very accurate, with the husband smoking a varying proportion of his cigarettes at home and the wife exposed to other sources of exposure besides the husband.

What is most surprising, however, is the sheer magnitude of the association. The two-fold increased risk in wives of heavier smokers is similar to that found by Hirayama for women actively smoking about 5 cigarettes a day, whilst it was stated that the heavy smokers smoked on average only 8.4 cigarettes a day at home and these presumably not at all in the direct presence of the wife.  If this is so, the study seems to be suggesting that one actively smoked cigarette is not so very different from one passively smoked one, which seems completely inconsistent with the dosimetry, especially when one realises that an active smoker will have greater passive smoke exposure than a passively exposed non-smoker.

The possibility that the association arises from so far unidentified confounding or biassing factors must therefore be seriously considered.  It is well-known that epidemiological observations are particularly susceptible to bias when, as here, a relatively small relative risk is involved.

**Greek study**

In contrast to the Japanese study, the small Greek case-control study of Trichopolous et al (31) is relatively lightweight, being based on only 40 lung cancer cases seen in non-smoking women.  However, their results (Table 6), though having quite wide confidence limits, agree well with those of Hirayama.  Taking into account a number of possible confounding factors (age, duration of marriage, occupation, schooling, residence) did not affect the general picture.

Although the trend is statistically significant, the limitations pointed out by the authors - the small number of cases, 35% of which were not cytologically confirmed, and the cases and controls being taken from different hospitals - would have meant that no great weight would have been attached to the results, had they not appeared at the same time as the first Hirayama paper (30).  It is an interesting observation, in comparison with the Japanese study, that the Greek study specifically excluded adenocarcinomas from their cases, since presumably it was assumed that this type of lung cancer was not smoking-associated.

**U.S. study**

Even taken together the Japanese and Greek studies are by no means totally convincing.  Doubts as to whether such a large effect on lung cancer incidence could possible be due to such an apparently small dose of tobacco smoke were supported by a further paper, Garfinkel (32), based on combined results from to veyr large prospective studies, the American Cancer Society's million person study and the US Veterans Study.  Two analyses were carried out.  The first (Table 7) showed no

Table 6.  Smoking Habits of Husbands of Greek Non-Smoking Women
With Lung Cancer and of Non-Smoking Control Women*

| Smoking habit of husband | | Lung cancer cases | Controls | Relative risk | Significance of trend |
|---|---|---|---|---|---|
| Non-smokers | | 11 | 71 | 1.0 | |
| Ex-smokers | | 6 | 22 | 1.8 | |
| Smokers (cigarettes/day): | 1-10 | 2 | 9 | } 2.4 | |
| | 11-20 | 13 | 32 | | $\chi^2$ = 6.45 |
| | 21-30 | 4 | 6 | } 3.4 | |
| | > 30 | 4 | 9 | | p = <0.02 |
| Total...... | | 40 | 149 | | |

* Data from Trichopoulos, Kalandidi, Sparros & MacMahon [31]

Table 7.  Lung Cancer Deaths Amongst Non-Smoking Women in the USA*

| Smoking habit of husband | Number of lung cancer deaths | Age standardized analysis mortality ratio | Matched group analysis mortality ratio |
|---|---|---|---|
| Non-smoker | 65 | 1.00 | 1.00 |
| Smoker: < 20 cigarettes/day | 39 | 1.27 | 1.37 |
| Smoker: ≥ 20 cigarettes/day | 49 | 1.10 | 1.04 |

* Data from Garfinkel [32]

significant relationship between lung cancer risk and the smoking habit of the husband. Indeed, after matching for age, occupation, education, race, urban/rural residence and absence of serious disease at the start of the study, non-smoking women married to smokers of 20 or more cigarettes a day had an estimated risk of lung cancer virtually identical to that of non-smoking women married to non-smokers.

The second analysis found no evidence of any trend in lung cancer rates in non-smokers over the period of either study. As death rates of smokers had increased substantially over the period one might have expected a similar rise to be seen in non-smokers, had passive smoking been a material cause of lung cancer risk in non-smokers.

It had been suggested that one reason why no effect of passive smoking was seen in the USA was that women spend more time out of the home and marry more often, Garfinkel recording no data on smoking habits of ex-husbands or passive smoke exposure from other sources. However, while inadequate measurement of the exposure variable is a weakness of this study (and indeed of all the other studies), Garfinkel made an interesting further point why passive smoking seems unlikely to play more than a small role in the development of lung cancer. he noted that in the study by Auerbach, Garfinkel and Hammond (37) of histological changes in bronchial epithelium taken from autopsy material, lesions frequently seen in cigarette smokers (such as atypical nuclei and lesions similar to carcinoma in situ) have very rarely been found in people who have never smoked.

### Hong Kong study

Table 8 summarizes the data presented by Chan (33) in a short letter in the Münchner Medizinische Wochenschrift. The letter gives insufficient detail to allow evaluation of the study, but it does not appear to support an association between passive smoking and lung cancer risk, there being a slightly smaller proportion of non-smoking lung cancer cases than controls married to smokers. Hong Kong is one of the areas where non-smoking women have a particularly high lung cancer risk and the figures come from a series of studies aimed at elucidating the cause of this.

### Louisiana study

Less than a month ago Correa (34) published a paper in the Lancet based on a case-control study involving 1338 lung cancer patients and 1393 comparison subjects in Louisiana,

Table 8.  Smoking Habits of Husbands of Hong Kong Non-Smoking Women
          with Lung Cancer and of Non-Smoking Control Women*

| Smoking habit of husband | Lung Cancer Cases | Controls | Relative risk[+] (95% limits) |
|---|---|---|---|
| Non-smoker | 50 | 73 | 1 |
| Smoker | 34 | 66 | 0.75 (0.43 - 1.30) |

* Data from Chan [33]

+ Not age-standardised

U.S.A., the main results of which are presented in Table 9. Although the number of non-smoking lung cancer cases is very small, only 8 males and 22 females with spouse smoking information, a higher proportion of spouses who smoked was seen among non-smoking lung cancer cases than among non-smoking controls. The overally relative risk was 2-fold which was not statistically significant, but when the comparison was restricted to spouses who had over 40 pack-years smoking compared with those who had never smoked a significant 3-fold relative risk was seen in the female subjects. The authors note that similar tabulations for smoking subjects did not show an increased risk but unfortunately do not present detailed figures. The study appears to suffer from a number of deficiencies:

(i)     the controls included a number of diseases, including ischaemic heart disease and peripheral vascular diseases, known to be smoking-associated,

(ii)    no information was recorded on time of exposure to the spouse's cigarettes,

(iii)   no mention is made of pipe and/or cigar smoking,

(iv)    no details were obtained on earlier spouses, nor length of marriage to the present one and

(v)     the use of pack-years is a poor way of separating smokers according to amount smoked, as age is also an important determinant of the variable.

In a separate tabulation, the authors show that matenal (but not paternal) smoking is positively related to lung cancer risk in smokers. A crude relative risk of 1.66 reduced to 1.36 after adjustment for active smoking. This suggests that were it possible to measure all aspects of active smoking exactly, the relative risk might reduce further. As the authors note "it is not clear whether the results reflect a biological effect associated with maternal smoking or the inability to control adequately for confounding factors related to active smoking." From the point of view that lung cancer risk is very strongly related to duration of smoking and that passive smoke exposure from the mother in childhood may extend duration, albeit at a lower exposure level, the possibility that a true effect of maternal smoking might exist merits further study.

**German study**

Knoth, Bohn and Schmidt (35) found that among 39 non-

Table 9.  Smoking Habits of Spouses of New Orleans Non-Smoking Men and Women
          with Lung Cancer and with Control Diseases*

| Sex | Cigarettes smoked by spouse (pack-years) | Lung Cancer Cases | Controls | Relative Risk |
|---|---|---|---|---|
| Male | None | 6 | 154 | 1 |
|  | 1 - 40 | 2 | 20 | } 2.0 |
|  | ⩾ 41 | 0 | 6 |  |
| Female | None | 8 | 72 | 1 |
|  | 1 - 40 | 5 | 38 | 1.18 |
|  | ⩾ 41 | 9 | 23 | 3.52[+] |
| Sexes combined | None | 14 | 226 | 1 |
|  | 1 - 40 | 7 | 58 | 1.48 |
|  | ⩾ 41 | 9 | 29 | 3.11[+] |

* Data from Correa et al [34]

+ $p < 0.05$

smoking female patients with lung cancer, 61.5% lived together
with smokers, the percentage being higher than expected on the
basis of the last microcensus which gave the percentage of male
smokers in West Germany as only 38.6% and even lower, 22.4%, in
those aged 50-69. The authors concluded that passive smoke
exposure is the most evident explanation of the development of
lung cancer in the non-smoking wives.

The conclusion seems scarcely justified. Not only is
the sample of patients very small, but the main comparison is
obviously invalid because people frequently live in households
with more than one other person, thus increasing the chance to
live with a smoker. Proper control data are sadly lacking.

**Total mortality**

Miller (38) has reported that non-smoking wives whose
husbands did not smoke cigarettes lived 4 years longer than
those whose husbands smoked cigarettes. This observation was
based on a study carried out in Erie County, Pennsylvania, in
which spouses of decedents were interviewed about their own
smoking habits and those of their spouse, conclusions being
based on average age at death measured in this way is
essentially meaningless and the results should be ignored. In
order to make inferences about relative risk of death, one
needs to have data about the smoking habits of the living
population and this was not obtained. Probably the result is
only a reflection of the fact that in the U.S.A. smokers, and
their spouses, tend to be younger than non-smokers.

**CONCLUSION**

Much of the data presented suggesting that passive
smoking is associated with more serious health hazards than
previously thought is open to considerable criticism, and none
is by any means completely convincing. More research is
certainly needed but at present the view that passive smoking
does not result in any material risk of serious disease for the
healthy non-smoker remains a reasonable one.

**REFERENCES**

1. SPEER, F. (1968).
   Tobacco and the nonsmoker.  A study of subjective
   symptoms.
   Archives of Enviromental Health, 16, 443.

2. WEBER, A., Jermini, C. & Grandjean, E. (1976).
   Irritating effects on man of air pollution due to
   cigarette smoke.
   American Journal of Public Health, 66, 672.

3. HUGOD, C., Hawkins, L.H. & Astrup, P. (1978).
   Exposure of passive smokers to tobacco smoke
   constituents.
   International Archives of Occupational and Enviromental
   Health, 42, 21.

4. U.S. PUBLIC HEALTH SERVICE (1979).
   Smoking and Health.  A Report of the Surgeon-General.
   U.S. Department of Health Education and Welfare.  DHEW
   Publ. No. (PHS) 79-50066.

5. SUDAN, B.J.L., Sterboul, J. (1979).
   Sensibilisation tabagique.  Le tabac: un allergen, la
   nicotine: un haptene.  Diagnostic par le test de
   degranulation des basophiles.
   La Nouvelle Presse Medicale, 8, 3563.

6. COLLEY, J.R.T., Holland, W.W. & Corkhill, R.T. (1974).
   Influence of passive smoking and parental phlegm on
   pneumonia and bronchitis in early childhood.
   Lancet, 2, 1031.

7. LEEDER, S.R., Corkhill, R., Irving, L.M., Holland, W.W. &
   Colley, J.R.T. (1976).
   Influence of family factors on the incidence of lower
   respiratory illness during the first year of life.
   British Journal of Preventive and Social Medicine, 30,
   203.

8. FERGUSON, D.M., Harwood, L.T., Shannon, F.T. & Taylor, B.
   (1981).
   Parental smoking and lower respiratory illness in the
   first three years of life.
   Journal of Epidemiology and Community Health, 35, 180.

9.  TAGER, I.B., Weiss, S.T., Rosner, B. & Speizer, F.E.
    (1979).
    Effect of parental cigarette smoking on the pulmonary
    function of children.
    American Journal of Epidemiology, 110, 15.

10. HASSELBLAD, V., Humble, C.G., Graham, M.G. & Anderson, H.S.
    (1981).
    Indoor environmental determinants of lung function in
    children.
    American Review of Respiratory Disease, 123, 479.

11. MELIA, R.J.W., Florey, C. du V., Altman, D.G.B., Swan, A.V.
    (1977).
    Association between gas cooking and respiratory disease
    in children.
    British Medical Journal, 2, 149.

12. GUILLERM, R., Radziszewski, E. & Caille, J.E. (1978).
    Effects of carbon monoxide on performance in a vigilance
    task (automobile driving).
    In Smoking Behaviour: Physiological and Psychological
    Influences. Edited by R.E. Thornton. p.148. Churchill
    Livingstone, Edinburgh.

13. ARONOW, W.S., (1978).
    Effect of passive smoking on angina pectoris.
    New England Journal of Medicine, 299, 21.

14. BUDIANSKY, S. (1983).
    Food and drug data fudged.
    Nature, 302, 560.

15. BRITISH MEDICAL JOURNAL (1978).
    Breathing other people's smoke.
    British Medical Journal, 2, 453.

16. REPACE, J.L. & Lowrey, A.H. (1980).
    Indoor air pollution, tobacco smoke and public health.
    Science, 208, 464.

17. BOCK, F.G.
    Nonsmokers and cigarette smoke: A modified perception of
    risk.
    Science, 215, 197.

18. LABORATORY OF THE GOVERNMENT CHEMIST (1980).
    Report of the Government Chemist 1979.
    Deparment of Industry. HMSO, London.

19.  RUSSELL, M.A.H. & Feyerabend, C. (1975).
     Blood and urinary nicotine in non-smokers.
     Lancet, $\underline{1}$, 179.

20.  FEYERABEND, C., Higenbottam, T. & Russell, M.A.H. (1982).
     Nicotine concentrations in urine and saliva of smokers and
     non-smokers.
     British Medical Journal, $\underline{1}$, 1002.

21.  RUSSELL, M.A.H., Cole, P.V. & Brown, E. (1973).
     Absorption by non-smokers of carbon monoxide from room
     air polluted by tobacco smoke.
     Lancet, $\underline{1}$, 576.

22.  ANDERSON, E.W., Andelman, R.J., Strauch, J.M., Fortuin,
     N.J. & Knelson, J.H. (1973).
     Effect of low-level carbon monoxide exposure on onset and
     duration of angina pectoris.  A study of ten patients
     with ischaemic heart disease.
     Annals of Internal Medicine, $\underline{79}$, 46.

23.  ARONOW, W.S., Stemmer, E.A. and Isbell, M.W. (1974).
     Effect   of carbon monoxide exposure on intermittent
     claudication.
     Circulation, $\underline{49}$, 415.

24.  BRUNNEMANN, K.D., Adams, J.D., Ho, D.P.S. & Hoffmann, D.
     (1978).
     The influence of tobacco smoke on indoor atmospheres.
     II.  Volatile and tobacco smoke nitrosamines in main and
     sidestream smoke and their contribution to indoor
     pollution.
     Proceedings, 4th Joint Conference of Sensing of
     Environmental Pollutants, New Orleans, Louisiana, 1977.
     p.876. American Chemical Society.

25.  STEHLIK, G., Richter, O., Altmann, H. (1983).
     Concentration of dimethylnitrosamine in the air of smoke-
     filled rooms.
     Ecotoxicology and environmental Safety, In The Press.

26.  FRIEDMAN, G.D., Pettiti, D. and Bawol, R.D.
     Prevalence and correlates of passive smoking.
     American Journal of Public Health, $\underline{73}$, 401.

27.  WHITE, J.R. & Froeb, H.F., (1980).
     Small-airways dysfunction in nonsmokers chronically
     exposed to tobacco smoke.
     New England Journal of Medicine, $\underline{302}$, 720.

28.  KAUFFMANN, F., Tessier, J-F. and Oriol,P. (1983).
     Adult passive smoking in the home environment: a risk
     factor for chronic airflow limitation.
     American Journal of Epidemiology, 117, 269.

29.  COMSTOCK, G.W., Meyer, M.B., Helsing, K.J. and Tockman,
     M.S. (1981).
     Respiratory effects of household exposure to tobacco
     smoke and gas cooking.
     American Review of Respiratory Diseases, 124, 143.

30.  HIRAYAMA, T. (1981a).
     Non-smoking wives of heavy smokers have a higher risk of
     lung cancer: a study from Japan.
     British Medical Journal, 282, 183.

31.  TRICHOPOULOS, D., Kalandidi, A., Sparros, L. & MacMahon, B.
     (1981).
     Lung cancer and passive smoking.
     International Journal of Cancer, 27, 1.

32.  GARFINKEL, L. (1981).
     Time trends in lung cancer mortality among non-smokers
     and a note on passive smoking.
     Journal of the National Cancer Institute, 66, 1061.

33.  CHAN, W.C. (1982).
     Zahlen aus Hongkong.
     Munchner Medizinische Wochenschrift, 124, 16.

34.  CORREA, P., Pickle, L.W., Fontham, E., Lin, Y. and
     Haenszel, W. (1983).
     Passive Smoking and Lung Cancer.
     Lancet, 2, 595.

35.  KNOTH, A., Bohn, H. and Schmidt, F. (1983).
     Passive smoking as cause of lung cancer in female
     smokers.
     Medizinische Klinik, 78, 54-59.

36.  HIRAYAMA, T. (1983).
     Lung Cancer in Japan.  Effects of Nutrition and Passive
     Smoking.
     Paper presented at International Conference  on Lung
     Cancer, New Orleans.

37.   AUERBACH, O., Garfinkel, L. & Hammond, E.C. (1979).
      Change in bronchial epithelium in relation to cigarette
      smoking 1955–1960 vs 1970–1977.
      New England Journal of Medicine, <u>300</u>, 381.

38.   MILLER, G.H. (1978).
      The Pennsylvania Study on Passive Smoking.
      Journal of Breathing, <u>41</u>, 5.

DISCUSSION

LECTURER: Lee                                    CHAIRMAN: Cumming

ASSENATO:    An article in the New England Journal of
             Medicine studied the development of pulmonary
             function in two populations of children.  In
             one group the mother was a smoker, and in the
             other a non-smoker, and the groups were
             observed prospectively for seven years.  After
             appropriate corrections and multi-variate
             analysis the $FEV_{1.0}$ in the smoking group was
             28, 51 and 101 ml over 1, 2 and 5 years
             respectively, or in percentage terms 7, 9.5 and
             7.0 less than the expected increase.  This
             suggests that passive smoking may have
             important effects on the development of
             pulmonary function in children.

             You mentioned the Registrar-General's statement
             of 1979, but omitted that of 1982 which stated
             "Although the available evidence is not
             sufficient to conclude that passive smoking
             causes lung cancer in non-smokers the evidence
             does raise concern about the possible serious
             public health problem"  I agree with your
             comments about the press and would suggest that
             reading the Daily Mail is less hazardous than
             passive smoking.

LEE:         I was aware of the study by Karger and found it
             difficult to evaluate.  There is a difficulty
             in assigning social class, the mothers
             education being the only variable available and
             I thought this somewhat weak.  It was difficult
             to attribute any possible effect to the smoke,
             to cross infection, or smoking during pregancy.
             I have not yet had the opportunity to look in
             detail at the mathematics, and would ideally
             like to discus it with the authors.

CUMMING:     This whole area is bedevilled by the problem
             identified by Jacques Chretien earlier, that as
             we are all passive smokers, the concept that
             non-smokers may be allocated to two groups one
             which is a passive smoking group and the other
             which is not, appears to me to be an incredible
             idea, and any paper based on that notion is
             unsound.  We need much more work, it is clearly

a quantitative and not a qualitative problem,
and I think we will have to leave the matter
there for this evening.

# NICOTINE AND THE CONTROL OF SMOKING BEHAVIOUR

D.M. Warburton, K. Wesnes and A.D. Revell

University of Reading, United Kingdom

## INTRODUCTION

It has been shown that a dose-response relationship exists between the number of cigarettes smoked each day and the risk of developing smoke-related diseases (U.S. Surgeon General, 1979). Accordingly, the weight of medical opinion over the last 20 years has argued for total abstinence, decreased consumption of cigarettes and certainly the use of products of a nominally lower yield of smoke constituents, as determined by a standard smoking machine. For example, The 1974 WHO Expert Committee on Smoking and its Effects on Health stated that "Those who are unable to stop smoking should try to reduce their exposure to such harmful substances in smoke as tar, nicotine, and carbon monoxide." The new recommendations of the 1978 WHO Expert Committee on Smoking Control has strengthened and reinforced these statements by saying that "Upper limits should be established for appropriate emission products of cigarettes. These limits (currently for tar, nicotine, and carbon monoxide) should be progressively lowered as rapidly as possible." (WHO, 1979).

This recommendation of decreasing nominal yields seems to have reduced the health risks from cigarette smoking as the yield levels have decreased from 40 mg of tar and 3.0 mg of nicotine to 18 mg of tar and 1.3 mg of nicotine (US Public Health Services, 1979). Accordingly, these pressures on smokers to reduce yields of their cigarettes continue, but it is the thesis of this paper that this attitude is mistaken and that a different aproach to minimising risk to smokers is now required. This view is based on the fact that smokers, when

217

given cigarettes with low yields, smoke more in order to try to compensate for the reduction. This compensation takes the form of increased cigarette consumption, increased smoke generation (including puffing) and smoke manipulation (including inhaling). Thus smoking lower yield brands will not necessarily be safer, because compensation will result in levels of exposure to smoke which the smoker had not anticipated.

In our work, we have examined the magnitude of compensation by smoke generation and smoke manipulation and, from this sort of data, it is possible to make an estimate of the additional exposure to smoke constituents that could result from compensation. In this chapter, we will discuss only compensation i.e. the increases in smoking in response to lower nominal delivery of products and not the general issue of titration upwards and downwards.

## CONSUMPTION

Increasing the number smoked is the most obvious way for smokers to compensate for lower yield. In a study of M.H.A. Russell (Russell, Wilson, Patel, Cole and Feyerabend, 1973), smokers were switched from their usual brand (1.34 mg of nicotine and 18 mg tar average) to cigarettes of less than 0.3 mg nicotine and 4 mg of tar. The consumption of the group increased by 17% after the switch, which delivered 0.68 mg of tar on the basis of the nominal deliveries of the products. However, in terms of the cigarette yield, subjects would have needed to smoke five to ten times as many to compensate completely for the lower nicotine i.e. an increase from 10.7 cigarettes to over 53.5 if they did not change their smoking behaviour in any way. Thus consumption is not a major method of compensation.

This finding is typical of several that show that subjects can compensate by increasing the number of cigarettes that are smoked. However, many negative studies can be explained in terms of changes in either smoke generation (including puffing) or smoke manipulation (including inhalation) and these will be considered in the next two sections.

## SMOKE GENERATION

Changes in smoke generation occur by modifying the flow of smoke into the mouth. In the last section, cigarette consumption was related to the nicotine yield of the cigarettes. These nicotine levels were obtained from standard smoking machines which puff according to the standard set of parameters. This method enables comparisons of cigarettes to be made, but only produces an approximation of human smoking.

One of the most obvious methods of changing smoke generation is intentional or inadvertant hole blocking. Cigarettes in the low tar range achieve this rating on a smoking machine by perforations in the filters so that an inflow of air dilutes the smoke of each puff. Smokers can achieve much higher deliveries by blocking the vent holds with their fingers or lips. By use of these techniques, an ultra-low yield brand with an estimated 1 mg on a smoking machine could be equivalent to a medium yielding brand with 13 mg of tar in the hands of the smoker (Koslowski, Rickert, Pope, Robinson and Frecker, 1982).

In a series of studies, Creighton and Lewis (1978) recorded the pattern of smoking in terms of number of puffs, puff interval, puff volume and puff shape. They found that there were marked inter-individual differences and the patterns did not conform to the standard variables of the smoking machine. The consequence of these variations in smoking pattern, in terms of nicotine deliveries, was examined using a puff duplicator (Creighton and Lewis, 1978). The divergence of the actual deliveries from the smoking machine yields was revealed when a group of male smokers derived an average of 2.25 mg of nicotine and a group of female smokers obtained an average of 1.4 mg of nicotine.

Tar levels were correspondingly higher with males achieving a delivery to the mouth of 33.4 mg and females getting 24.3 mg from a product with a nominal yield of 23.3 mg i.e. an average puffing intensity of 143% for males and 104% for females. The coefficient of variation between nicotine deliveries for different subjects who smoked the same brand ranged from 24% to 38% with a mean of around 30%. Puff volumes, which range from 23 - 79 ml, can result in a change of deliveries from 2.09 to 3.19 mg of nicotine, 25.8 to 60 mg of tar and from 21.6 to 53.8 mg of carbon monoxide. An increase in puff number from 10 puffs to 14 puffs could result in a cigarette with a standard machine yield of 1.32 mg of nicotine and 13.9 mg of tar delivering 1.84 mg of nicotine and 30 mg of tar, an increase in mouth tar of 216%.

Estimates of nicotine entering the smoker's mouth can also be made from the nicotine deposited in the cigarette filter and the latter's filtration efficiency. In our own studies (Warburton and Wesnes, 1978), we found changes in the estimated delivery depending on the strength. Low to middle tar smokers smoked both a 0.3 mg nicotine cigarette with 4 mg of tar and a 0.7 mg nicotine cigarette with 9 mg of tar more intensely obtaining mean nicotine deliveries of 0.68 mg and 1.3 mg of nicotine respectively. Titration had occurred by smoking the lower yield cigarettes more intensely. As a consequence, the delivery of tar to the smoker would have doubled in each case giving mouth levels of 9 mg and 16.7 mg of tar respectively. This calculation assumes that the ratio of tar to nicotine remains constant but both the tar to nicotine ratio and carbon monoxide to nicotine ratio increase with increased puffing (Creighton and Lewis, 1978), so that smokers who are puffing harder to obtain more nicotine will get proportionally larger quantities of tar and carbon monoxide.

Clearly the machine-estimated yield can give a misleading picture of the smoke constituents entering a smokers mouth. With low yield products, smokers are obtaining more tar and carbon monoxide as they attempt to compensate by smoke generation. Therefore, it is important that we understand the control mechanism for this process.

## CONTROL OF SMOKE GENERATION

One approach to understanding the mechanism or mechanisms that are involved in a particular process is to vary systematically the variable under consideration and examine the concomitant variation in the measures of the process. We have presented a series of cigarettes of covarying yield and examined the changes in nicotine presented to the smokers mouth. This dependent variable was estimated from the analysis of the nicotine that was deposited in the cigarette butt. We found that nicotine presented changed systematically when smokers were presented with cigarettes of covarying yield. As yield increased nicotine presented also increased but the relation was not a linear one. The best fitting curve was exponential with an exponent of 0.50 and a constant of − 0.024 i.e. virtually zero so that it can be assumed that the curve passed through the origin. An exponent less than one means that puffing increased much more for increments in yield at low levels than at high.

Exponential curves with an exponent less than one are common in sensory psychology and describe the relation between subjective sensory experience and the physical intensity of the sensory stimulus. In sensory experience and the physical intensity of the sensory stimulus. In sensory physiology the same exponent is found for the relation between frequency of firing of the sensory neurone and physical intensity as that for the relation between subjective experience and physical intensity of the stimulus for a specific modality i.e. subjective experience is mapping onto sensory neurone activity. Exponents in the range of 0.5 - 0.8 are found for the relation between subjective experience and the physical intensity of taste and odour stimuli.  It seemed a reasonable hypothesis that puffing intensity could be controlled by the flavour impact of the cigarette smoke.

In order to get more support for this idea, we presented a series of cigarettes of differing nicotine yields to smokers and obtained their subjective estimates of the flavour of the cigarette smoke after four puffs by the technique of magnitude estimation.  By means of a standard cigarette, the scales were adjusted so that it was possible to plot the subjective estimates of flavour against machine-smoked nicotine yield.  It should be noted that nicotine is only used as an index of the set of components of smoke.  The best fitting curve for the subjective sensory experience of the cigarette flavour was exponential with an exponent of 0.51.

Thus there was an analogous exponential relation between subjective strength and nicotine yield and between puffing intensity and nicotine yield.  The similarity of the function suggests that cigarette puffing is under the control of the taste receptors in the mouth which are sensitive to the smoke constituents, including nicotine.  When flavour impact is decreased, puffing increases to compensate for the decreased yield and more smoke is generated.  If, as seems likely, nicotine is one of the most important ingredients of cigarettes for the smoker, increased smoke generation for nicotine results in proportionately more tar and carbon monoxide being drawn into the mouth and potentially an increased risk to health.  It seems from this data that the cigarettes of the future must maintain the flavour level to prevent increased puffing and the consequent higher tar and carbon monoxide levels in the mouth prior to inhalation.

## SMOKE MANIPULATION

A crucial part of cigarette smoking is further manipulation of the smoke which usually includes inhalation. About 90% of cigarette smokers say that they inhale to some extent and 77% say that they usually inhale "a lot" or "a fair amount" (US Public Health Services, 1979).  Since nicotine from tobacco smoke is absorbed from the lungs, it follows that a high percentage of the mouth nicotine is made available for absorption by inhalation.  Smoke inhalation results in very efficient absorption of nicotine and the large percentage of

smokers who do inhale provides evidence that an aim of smoking
is to obtain nicotine. Thus it is clear that we need a measure
of compensation by inhalation.

One index of smoke uptake is the amount of carbon
monoxide exhaled after a cigarette. Carbon monoxide is
absorbed into the bloodstream from the alveoli and not the
mouth. When the residual smoke has been expelled from the
lungs after smoking, carbon monoxide exchange from the
carboxyhaemoglobin in the blood to the lungs will occur so that
the level of exhaled, end-tidal carbon monoxide provides an
index of smoke uptake. In addition, the exhaled carbon
monoxide can be used to estimate the exposure of the lungs to
tar in the absence of any other measure. The advantage of this
method is that it is non invasive. However, there are
disadvantages in using these estimations (Ashton, Stepney and
Thompson, 1981).

Firstly, we might not expect that smoke constituents
would be correlated because tar is is the particulate phase
while carbon monoxide is part of the vapour phase and nicotine
is in both phases. Creighton's data on smoke generation with
the puff analyser (Creighton and Lewis, 1978) enable a test of
the first criticism. When the carbon monoxide deliveries and
tar deliveries are correlated, the correlation coefficient is
0.755, significant well beyond the 0.001 level. Thus despite
the widely varying range of puff shapes, puff volumes and puff
intervals used by Creighton and Lewis (1978), there was still a
highly significant correlation between the two measures which
gives confidence in estimates for groups of smokers.

Secondly, while only 60% of the carbon monoxide is
absorbed from the lungs, 60-80% of the tar is retained by
deposition and over 90% of the nicotine is absorbed after
inhalation (Creighton, 1973, cited by Ashton et al, 1981).
However, this only means that estimates of exposure of tar from
the exhaled carbon monoxide will be conservative. Higher
values are obtained with deeper inhalation and so a value which
is relatively higher for one cigarette than another will be
even higher in absolute terms.

A good example of compensation by changes in inhalation
is an eleven week crossover study in which smokers switched
from their own 1.4 mg nicotine brand with 18 mg of tar to 0.6
mg cigarettes with 6.5 mg of tar i.e. a decrease of 64% in
machine-estimated tar yield (Ashton, Stepney and Thompson,
1979). They found that the plasms carboxyhaemoglobin (COHb)
levels were only 19% lower after smoking a 0.6 mg cigarette
than after smoking their own brand. From their data on the

rise in COHb, we have computed a relative smoking intensity index by dividing the rise in COHb level by the machine-estimated delivery of carbon monoxide.  For their own cigarette, the index was 0.078 compared with 0.11 with the lower yielding product i.e. 143% greater smoke uptake with the lower yield cigarette.  A further revealing index is the initial tar exposure index which we obtained from the product of the smoke uptake index and the machine-estimated yield of tar.  This index was 1.41 for their own branch and 0.73 for the lower yielding cigarette i.e. a 49% reduction in initial tar exposure.

This saving in exposure, while not as much as the 64% that would be predicted from the nominal yield, would be important if the smokers did not increase their consumption. However, consumption also increased significantly and 24 hr values for urinary nicotine and cotinine excretion show that the subjects were obtaining just as much nicotine with the reduced yield cigarette as from their own. Another smoking intensity index can be computed from the presmoking COHb levels, a measure of the amount of smoking outside the laboratory.  The presmoking COHb levels on the lower yield cigarette were 98% of those with smoking their own brand, so that there was no saving in carbon monoxide exposure at all. The values for the relative inhalation index were 0.45 with their own cigarette and 0.79 (i.e. 76% higher) with the nominally lower yield brand.  The tar exposure indices were 8.24 and 5.1 respectively i.e. only a 38% saving in tar exposure compared with the expected 64% reduction.

In summary, smokers compensated by smoke manipulation, as well as consumption and the consequence was no reduction in daily nicotine absorption and concomitantly no reduction in carbon monoxide exposure and some reduction in tar exposure. Clearly the smokers are sensitive in some way to nicotine yield and control inhalation accordingly.

## CONTROL OF INHALATION

As we pointed out earlier there was evidence that flavour impact is an important cue for smoke generation and so it would be unsurprising if oral cues were also part of the mechanism that controlled smoke manipulation, especially inhalation.  In order to investigate the involvement of mouth cues in judgements of cigarette strength, a series of cigarettes with varying nicotine, but constant tar and carbon monoxide was prepared with and without menthol. Strong mentholation desensitizes the taste receptors in the mouth and so minimizes oral cues. These two sets of cigarettes were compared in two conditions, with and without inhalation. The non-menthol cigarettes which were puffed and the smoke inhaled gave the smoker oral cues and cues from the physiological effects of the nicotine that was absorbed from the lungs and to a much lesser extent from the mouth. The non-menthol cigarettes which were puffed, but the smoke was not inhaled, gave the smoker oral cues and weak physiological cues from any nicotine which was absorbed orally. The menthol cigarettes which were puffed and the smoke inhaled still gave physiological cues from the absorbed nicotine but diminished oral cues. The menthol cigarettes which were puffed but the smoke was not inhaled gave minimal cues for the judgement of cigarette strength. The subjects ranked the three cigarettes in the four different conditions for strength on four separate occasions. Tests of exhaled carbon monoxide confirmed that subjects had complied with the inhalation instructions.

There was clear evidence that subjects could distinguish the cigarettes when they were given both oral and physiological cues from the inhaled nicotine. There was clearly no differentation in the menthol and non-menthol, non-inhale conditions, showing that oral cues are not sufficient for differentiating the strength of the cigarettes. These two pieces of evidence point strongly to physiological effects of absorbed nicotine being the cues for strength discrimination. There was some strength discrimination in the menthol, inhale condition which supports the idea of differentiation by the effects of nicotine after inhalation from the mouth. However the discrimination was not as good as with non-mentholated products which suggests that nicotine was producing some sensory cues in the respiratory tract between the mouth and the lungs, perhaps due to stimulation of the trigeminal nerve receptors. Menthol would block these receptors in the menthol -inhale condition and allow only internal physiological cues to be used for judging strength.

The nature of this internal mechanism which we will call the "nicotinostat" will be considered next. It is obviously that a necessary condition for controlled, self-administration of a drug is the ability to discriminate the level of drug in the body. In a study which was designed in order to gather direct evidence for a nicotinostat, we gave subjects nicotine tablets to preload them with nicotine and stimulate the "nicotinostat", prior to smoking a 0.6 mg nicotine cigarette (Wesnes, Pitkethley and Warburton, in preparation). Unpublished work by Dr M A H Russell and Dr K Wesnes had shown that oral absorption from tablets containing

1.5 mg nicotine gave venous levels of 6.0 ng/ml at pH 6 and 10.5 ng/ml at pH 9. Puffing behaviour, butt nicotine and exhaled carbon monoxide were measured. No differences were seen in puffing variables or butt nicotine levels for nicotine and placebo conditions. However, there was a significant reduction of exhaled carbon monoxide after the subjects had received a nicotine tablet indicating reduced inhalation. Clearly the smokers were reducing their nicotine intake according to a nicotinostat sensitive to plasma nicotine levels.

The interesting question is how this mechanism operates and where it is located. There is good evidence from drug discrimination studies that the cue is based on cholinergic action (Rosecrans and Chance, 1977). Specifically nicotine acts via nicotinic cholinergic receptors, as shown by agonists and antagonists (Romano, Goldstein and Jewell, 1981). Thus there must be some cholinergic control mechanism in the body which is sensitive to the absorbed nicotine.

During inhalation, the smoker aerosol passes down the bronchi into the alveoli and absorption occurs into the pulmonary capillaries. After absorption into the pulmonary capillaries, the nicotine-loaded blood leaves the lungs via the pulmonary veins into the heart. From there, the nicotine is pumped out into the aorta from which the large arteries branch off to distribute blood around the body. Nicotine in smoking doses acts on the heart and brain. Information from one or more of these organs could be combined to control the pattern of smoking throughout the day and of smoking a single cigarette.

## (1)  Cardiovascular Control

The acute cardiovascular responses to tobacco and nicotine have been summarized in the Surgeon General's report (1979). The acute changes after smoking are increased heart rate (10 to 25 per minute, (blood pressure (10 to 20 mm Hg systolic, 5 to 15 MM Hg diastolic) and cardiac output (0.5 l/min/m2) typically occur in smokers after smoking one or two cigarettes. These effects are assumed to be mediated mainly by sympathetic excitation together with some due to catecholamine release from the adrenal medulla and chromaffin tissue. It is now established that people can learn to discriminate their own changes in heart rate (DiCara, Barber, Kamiya, Miller, Shapiro and Stoyva, 1975), so that could be a cue for nicotine control, but the evidence that we shall discuss next suggests that a brain system is involved.

## (2) Brain Control

The carotid artery from the aorta takes blood directly
to the brain and about one fifth of the absorbed nicotine i.e.
250 ug passes to the brain within 10 secs. Nicotine passes
easily into the brain and well over ninety per cent is taken up
on the first pass through the brain (Oldendorf, 1977).
Microautoradiograms reveal radioactive nicotine in cortical
cells, high levels in the hippocampus, the cerebellum, and
nuclei of the hypothalamus and brain stem (Schmiterlow, Hanson,
Applegren and Hoffman, 1967).

The study of Romano et al (1981), that was mentioned
earlier, found that the nicotine discrimination was due to the
cholinergic action of the drug on the brain because a) the cue
could be produced by an injection into the cerebral ventricles
of a nicotine does which was one sixth of the subcutaneous
dose; b) a nicotine agonist, which passes the blood-brain
barrier poorly, had no effect systemically but had cue
properties when injected intraventricularly; and c) a nicotine
antagonist, which passes across the barrier, blocked the
nicotine discrimination but another which does not pass across
the barrier, was ineffective (Romano et al, 1981).

Studies with iontophoresis have revealed that nicotine
acts on acetylcholine receptors in the thalamic nuclei,
hippocampus and reticular formation nuclei, but not those on
cortical cells and caudate nucleus cells, which gives three
possible sites of action. An important discovery was that the
effective intraventricular dose of nicotine was only one sixth
of the effective systemic dose (Romano et al, 1981) which
indicates that the site of action of nicotine is not easily
reached from the ventricular system, or, at least, is some
distance from the lateral ventricles. If this inference is
true it would rule out the hippocampal formation and thalamus,
but not the mesencephalic reticular nuclei.

Cortical acetylcholine release and cortical excitation
can be produced by stimulation of the mesencephalic reticular
information and this phenomenon can be reduced in one
hemisphere by destruction of this region ipsilaterally (Celesia
and Jasper, 1966). Kawamura and Domino (1969) found that
"smoking" doses of nicotine did not produce cortical
desynchronization in cats after bilateral lesions in the
tegmental region of the midbrain. The tegmental region in the
mesencephalic reticular formation is the origin of a
cholinergic pathway which terminates on cells at the sensory

cortex and produces electrocortical arousal (see review by Warburton, 1981).

Thus it seems that "smoking" doses of nicotine ascend in the carotid artery and excite nicotinic receptors on the tegmental neocortical cholinergic pathway in the midbrain. The outcome of activation of the pathway at the midbrain is release of acetylcholine at the cortex. Many human studies have shown that smoking increases the amount of cortical desynchronization in the form of an upward shift in dominant alpha frequency (See extensive review in Edwards and Warburton, 1983). This cortical desynchronization is similar to the alert pattern in a person who is concentrating. There is good evidence from research on biofeedback that people can learn to discriminate changes in their states of electrocortical activity (DiCara et al, 1975) and it is a reasonable hypothesis that nicotine by its action on nicotinic cholinergic receptors in the mesencephalic reticular formation changes electrocortical activity and that people learn to control their nicotine intake on the basis of these changes.

## TOWARDS A SAFER CIGARETTE

Smokers control their nicotine intake by varying their cigarette consumption, their strength of puffing and their inhalation, in order to obtain an optimal dose of nicotine which will act on the body and satisfy the sorts of needs such as mental activation and relief from stress that were discussed in our companion chapter (Wesnes and Warburton, 1983). Nicotine in "smoking" doses seem to be safe for normal healthy adults in comparison with other stimulant and sedative substances. Examination of the evidence (US Public Health Services, 1979) suggests that any health problems, that are associated with smoking, are more likely to be related to the tar of the smoke. It follows that a less hazardous cigarette will not be one with low nicotine, low tar and low carbon monoxide because such non-selective reductions in smoke yield ignore the smoker's needs. Instead a more sensible cigarette would be a product with the reduction of some smoke constituents but sufficient nicotine to satisfy the smoker, and so prevent compensation. This type of product would provide the benefits of smoking with minimum risk. the idea is not new but was proposed 10 years ago (Russell et al, 1973) but only now are the scientific thinking and technology ready for this sort of innovation.

Obviously such an innovatory cigarette must be acceptable to smokers. So far manipulations of smoke delivery have not been satisfactory because they remove important flavours. It is therefore essential that the product must be acceptably flavoured so that it would not need to be puffed more intensively in order to obtain the full flavour impact for the smoker. Clearly, future progress must be in the direction of reduction of specific smoke constituents to reduce risk but adding flavouring in order to satisfy taste requirements of the smoker.

## REFERENCES

1. ASHTON, H., Stepney, R. and Thompson, J.W. (1979).
   Self titration by cigarette smokers.
   British Medical Journal, 2, 357-360.

2. ASHTON, H., Stepney, R. and Thompson, J.W. (1981).
   Should intake of carbon monoxide be used as a guide to intake of other smoke constituents?
   British Medical Journal, 282, 10-13.

3. CELESIA, G.G. and Jasper, H.H. (1966).
   Acetylcholine released from the cerebral cortex in relation to state of activation.
   Neurology, 16, 1053-1063.

4. CREIGHTON, D.E. and Lewis, P.H. (1978).
   The effect of smoking pattern on smoke deliveries.
   In Smoking Behaviour, ed. by R.E. Thornton, 289-314, Churchill Livingstone, Edinburgh.

5. DOLL, R. and Hill, A.B. (1964).
   Mortality in relation to smoking: Ten years' observations of British doctors.
   British Medical Journal, 1, 1399-1410 and 1460-1467.

6. DICARA, L., Barber, T.X.D., Kamiya, J., Miller, N.E., Shapiro, D. and Stoyva, J. (1975).
   Biofeedback and Self-Control.
   Aldine, Chicago.

7. EDWARDS, J. and Warburton, D.M., (1983).
   Smoking, nicotine and electrocortical activity.
   Pharmacology and Therapeutics, 19, 147-164.

8. KAWAMURA, H., and Domino, E.F. (1969).
   Differential actions of m and n cholinergic agonists on the brainstem activating system.
   International Journal of Neuropharmacology, 8, 105-115.

9.  KOSLOWSKI, L.T., Rickert, W.S., Pope, M.A., Robinson, J.C.
    and Frecker, R.C. (1982).
    Estimating the yield to smokers of tar, nicotine and
    carbon monoxide from the 'lowest yield' ventilated
    filter-cigarettes.
    British Journal of the Addictions, 77, 159-165.

10. OLDENDORF, W.H. (1977).
    Distribution of drugs in the brain.
    In Psychopharmacology in the Practice of Medicine, ed. by
    M.E. Jarvik, pp. 167-175, Appleton-Century-Crofts, New
    York.

11. ROMANO, C., Goldstein, A., Jewell, N.P. (1981).
    Characterization of the receptor mediating the nicotine
    discriminative stimulus.
    Psychopharmacology, 74, 310-315.

12. ROSECRANS, J.A. and Chance, W.T. (1977).
    Cholinergic and non-cholinergic aspects of the
    discriminative stimulus properties of nicotine.
    Advances in Behavioral Biology, 22, 155-185.

13. RUSSELL, M.A.H. (1976).
    Tobacco smoking and nicotine dependence.
    In Research Advances in Alcohol and Drug Problems, Vol 3,
    ed. by R.J. Gibbins, Y. Israel, H. Kalant, R.E. Popham,
    W. Schmidt, and R.G. Smith, pp. 1-48, Wiley, New York.

14. RUSSELL, M.A.H., Wilson, C., Patel, U.A., Cole, P.V. and
    Feyerabend, C. (1973).
    Comparison of effect on tobacco consumption and carbon
    monoxide absorption of changing to high and low nicotine
    cigarettes.
    British Medical Journal, 4, 512-516.

15. SCHMITERLOW, C,.G., Hansson, E., Andersson, G., Appelgren,
    L.E., and Hoffman, P.C. (1967).
    Distribution of nicotine in the central nervous system.
    Annals of the New York Academy of Sciences, 142, 2-14.

16. US PUBLIC HEALTH SERVICES: Smoking and Health, (1979).
    A report to the Surgeon General: 1979.
    US Department of Health, Education and Welfare,
    Washington DC.

17. WARBURTON, D.M. (1981).
    Neurochemical Bases of Behaviour.
    British Medical Bulletin, 37, 121-126.

18.  WARBURTON, D.M. and Wesnes, K. (1978).
     Individual differences in smoking and attentional
     performance.
     In Thornton, R.E. (ed), Smoking Behaviour: physiological
     and psychological influences. London: Churchill-
     Livingstone.  pp. 19-43.

19.  WESNES, K. and Warburton, D.M. (1983).
     Smoking, nicotine and human performance.
     Pharmacology and Therapeutics, 21, 189-208.

20.  WHO (1979).
     Controlling the Smoking Epidemic.
     Geneva: World Health Organization.

DISCUSSION

LECTURER: Warburton                                CHAIRMAN: Cumming

FLETCHER:         There have been experiments with Naloxone in
                  relation to endogenous endorphins. Does this
                  fit into your hypothesis?

WARBURTON:        It does not relate specifically to the work we
                  have described. Some work has been done
                  relating nicotine to endogenous opiates, and
                  there is some evidence that smoking exposure
                  and changes in the chest are related to
                  endorphins, but I know of no evidence
                  suggesting that nicotine acts on the endogenous
                  opiates in the brain, but if it could be
                  demonstrated it would be powerful evidence for
                  the addictive nature of nicotine. the work at
                  the Salk Institute in this area produced no
                  positive result.

CUMMING:          Does any member of the audience know of any
                  work relating nicotine to endorphin secretion?
                  No.

WARBURTON:        The pleasure pathways appear to be coded wih
                  catecholamines and the evidence that nicotine
                  acts on catecholamine pathways is very poor.
                  Nicotine in smoking does appear to have no
                  effect, but experiments using very high doses
                  at frequency intervals have shown some effect.
                  With our nicotine tablets we have been unable
                  to find any pleasurable effect.

BAHKLE:           Although nicotine is not the sole motivation it
                  does seem to be a strong motivation. What is
                  the success of chewing nicotine chewing gum in
                  helping people to give up smoking.

WARBURTON:        There are successes in the use of this chewing
                  gum, I don't remember the figures but they are
                  in excess of the 20-25% by conventional
                  methods. There are a number of problems — first
                  the nicotine must be bound to the gum so that
                  accidental ingestion by children does not
                  produce poisoning, so that it can only be
                  produced by half an hour of hard chewing.
                  Gastric absorption is also poor, so it is not

as affective as a cigarette.

CHRETIEN:        Do you have any information on the circadian
                 rhythm of nicotine?

WARBURTON:       We have some laboratory information, we looked
                 at the effects of smoking on light and heavy
                 smokers and found that a cigarette which worked
                 well for heavy smokers did not produce a
                 similar effect in light smokers when they
                 smoked in the morning. This was a surprise, so
                 we asked them how they found the cigarette and
                 they replied that it made them sick. On giving
                 them the same cigarette in the afternoon we
                 found it to be perfectly effective. If there
                 is a duirnal shift in chemicals in the brain
                 maybe the light smoker does not need to smoke
                 in the mornings but when he does he takes too
                 much nicotine, with a resultant toxic effect.
                 This may be why a light smoker is a light
                 smoker.

CUMMING:         After coffee Michael Sleigh will show two films
                 demonstrating ciliary activity.

# WORK AND STRESS AS MOTIVES FOR SMOKING

K. Wesnes, A. Revell, and D.M. Warburton

Department of Psychology
University of Reading
Earley Gate, Reading, Berks

In this paper we describe some findings from our research into smoking behaviour. In the first two sections evidence will be presented that smokers feel that smoking helps them both to concentrate and to relax, cite these types of help as motives for smoking, and smoke more during times of work and stress. In the following two sections studies will be described which have shown that smoking has a beneficial effect on mental efficiency and that this effect is due to the action of nicotine on the central nervous system. Finally we shall consider the role of nicotine in smoking motivation, and present evidence that the availability of nicotine influences smoking behaviour in work situations, particularly by altering inhalation patterns.

## SELF-REPORTED EFFECTS OF SMOKING AND SMOKING MOTIVATION

Major surveys carried out in this and other laboratories have indicated that the vast majority of smokers respond positively to the questions "Smoking helps me to relax" and "Smoking helps me to think and concentrate" (Warburton and Wesnes, 1978). While other surveys have further deduced that smokers report that they smoke in order to relax, to our knowledge none has asked smokers whether they smoke in order to concentrate. We have therefore carried out a questionnaire survey on a student population concerning their smoking habits and motivation, in which we specifically asked them whether smoking helped them concentrate, and if so, whether this was a motive for smoking. It is clear from Table 1 that as has been previously found, a large proportion of smokers report that smoking helps them to concentrate, but more importantly 74% of these smokers report that this help is a motivation to smoke.

233

**TABLE 1:**     **Questionnaire responses of 378 male and female student smokers**

Q.   Does smoking help you concentrate?

|  |  |
|---|---|
| not at all | 16% |
| a little | 35% |
| quite a lot | 35% |
| very much so | 10% |

Q.   If smoking helps you concentrate, do you smoke to help you concentrate?

|  |  |
|---|---|
| not at all | 10% |
| partly | 66% |
| completely | 8% |

## SELF-REPORTED SMOKING RATES DURING WORKING AND STRESS

From the previous section it is clear that smokers report that smoking helps them both to concentrate and to relax, and also claim that these effects are motives for smoking. Consequently, if these are major motives, it would be expected that smokers would smoke more during times of work and stress than during other times. Evidence in support of this comes from two studies.

A survey of smoking at work by Meade and Wald (1977) showed that cigarette smoking was frequent during the working day. When there were no restrictions on smoking at work, male and female office workers smoked 56% of their cigarettes during working hours, no matter whether they were light or heavy smokers. The highest smoking rate per hour (about 30% greater than the average hourly rate for the whole day) occurred at work and so did the second highest hourly rate. This occurrence of peak smoking rates during the working day suggests that smoking may be used by people to help cope with the demands of work, especially work which involves thinking and concentration.

University examinations are a time at which students commonly report that they are under stress, and also one in which it is generally true to say that students tend to study harder than during others. In a recent study (Warburton, Wesnes & Revell, 1983) 48 first year undergraduates kept a detailed diary of their smoking habits during an examination week and then later during a quiet period of a summer term. From Table 2 it can be seen that the students smoked more throughout the day during the examination period, than during the equivalent times in the non-examination period. Further, during the examination period they smoked more cigarettes on those mornings preceding an afternoon examination, this figure representing an 80% increase on the smoking rate of the non-examination period. The differences between the two periods were highly statistically significant, and the trends highly consistent throughout the group, with for example only two of the 48 subjects not increasing their morning smoking during the examination period.

An identical pattern occurred for subjective estimates of the strength and depth of the inhalation of the cigarette smoke during the two periods. Throughout the examination period the students reported that they inhaled more strongly, and to a greater depth, than during the non-examination period, the differences again being highly significant and extremely consistent across subjects. The students also kept butts from the cigarettes they smoked during both periods and these were analysed for nicotine content. Intriguingly, despite the subjective reports of increased strength and depth of inhalation during the examination period, the nicotine retained in the butts was actually slightly lower during this period, indicating that the smokers had generated less smoke from the cigarettes. Thus during the examination period, the students smoked more cigarettes and also inhaled more of the smoke they generated, while they possibly generated less smoke from the

**TABLE 2: Number of cigarettes smoked during an examination period and a non-examination period**

| | | |
|---|---|---|
| EXAM | Mornings | 3.07 |
| PERIOD | Afternoons | 4.15 |
| | Evenings | 7.1 |
| | Mornings before afternoon exams | 3.95 |
| NON-EXAM | Mornings | 2.2 |
| PERIOD | Afternoons | 3.5 |
| | Evenings | 5.97 |

cigarettes. This increased self-reported inhalation can be interpreted as an attempt to maximize nicotine absorption during the stressful period. Certainly, if it were the oral-manipulative aspects of smoking which the smokers were seeking, or the flavour of the smoke, an increase in smoke generation would have been expected as opposed to an increase in inhalation.

## EFFECTS OF SMOKING ON PERFORMANCE

From the evidence presented so far it is clear that smokers claim firstly that smoking helps them to concentrate and relax, and secondly that they smoke for these effects. Further, in the previous section, evidence was presented that they smoke more when working and in times of stress. In this section we shall consider whether it can be objectively determined that smoking has beneficial effects on performance in work-like situations. Over the last decade we have carried out an extensive series of studies of the effects of smoking on a variety of tasks requiring high levels of concentration. These experiments have been reviewed recently (Wesnes and Warburton, 1983a) and the consistent finding has been that smoking has a beneficial effect on performance. In our early studies, smoking enabled smokers to maintain their concentration over 80 minute vigilance tasks, whereas efficiency declined over time in non-smoking conditions.

In an extended series of studies using a task requiring high levels of cognitive processing, smoking has been found not only to prevent the decline in efficiency which occurs over time, but also to improve performance above rested baseline levels. These improvements have been determined using a rapid visual information task in which subjects monitor rapidly presented digits for sequences of three consecutive odd or even digits. They perform the task for a ten minute period and then smoke a cigarette or rest according to the experimental condition. Following this they then perform the task again, this time for 20 minutes. In non-smoking conditions performance drops from the first to the second period, whereas in smoking conditions, performance is actually better for the first 10 minutes after smoking than before smoking.

## ROLE OF NICOTINE IN THE EFFECTS OF SMOKING ON PERFORMANCE

Cigarette smoking is a highly efficient technique for delivering nicotine to the central nervous system. Following the inhalation of cigarette smoke nicotine immediately enters the bloodstream and part of it is carried within seconds to the brain, where it increases electrocortical activity by its action on the ascending cholinergic pathways of the midbrain reticular formation (Warburton, 1981). On the basis of these electrocortical and neurochemical effects, it would be expected that nicotine would have a favourable effect on human information processing.

We believe that it is the nicotine absorbed during smoking which is responsible for the improvements in performance described in the previous section, and in order to test this we have studied the effects of nicotine administered in tablet form. In a visual vigilance task nicotine helped reduce the vigilance decrement which occurred over time in the placebo condition (Wesnes, Warburton and Matz, 1983). Further, the tablets produced the same effects in non-smokers, light smokers and heavy smokers. In another study nicotine tablets improved performance by reducing the Stroop effect, equivalent improvements again occurring in both smokers and non-smokers (Wesnes and Warburton, 1978). In a third study (Wesnes, 1979) in which non-smokers were given nicotine tablets while performing the rapid information processing task described earlier, nicotine tablets had comparable effects to those of cigarette smoking (e.g. Wesnes and Warburton, 1983b). These studies therefore provide good evidence that nicotine is involved in the beneficial effects of smoking on concentration.

## ROLE OF NICOTINE IN SMOKING MOTIVATION

On the basis of the evidence presented so far it is reasonable to propose that smokers smoke in work situations in order to obtain the beneficial effects of nicotine. Further, in previous studies the improvements following smoking are related to the nicotine yield of the cigarettes, although the relationship is not simple. Generally, cigarettes having nicotine yields in the range 1.3 - 1.5 mg produce the greatest improvements, while cigarettes which yield either more or less nicotine than this produce smaller improvements. On this basis we would predict that in work situations, smokers are seeking a particular level of nicotine, and if given cigarettes which have standard yields which do not produce these levels, then they will alter their smoking behaviour accordingly.

This is precisely what happened in one study of the effects of smoking on the performance of the rapid information processing test, in which we retained the cigarette butts for nicotine analysis (Wesnes and Warburton, 1983b). By measuring the amount of nicotine retained in the cigarette filter, it is possible to estimate how much nicotine entered the smoker's mouth. This figure can then be compared to the smoking machine yield of the cigarette, which is the amount of nicotine drawn

from the cigarette when a 35 ml puff lasting two seconds is
taken every minute until the cigarette is smoked to a few mm
from the filter. In this study the smokers managed to obtain
2.25 times more nicotine than the machine smoked yield of one
cigarette of 0.28 mg, whereas they obtained only 1.03 times
more nicotine from a cigarette having a yield of 1.65 mg.

In another experiment we additionally measured the
levels of exhaled carbon monoxide before and after smoking the
experimental cigarettes, in order to make inferences about the
degrees to which the smoke was taken up from the various
cigarettes. From Table 3 it can be seen that the four
cigarettes had a wide range of machine smoked nicotine yields.
When we consider the deliveries of nicotine to the smokers,
which were inferred from the nicotine contents of the cigarette
butts, it is clear that they generated significantly more
nicotine from the three lower yield cigarettes than would be
predicted from machine smoking, and less from the highest yield
cigarette. The smoking intensity index presented in Table 3 is
obtained by simply dividing the increase in exhaled carbon
monoxide produced by smoking the cigarette, by the machine
smoked carbon monoxide yield. Statistical testing revealed
that the smoking intensity index of the cigarette having the
lowest nicotine yield was significantly higher than that of the
other three, between which there were no significant
differences. This indicates that when the machine smoked yield
is taken into account, the subjects absorbed proportionally
more carbon monoxide from this cigarette than from the other
three cigarettes. As carbon monoxide is only taken up during
smoking by alveolar absorption, and as the puffing intensity
figures for the three lower delivery cigarettes were not
significantly different, the subjects must have altered their
pattern of inhalation when smoking the lowest nicotine
cigarette. This alteration could have been achieved by
inhaling more of the smoke, and/or inhaling it to a greater
depth and/or holding it longer in the lungs. However, as the
precise way in which subjects alter their inhalation patterns
cannot be determined by measuring the increases in exhaled
alveolar carbon monoxide resulting from smoking various
cigarettes, for the purposes of this paper the expression
'smoking intensive' will be used to describe the overall
inhalation strategy. Thus in this experiment the subjects
increased their smoking intensity when smoking the lowest yield
cigarette. Because nicotine is primarily absorbed by the
inhalation of tobacco smoke, we interpret this increased
smoking intensity to represent an attempt to maximize the
nicotine absorption from this cigarette. This is consistent
with the finding in this experiment that immediately after
smoking, the performance improvement with the lowest yield

**TABLE 3.** **Puffing intensities and smoking intensities of four cigarettes smoked during a performance experiment**

| | | | | |
|---|---|---|---|---|
| Machine smoked Nicotine yield (mg) | 0.8 | 1.2 | 1.3 | 1.7 |
| Puffing Intensity (%) | 115.1 | 116.6 | 112.3 | 87.7 |
| Smoking Intensity | 0.75 | 0.51 | 0.41 | 0.49 |

cigarette was comparable to those obtained with the higher yield cigarettes.

Although we have argued that in the previous two studies smokers altered their smoking habits in order to obtain optimal levels of nicotine, it must be noted that the cigarettes used in these studies had tar and carbon monoxide yields which covaried with the nicotine yield. Thus alternatively it could be argued that the altered smoking patterns were attempts to obtain optimal levels of other constituents of cigarette smoke besides nicotine. However, evidence against this alternative proposal comes from another study in which we looked at the effects of nicotine tablets on smoking behaviour in a work situation. In this experiment we studied the smoking behaviour of a group of smokers who all smoked a popular brand of cigarettes. The subjects were told that we were looking at the combined effects of smoking and nicotine tablets on the performance of a mirror drawing task. To perform the task the subjects sat in front of a mirror and were instructed to guide a pen along a narrow star shaped track. A screen prevented the subjects from seeing either their hand or the track directly and this forced them to use the mirror. The sessions were arranged so that after performing the task once they were given a tablet which they held in their mouths for 5 minutes before swallowing the remains. They then smoked an experimental cigarette after which they performed the task for the second time.

However, the purpose of the experiment was not to study the effects of nicotine and smoking on performance, but instead to study the changes in smoking behaviour resulting, firstly, from giving the subjects a cigarette of lower yield than their normal brand and, secondly, giving them nicotine prior to smoking such cigarettes. In order to conceal the true purpose of the experiment from the subjects, they were not informed that the mirror which they used to perform the task was two-way, nor that they were being filmed through it.  The smoking behaviour of the subjects was measured in three ways:  (1) the cigarette butts were retained for nicotine analysis;  (2) exhaled end-tidal carbon monoxide was measured before and shortly after smoking and (3) a video film was made of the smokers when they smoked the cigarettes.  In one condition they were given a placebo tablet followed by their own brand of cigarette, which was specially prepared without brand markings. In a second condition they again received a placebo tablet followed by an identical looking cigarette, which was a lower yield version of their own cigarette, commercially described as 'mild'.  In the third condition they received a tablet containing 1.0 mg. nicotine followed by the low yield cigarette.

The results of the study are summarized in Table 4.  A comparison of the two conditions in which they received placebo tablets, showed that there were no significant differences in any of the smoking variables derived from analysis of the video records of the subjects smoking the cigarettes.  Thus there were no differences in the number of puffs taken, the duration of the puffs, the time taken to smoke the cigarettes and the inter-puff intervals, and therefore it would be expected that the subjects generated equivalent amounts of smoke relative to that available from the two cigarettes, and this was confirmed by the puffing intensity values which were not significantly different.  However, when the amounts of carbon monoxide taken up from the two cigarettes are adjusted for the machine smoked carbon monoxide yields, it is clear from the resulting smoking intensity indices that the smokers obtained proportionally more from the lower yield cigarette.  Statistical testing indicated that the differences in the indices were significant.  Thus the mechanism of compensation that the smokers employed when smoking a lower yield cigarette in this study was to increase the smoking intensity, which can be interpreted as an attempt to maximize the nicotine absorption from the smoke.

Nonetheless, as has been pointed out earlier, this is not the only possible explanation of the data as the subjects may have been compensating for some of the other constituents of the smoke which covaried in these cigarettes.  However, these alternative explanations cannot be applied to the differences in inhalation patterns which occurred between the two conditions in which the subjects smoked the lower yield cigarette.  Again, there were no differences in any of the smoke generation parameters when smoking the low yield cigarette following either the placebo or the nicotine pre-load, but when given nicotine prior to smoking, the subjects

TABLE 4:    The measures of smoking behaviour for each of the
three conditions of the nicotine pre-load study.

| | TABLET | Placebo | Placebo | 1mg nicotine |
|---|---|---|---|---|
| | CIGARETTE | 1.4 mg Nicotine | 0.9 mg Nicotine | 0.9 mg Nicotine |
| Time taken to smoke cigarette (secs) | | 370 | 337 | 342 |
| Number of puffs taken | | 16.2 | 16.5 | 16.8 |
| Puff duration (secs) | | 1.8 | 1.8 | 1.8 |
| Interpuff interval (secs) | | 25.5 | 22.8 | 22.4 |
| Puffing Intensity (%) | | 133 | 117 | 122 |
| Smoking Intensity | | 0.43 | 0.61 | 0.43 |

significantly reduced their smoking intensity to the same level
as when smoking the higher yield cigarette. This reversal can
of course only be explained in terms of nicotine because the
cigarettes were identical in the two conditions.

The results of this study provide strong evidence that
.alterations in inhalation patterns reflect nicotine-seeking
behaviour, and have implications for some of the work described
earlier. For example, this finding considerably strengthens
the argument that in the performance experiment, the basis for
the markedly increased smoking intensity observed with the
cigarette having the lowest nicotine yield was an attempt to
maximize the nicotine absorption from the inhaled smoke. Also
in the study of the smoking behaviour of students taking
examinations, the increased self-reported inhalation could
represent an attempt to obtain more nicotine to help them cope
with the situation.

## CONCLUSIONS

Smokers report that smoking helps them to concentrate
and to relax, and claim that these effects are motives for
smoking. Further they smoke more cigarettes at work and in
times of stress. These findings indicate that work and stress
can be considered as motives for smoking. We believe that
nicotine is responsible for these effects and thus the basis of
these motives for smoking. In a time of stress the rise in the
number of cigarettes smoked and the increase in inhalation
could reflect an increased need for nicotine, as opposed to
some other aspect of smoking. In laboratory tests we have
found that smoking improves mental efficiency and have found
equivalent effects with nicotine tablets. Furthermore, in
laboratory work situations, smokers adjust their smoking
patterns when given cigarettes having low yields of nicotine,
in a manner ideal for increasing nicotine absorption.
Increased inhalation appears to be an important technique by
which smokers achieve this compensation, and in the last study
reported there was firm evidence that the absorption of
nicotine was the reason for this compensation.

## REFERENCES

MEADE, T.W. & Walde, N.J. (1977).
Cigarette smoking patterns during the working day.
Br. J. prev. soc. Med., 31, 25-29.


WARBURTON, D.M. (1981).
Neurochemical bases of behaviour.
Br. Med. Bull., 37, 121-125.

WARBURTON, D.M. & Wesnes, K. (1978).
Individual differences in smoking and attentional
performance.
In: Thornton, R.E. (ed), Smoking behaviour: physiological
and psychonogical influences.  London: Churchill-Livingston.
pp.19-43.

WARBURTON, D.M., Wesnes, K. & Revell, A. (1983).
Personality factors in self-medication by smoking.
In: Janke, W. (ed) Response Variability to Psychotropic
Drugs.  London: Pergamon.  In Press.

WESNES, K. (1979).
The effects of nicotine and scopolamine on human attention.
Unpublished Doctoral Thesis.  Reading University.

WESNES, K. & Warburton, D.M. (1978).
The effects of cigarette smoking and nicotine tablets upon
human attention.
In: Thornton, R.E. (ed), Smoking behaviour: physiological
and psychological influences. London: Churchill-Livingston.
pp. 131-147.

WESNES, K. & Warburton, D.M. (1983a)
Smoking, nicotine and human performance.
Pharmacology and Therapeutics, 21: 189-208.

WESNES, K. & Warburton, D.M. (1983b)
Effects of cigarette smoking on rapid information processing
performance.
Neuropsychobiology, 9: 223-229.

WESNES, K., Warburton, D.M., & Matz, B. (1983).
Effects of nicotine on stimulus sensitivity and response
bias in a visual vigilance task.
Neuropsychobiology, 9: 41-44.

DISCUSSION

LECTURER: Wesnes                              CHAIRMAN: Cumming

HEATH:              You have related the smoking habit to working
                    stress and the role played by nicotine.  Not
                    all socities use nicotine, and in the Andes
                    region of South America all the Indians chew
                    coca leaves, these are not cocaine addicts such
                    as one might meet in London or Los Angeles.
                    There are interesting parallels because in Peru
                    it is said that they all chew coca because it
                    makes them concentrate more and work better.
                    The leaves are freely available everywhere, and
                    although supposed to increase work the typical
                    picture is of the Indians lounging against a
                    wall chewing this bolus of coca leaf.  It has
                    always seemed to me that the easy and cheap
                    availability of this substance, habit, boredom
                    and convention go together.  This habit is
                    taken up by children of five or six years of
                    age.  In Merseyside there is a high rate of
                    unemployment, and the characteristic of
                    Merseyside life is a group of youths lounging
                    up against the way like the Indians, only with
                    them it is not coca leaves, but the cigarette.
                    It seems to me that the determinant is the easy
                    and cheap availability of the substance, habit,
                    boredom and convention.  Do you think there is
                    a parallel between the chewing of coca leaves
                    and the smoking of cigarettes.

WESNES:             The parallel seems evident.  I am not sure of
                    the neurochemical effects of coca, but boredom
                    usually creates the demand for a stimulus.

CUMMING:            Donald Heath is, like myself, biased in favour
                    of a quantitative approach, and the observation
                    of lounging against walls is not very
                    quantitative.  However if one measures the
                    output of tin mines in Boliva where inflation
                    is driving the cost of coca leaves beyond the
                    pocket of the workers, the mine-owners are
                    complaining the productivity is falling pari
                    passive with the unavailability of coca leaves.
                    Since it improves productivity in tin mines it
                    may have affects other than on boredom.

RAWBONE:    You have presented evidence of the role of nicotine in stressful situations where the act of smoking might be cured by that situation. What do you suppose is its role in the maintenance of the smoking habit, its role as an addictive drug?  My second question relates to the observation that a higher nicotine cigarette improves work performance, which does not seem to accord with the fact that when such a cigarette is smoked the smoking intensity is adjusted, apparently to give the same nicotine dose.

My third question relates to the use of carbon monoxide boost divided by machine carbon monoxide as an indicator of inhalation.  A carbon monoxide boost can stem from an increased quantity delivered at the mouth, or to an increased inhalation.  I believe the former mechanism to be dominant.  A better indicator would be to divide the mouth intake by the boost.

WESNES:     I am not certain whether carbon monoxide is a good indicator of nicotine uptake, but with the equipment in our laboratory we would have difficulty in measuring the carbon monoxide delivery at the mouth.  I am at a loss to explain why more carbon monoxide should be taken up when the puffing intensity, as indicated by butt nicotine, is unchanged.  The velocity of inspiration might increase, and this might diminish the filtration efficiency of the filter but this is not a convincing explantion – I understand your reservations and have noted them.  Turning to the interesting point that a cigarette containing 1.7 mgm is undersmoked, whilst one containing 1.3 mgm is oversmoked to obtain roughly the same amount of nicotine and yet manifest different levels of performance.

There are two points – nicotine delivery is not the only criterion of performance improvement, another being the enjoyment and we found that the 1.7 mgm cigarette was not enjoyed so that appreciation of the smoke might be an additional factor.  Also the mouth delivery may not be directly related to lung absorption.

The role of nicotine as an addictive agent differs from other alkaloids in that it produces psychological dependence, there is no evidence that there is an increased consumption associated with long continuance of the habit. Nicotine plays a major role in reinforcing smoking which would probably disappear if cigarettes contained none.

RAWBONE: Delivery and uptake of nicotine is a good smoke marker since 95% of the nicotine delivered to the mouth is retained and does not depend on the method of inhalation. The nicotine/carbon monoxide rates is heavily dependent upon the flow profile and machine values are misleading. We did some experiments on inspiration of 50% carbon monoxide mixtures and varied puff volume, depth and duration of inspiration and found that depth had no effect, duration only a marginal effect, but that the volume of CO had a strong correlation with CO increment.

RICHARDSON: You have interpreted arousal entirely in terms of absorbed nicotine stimulating the reticular system. Do you think there is another mechanism, that of tar and nicotine stimulating endings in the respiratory tract causing arousal through stimulation of afferent nerves.

WESNES: It might contribute in part. The application of nicotine to the reticular system does produce arousal in exactly the same way as does a cigarette. This arousal is blocked by cholinergic agents and not by adrenergic, and sensory neurones are not cholinergic.

RICHARDSON: Sensory nerves do have some cholinergic fibres in the central nervous system. You might get evidence on this from the time course of arousal – stimulation of afferent nerves would be quicker than arousal depending on absorption and then delivery by the blood.

WESNES: Where would these afferent neurones contact wih the cortex, and would they stimulate directly or via the reticular formation?

RICHARDSON: At the level of the medulla, and into the reticular formation.

WARBURTON:   It is a useful point about stimulation via nerves, but this would not explain the effect of the nicotine tablet which does not enter the respiratory tract.

GUYATT:   You speak of the depth of inhalation and the strength of inhalation and I cannot see the difference. Secondly in the experiments with different nicotine loads, and using tablets, were they carried out in a fixed order?

WESNES:   Each experiment was carried out on separate days and the order was randomised. When I designed the questionnaire I thought it possible to distinguish depth and strength, but the smokers gave equivalent answers to both.

HORSFIELD:   You say that smoking increases the performance of smokers, so that not smoking decreases the performance of smokers. Do you have any information on the baseline performance of non smokers as compared with smokers.

WESNES:   We have found no differences in the baseline levels of the two groups. You are arguing that smoke deprivation causes a diminution in performance and that a cigarette improves performance towards normality. Evidence against that is that non-smokers are improved by nicotine tablets and we found no difference in light, medium or heavy smokers on their baseline performance.

HIGENBOTTAM:   Areas of high unemployment also abuse alcohol and that may be in part responsible for the lethargy of Merseyside. The major deficit in the studies you have shown us is the absence of blood nicotine levels, with the reservation about the meaning of venous nicotine levels this would be a relevant measure. I was interested to note that you found that cigarette puffing was unaltered by the administration of a nicotine tablet, which suggest there may be local respiratory effects of nicotine in determining puffing pattern.

WESNES:   Smoking has such a rapid effect on performance that blood nicotine should be measured during smoking. We have not measured blood nicotine,

because of the need for venipuncture the stress
of which might interfere with the dependent
variables.    The blood nicotine data wiht
tablets shows the peak of blood level
corresponds wih the peak of performance.

CUMMING:        A significant problem in the determination of
blood nicotine levels is whether the mean blood
level in the systemic circulation is relevant,
or the peak blood level reaching the brain and
information on this will be necessary before we
can interpret the effects of nicotine by the
gastric route and the pulmonary route.

# PARTICULATE DEPOSITION IN SMOKING

V. Prodi and A. Mularoni

Dip di Fisica dell'Universita, Bologna, Italy

## Summary

Tobacco smoke is a condensation aerosol, produced by cooling of low temperature flameless combustion products. These are liquid at body and room temperature, therefore they are spherical in shape and tend to spread on the surface upon contact.

Their density is close to unity ($1 \text{ g cm}^{-3}$) since the most important component is water. Therefore the geometric diameter coincides with the aerodynamic diameter, for particles larger than 0.5 $\mu$m, and with the diffusive diameter for particles smaller than 0.5 $\mu$m.

The deposition mechanisms as a function of particle size and of other parameters are reviewed and their role in total deposition and in the regional detail is assessed.

The size distribution of the aerosol cannot be in general uniquely defined since the smoke is in continuous dynamic size evolution because of the progressive cooling through the tobacco mat and filter. In addition, there may be an important uptake of water in the passage through the airways, with a consequent difficulty in predicting the effective particle size and therefore the total and regional depositions.

The size distribution is measured, in situations of practical interest, by means of an inertial particle spectrometer in the range above 0.5 $\mu$m.

Regional and total deposition values obtained in vivo are reviewed.

## Particulate matter deposition in smoking

### Introduction

The deposition of airborne material in the airways is the result of the interaction of particles, with their intrinsic properties, with the airways, with their morphological and aerodynamics characteristics. This is why both have to be considered in detail. In the case of smoking, in addition, airborne particles can condense water and this adds to the complication of rapidly changing size. We shall consider aerosol properties, airway characteristics and stable particle deposition (1). Finally tobacco smoke will be considered, with size dynamics, together with specific measurement techniques of the size distribution and of total and regional deposition.

### Particle characteristics

Airborne particles subject to a constant force reach a constant velocity (2,3): the ratio of the velocity V to the force F is called mobility B

$$B = \frac{V}{F} = \frac{C(d)}{3\pi \eta d}$$

where d is the particle diameter, n is the viscosity of air, C(d) is the Cunningham slip correction that accounts for a lower air resistance for small particles. C(d) is 1.016 at 10 um, 1.16 at 1 um and 2.9 at 0.1 um. For an aerosol particle subject at t = 0 to a constant force it takes a time of the order of t (relaxation time) to reach dynamic equilibrium

$$\tau = \frac{d^2 \rho_p \, C(d)}{18 \eta}$$

where $\rho_p$ is the particle density. $\tau$ is 300, 3 and 0.09 usec respectively for 10, 1 and 0.1 um. particles.

For aerosol particles to deposit, it takes a force with a non-negligible component directed toward the airway wall. The main mechanisms responsible for this in the respiratory tract are: settling under gravity, inertia, Brownian diffusion and electrostatic forces.

**Settling**

The particle settling velocity is

$$V_s = \tau g$$

where g is the acceleration due to gravity.

The distance travelled by a particle during the time t is simply:

$$\Delta h = V_s t = \tau g t$$

which depends on the square of particle size and on the available time. In the airways the higher is h the higher is the deposition efficiency DE due to gravitational settling, i.e. DE increases with increasing size and with respiratory period (with decreasing respiratory frequency). The time available is always much greater than the relaxation time, therefore if the size is constant, also the settling velocity is constant. This is a time dependent mechanism.

**Inertia**

The flow lines in a bending conduit are forced to follow the curvature: aerosol particles carried by the gas tend to preserve their velocity in direction and value. Therefore they depart from their initial flow line by a distance proportional to the Stokes number:

$$Stk = \frac{U \tau}{R}$$

where U is the fluid velocity and R is the radius of curvature of the flow lines. For a given bend, the distance travelled by a particle is:

$$s = U \tau$$

proportional to the flow velocity and to the relaxation time (a measure of particle inertia); the displacement from the flow lines takes place in times of the order of the relaxation time, therefore it is called a "prompt" mechanism.

**Brownian diffusion**

Aerosol particles are constantly hit by gas molecules: the

momentum transferred to them is never perfectly balanced. This imbalance is the more likely the smaller in the particle. Therefore particles are displaced at random; the root mean square displacement in any direction during the time t is:

$$x^2 = 2 \, D_p \, t$$

where $D_p$ is the particle diffusion coefficient, which is proportional to particle mobility B: the displacement and therefore the deposition probability increases for decreasing size and for increasing available time. Therefore this also is time-dependent mechanism.

## Electrostatic forces

To the effect of particle behaviour, the airways can be considered a conductor and therefore the potential is constant. The only electrostatic effects can therefore come from the particles. If they have a unipolar charge and are in a high concentration, there may be an electrostatic repulsion and a scattering of the particle cloud as a whole, though this very seldom happens. Much more frequent are image forces between individually charged particles and airways walls: this can happen even if the aerosol as a whole is close to neutrality with particles charged of either polarity.

The force and therefore the displacement velocity depends on the particle to wall distance. During the residence time t in the airways all the particles initially at a distance smaller than $\Delta r$:

$$\Delta r = (\frac{3}{4} \, B \, q^2 \, t)^{1/3}$$

reach the wall and are captured, where q is the electrostatic charge carried by the particle of mobility B. This mechanism is more effective the longer the residence time and the smaller is the particle size.

Deposition probability for each particle depends on the combination of these mechanisms: the size range where Brownian diffusion prevails, below around 0.5 um, is called diffusive size range; the size range where settling and inertial impaction prevail is called aerodynamic range (above 0.5 um). In the aerodynamic range a concise parameter is the so called aerodynamic diameter, which is the diameter of the unit density (1 g cm$^{-3}$) sphere which has the same settling velocity as the unknown particle. The aerodynamic diameter is a generalisation to account for the differences in particle density and, to some extent, shape.

Brownian diffusion is independent of density: therefore the aerodynamic diameter is not an appropriate parameter to describe the behaviour of smaller particles. Instead, a diffusive diameter should be introduced, defined as the sphere having the same diffusion coefficient of the unknown particle.

## Air flow in the respiratory tract

Deposition depends strongly on the effective flow in the airways, because it can affect the time-dependent mechanisms through the mean residence time, MRT, and the "prompt" mechanisms (impaction) through the Volumetric Flow Rate (4), VFR.

Respiratory frequency and tidal volume govern the MRT and VFR. Mixing between tidal and residual air has a very important role in transferring inhaled particles into the residual volume and thereby markedly increasing their residence time.

The respiratory tract is generally modelled after Weibel (5) as series of tubes branching out in regular dichotomy in each generation from trachea (generation 0) down to the 23rd generation.

The air flow in the respiratory tract is very complex (6) characterised by a wide range of Reynolds numbers (Re). At a VFR of 1000 $cm^3$/sec, Re is larger than 2000 down to lobar bronchi inclusive, which means turbulent flow and good mixing in this section of the conductive structures. In addition the flow can be turbulent even at lower VFR because of the corrugated walls and of the short cylindrical sections (a few diameters only) which prevent a fully developed laminar flow to be reached: at a VFR of 1000 $cm^3$/sec there can be turbulence down to segmental bronchi.

A laminar regime in the whole respiratory tract occurs only at low VFR, corresponding to a minute volume of 3500 $cm^3$.

Even in laminar conditions at the bifurcations double vortices are established and propagate into the daughter tubes during inhalation, and conversely two double vortices during exhalation propagate into the parent tube. This can happen at Re down to several times unity, corresponding, at $1000^3$/sec VFR to $15^{th}$ generation (bronchioles without cartilage).

The general effect is that well developed laminar flow pattern is never reached and mixing is practically extended to the whole anatomical dead space.

Even below the 15th generation the velocity profile is fairly flat because the flow is not confined by smooth walls, but rather by alveolar openings: therefore at the opening, the velocity of the air is finite and it persists during the expansion of the alveolus; tidal air penetrates into alveoli like a tongue tangential to the wall. This pattern is not reversed during exhalation.

An important mixing mechanism, active in the whole respiration tract, is due to the non-uniformity, both geometric and dynamic, of the airways. Their diameter in each generation is considerably scattered; branching angles, also, are asymmetric; this produces a non-uniform flow in the following bifurcations which is not reversed in the exhalation.

This non-uniformity can be enhanced by the relatively rigid structures (blood vessels and bronchi) of the airways: during expansion and contraction, geometric similarity is not preserved. The mixing effect is enhanced at lower values of the functional residual capacity and this can account for the increasing deposition with decreasing expiratory reserve volume, (7,8) ERV.

In Table 1 the dynamic condition of the respiratory tract is summarised.

In addition, a considerable scatter of deposition data (8) is found among subjects inhaling in the same controlled conditions.

Unfortunately accurate morphometric data are not available for populations; it is not possible therefore to correlate the scatter of deposition values to any distribution of morphometric parameters, but there is little doubt that this scatter could be due to anatomical as well as physiological or pathological differences(6).

**Total Deposition**

Airborne particles are characterised by a probability of entering the airways: this is defined as inhalability. In the case of smoke, this probability is practically 1 since the flow is confined and as such completely enters the airway. Only side-stream smoke is not confined, but particle size in

ordinary conditions is small and inhalability is practically
unity.

Deposition is defined as the probability for an inhaled
particle to touch a surface of the respiratory tract and to
adhere to it.  Total deposition is intended as referring to the
entire respiratory tract, while for a more detailed picture of
the incorporation the respiratory tract is divided into three
regions and the deposition in each is defined as regional
deposition (4,1).

**Extrathoracic airways,** in which the deposition is mainly due to
inertia and particles are cleared within minutes either by
mechanical transport of particles or secretions:
**tracheobronchial airways** in which particles are deposited by
inertia and settling and from which particles are removed
within hours by mechanical transport of secretions; **alveolar
air spaces,** characterised by small particle-to-wall distances,
in which particles are deposited mainly by gravitational
settling and Brownian diffusion.  Removal takes months or even
years, mainly by phagocytosis and subsequent cell transport to
the mucociliary excalator or to the lymphatic system, or
solubility; **regional deposition** is the probability for an
inhaled particle to reach a surface of the given region and
adhere to it.  Total deposition is the sum of regional
depositions.  Deposited particles are cleared from the
respiratory tract or translocated to other body regions;
**retention** is the probability for a deposited particle to be
retained in the body, while regional retention is such a
probability referred to a given region.

## Total deposition of stable particles

At first the deposition will be considered for stable
particles, i.e. insoluble and not wettable;  for them the role
of various mechanisms will be briefly described.  Total
deposition is measured by measuring inhaled $C_I$ and exhaled $C_E$
concentrations of monodisperse aerosols (which have uniform
size distribution) and by means of the expression:

**Size**

$$C_I - C_E \, / \, C_I$$

The behaviour of DE as function of size is shown in Fig. 1
(from Heyder et al. (4) for mouth breathing.  The general trend
shows a minimum around 0.5 um and increases both for decreasing
and for increasing size.  Below 0.5 um the increase is due to
increasing diffusion coefficient.  Above 0.5 um deposition is
due to gravitation and impactation.

V. PRODI AND A. MULARONI

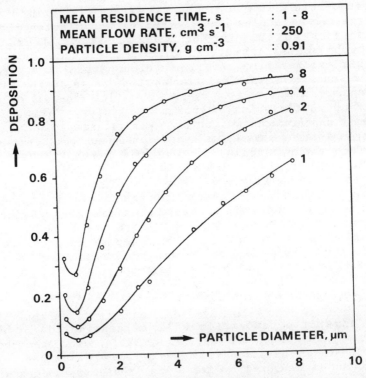

FIGURE 1: Effect of particle size and MRT on total deposition. (From Heyder et al 4).

**MRT**

Fig. 1 shows also the effect of MRT on total deposition (4): DE increases with increasing MRT because both diffusion and settling are time dependent. Impaction becomes important at higher sizes and this is shown by the smaller effect of MRT at higher VFR.

**VFR**

Total deposition is independent of VFR up to 1-1.5 um since impaction is not effective for the range of VFR encountered. At higher sizes VFR begins playing a more and more important role, as shown in Fig. 2 (from Heyder et al. 4). The importance of impaction is depicted in Fig. 3 where MRT and VFR are varied while keeping the Tidal Volume constant. There is a definite cross-over of the curves, which is even more dramatic for nose breathing, as shown in Fig. 4 (also from Heyder et al. 4), where it takes place around 1 um showing the contribution of impaction to nose deposition.

### Biological variability

It is now generally accepted (6) that even under strictly controlled breathing conditions and residual volumes there is definite intersubject variability of total deposition. An example of this is given (8) in Fig. 5 where DE is plotted as function of particle size for six volunteers between 0.3 and 1.5 um unit density spheres, breathing at 1000 cm$^3$ TV and 15 resp/min., each at his own expiratory reserve volume, ERV. This is interesting since in this range total deposition is also alveolar deposition.

The scatter of data reaches a factor of 2 and cannot be explained on the basis of respiratory parameters.

For each volunteer instead, with 0.6 um aerosols, a marked dependence on ERV is found. The relative DE can be expressed as a -1/3 power of ERV relative to normal, probably due to a stronger mixing with smaller volumes (9): to a 30% variation of ERV around the normal value, a 10% variation of DE corresponds (8).

### Electrostatic charge

Tobacco smoke particles are formed by vapour condensation: therefore the absolute value of the charge carried is very low

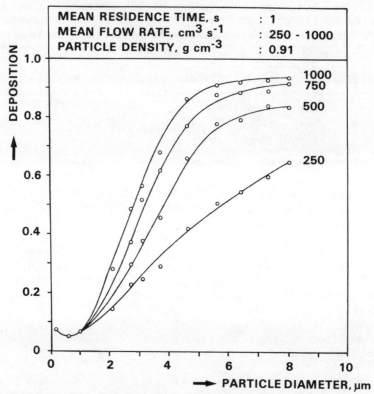

FIGURE 2:    Effect of particle size and VFR on total
deposition.  (From Heyder et al. 4).

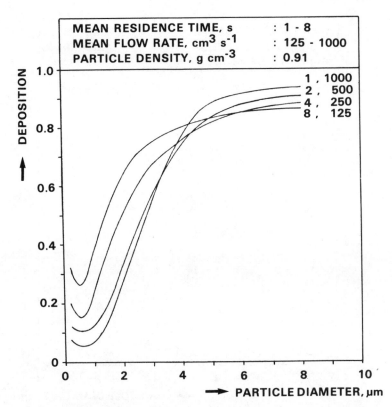

FIGURE 3:    Effect of particle size on total deposition for
mouth breathing at 1000 cm³ TV. (From Heyder et al
4).

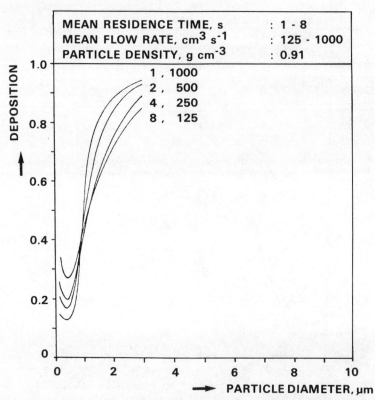

FIGURE 4:    Effect of particle size on total deposition for
             nose breathing at 1000 cm$^3$ TV. (From Heyder et al
             4).

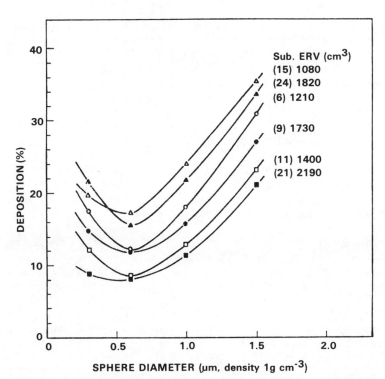

FIGURE 5:    Total deposition for a group of six volunteers, at 1000 cm$^3$ TV and 15 resp.min$^{-1}$. (From Tarroni et al. 8).

and electrostatic mechanisms have a negligible effect (10).

## REGIONAL DEPOSITION

### Nose deposition

In the case of tobacco smoke, the nose is not used as an entry pathway. It may be important for passive smoking and for active smoking during exhalation only.

Nose deposition is due to impaction and depends strongly on flow rate: at 30 litres per minute, it starts around 2 um and it is practically quantitative between 9 and 10 um (6). During exhalation, nose deposition has the same efficiency (11), but this applies only to the fraction transmitted.

### Mouth breathing

Extrathoracic deposition

Regional deposition in mouth breathing has been studied by Lippmann and Albert (12), Chan and Lippmann (13), and by Stalhofen et al (14) by external counting of labelled monodisperse particles deposited in airways.

Head deposition too can be linearly fitted with the $\lg \varepsilon_p D_p^2 F$ parameter. Lippmann's (6) data have been extrapolated to obtain the size for quantitative head deposition, that is around 17 um for a 500 $c^3 \sec^{-1}$ VFR.

Stahlhofen et al's (14) data show a slightly higher efficiency pointing to 100 percent deposition around 11 um at the same flow rate.

The inertial parameter is not fully representative of deposition since the geometry of the airways may be dependent on the flow.

In Fig. 6 the average extrathoracic deposition of three subjects (14) is reported for 2 flow rates together with Lippmann's curve (6).

In Fig. 7 the effective total and regional depositions are shown as the average of three subjects (14), for two breathing patterns: $V_t$ = 1500 $cm^3$, 15 resp. $min^{-1}$ MTR = 2 sec, VFR = 750 $cm^3 sec^{-1}$) and $V_t$ = 1000 $cm^3$, 7.5 resp. $min^{-1}$ (MTR = 4 sec, VFR

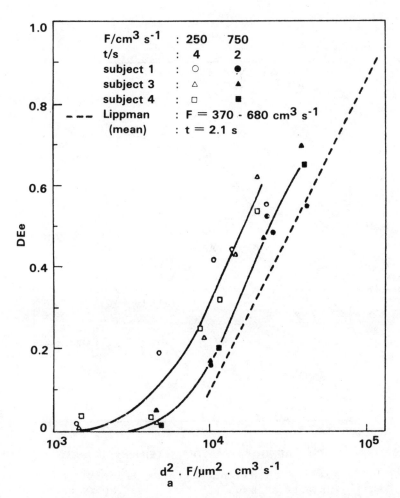

FIGURE 6: Head deposition during inhalation via the mouth vs impaction parameter. The solid lines and points are for three subjects at two VFR (From Stahlhofen et al.(14) while the broken line is Lippmann's (6) fitted curve.

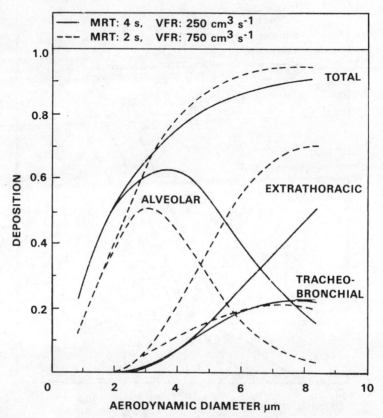

FIGURE 7:    Average total and regional deposition for three
             subjects at two different respiratory patterns and
             for mouth breathing (based on data of Stahlhofen
             et al. 14).

= 250 cm$^3$ sec$^{-1}$). For the extrathoracic deposition the curves are derived from the same data points of Fig. 6.

**Tracheobronchial deposition**

Tracheobronchial deposition has been studied more recently by Chan and Lippmann (13) both in hollow casts and in vivo and by Stahlhofen, Gebhart and Heyder (14) on three healthy subjects.

In addition, detailed studies have been performed on hollow casts of human bronchial tree by Chan, Schreck and Lippmann (15), that have pointed out the flow pattern in the trachea, and preferential deposition sites in connection with airflow and turbulence.

The studies of Chan and Lippmann (13), have shown a remarkable biological variability of deposition data even in the tracheobronchial tree.

Fig. 8 shows TB deposition expressed as a function of the aerosol entering the trachea. The straight lines represent the average and the scatter of the values found by Lippmann (6), while the data of Stahlhofen et al (14) are shown with the points. These show a smaller scatter of data and two distinct behaviours at two rates as well as values of deposition slightly lower than Lippmann's average. The actual tracheobronchial deposition is a bell-shaped curve that departs from zero around or slightly above 2 um aerodynamic size and reaches a maximum, according to the flow conditions, between 6 and 10 um.

In Fig. 7 the average actual TB deposition for three subjects (14) is plotted as a function of particle size for two respiratory patterns.

It has been pointed out that deposition depends strongly on health and increases for smokers and again for bronchitic patients. Chan and Lippmann (13) have proposed a parameter, called Bronchial Deposition Size, BDS, derived by expressing tracheobronchial deposition as a function of the Stockes number. This was found 1.20 cm for healthy lung disease and 0.6 for severely disabled patients.

This is an additional effect brought about by cigarette smoke, consisting probably in increased turbulence and therfore deposition for smokers.

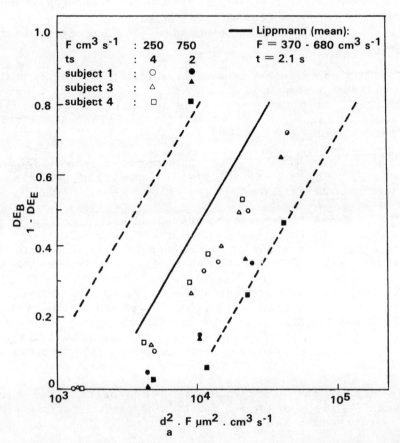

FIGURE 8:    Deposition in the ciliated tracheobronchial region
             during mouth breathing, in percent of aerosol
             entering the trachea.  The straight lines
             represent the average and the scatter of
             Lippmann's (6) data while the points are the
             values obtained by Stahlhofen et al (14).  (From
             Stahlhofen et al. 14).

## Alveolar deposition

The gas exchange region of the airways is characterised by a very large surface area and therefore by a small average particle-to-wall distance and a large cumulative cross-section. Therefore deposition in the aerodynamic size range is practically due to gravitational settling.

Alveolar deposition therefore increases with increasing MRT at constant VFR. Because of the behaviour of extrathoracic and tracheobronchial deposition, also alveolar deposition follows a bell-shaped curve: the relative maximum is around 3 um and can be shifted to smaller size both for increasing MRT at constant VFR and for increasing VFR at constant MRT. In the first case the alveolar deposition values increase since the extrathoracic and tracheobronchial deposition do not vary appreciably a gravitational deposition is more effection. Fig. 9 from Heyder et al (4) shows this effect in detail.

Increase in VFR causes a higher deposition by impaction in the higher regions and therefore transmits a lower fraction of large particles to the alveolar region and the effect of increased VFR takes over the effect of decreased MRT.

The data of Stahlhofen et al (14) for alveolar deposition are also summarised in Fig. 7 as the average of their three subjects. These are in good agreement with Lippmann and Albert's (1,2).

## Cigarette smoke size distribution

Smoke particles are liquid, in rapid dynamic evolution while passing through the tobacco mat and while mixing with air outside and within the respiratory tract.

The immediate appearance of the smoke displays a difference between sidestream and mainstream smoke.

Sidestream smoke is markedly blue in scattered light under white light illumination and is brown in transmitted light. This means a strong Rayleigh scattering, i.e. particles considerably below the wavelength of the incident light. In the environment these particles are difficult to measure because they are unstable. In this size range the instruments most widely used are diffusion batteries and electric mobility analysers.

The diffusion battery (3,2) is a system of high surface to

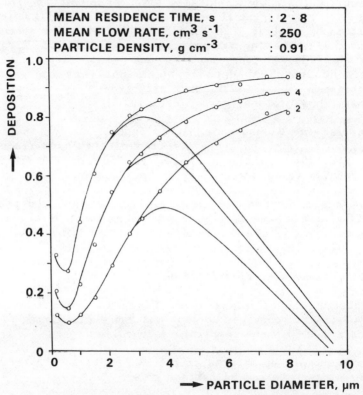

FIGURE 9:    Effect of particle size and MRT on total and
             alveolar deposition. (From Heyder et al.4)

volume ratio, therefore with a high probability of deposition by diffusion.

Stages of progressively increasing surface are placed in cascade: the penetration through all the stages upstream is measured at each stage and from the penetration as a function of the number of stages the aerosol diffusion coefficient is drawn. The penetration is generally measured by means of Condensation Nuclei Counter, CNC, in which a vapour supersaturation is produced, the particles grow to sizes easily detectable by optical means.

Alternatively electric mobility analysers are used. In these, particles are exposed to gaseous ions and charged. Their drift velocity in an electric field, if the charge is known, is a measure of the mechanical mobility and therefore of the size. In this case the detection is based on an electric current measurement which is equivalent to particle counting as with the CNC. Both these techniques require dilution of the aerosol and residence times within the equipment of seconds; therefore considerable changes by evaporation-condensation and coagulation may take place.

On the other hand the mass carried by particles in this size range is generally much smaller as compared with the mass carried by mainstream smoke.

Tar airborne material in the mainstream smoke is composed of bigger particles; aerodynamic techniques are then used for their characterisation. In the literature, values obtained (17) with the aid of an aerosol centrifuge after dilution of 10:1 with clean air have been reported. The mass median aerodynamic size is 0.45 with a geometric standard deviation around 1.5. The dispersion of size is relatively small and this can be explained by considering the condensation process leading to their generation.

More recently (18), mainstream particles have been measured by means of a cascade impactor and gave 0.85 um and 0.79 for low and middle tar cigarettes respectively; no indication on the size of distribution is given in this particular case.

In order to resolve the discrepancy an instrument, recently developed by the author, the inertial particle spectrometer (INSPEC) has been applied to size measurement of mainstream smoke.

The main advantage is that it does not need aerosol dilution and has a very short residence time.

In the INSPEC aerosol particles are injected into a clean
airstream in a curved channel (22). The particles persist in
their initial velocity by inertia and are then separated while
airborne according to their aerodynamic diameter. The
separation is preserved and magnified when the whole stream is
drawn through a filter;  on this, particles deposit at a
distance from the inlet which is a unique function of
aerodynamic size. For tobacco smoke particles the deposit is
apparent from the tar distribution and can be assessed either
by photometric scanning or by chemical analysis of the tar
content of sections of known size ranges. A bimodal
distribution has resulted, with a mode around 0.6 um and a
coarser mode around 1.5 um, with a relatively narrow size
distribution.

**Total deposition of tar particulate**

The size distribution of mainstream smoke, if compared with
the deposition of stable particles, should imply a deposition
efficiency of the order of 20%.

Total deposition measurements are not easy since it is
difficult to have a reliable reference source. A continuous
source is not realistic and on the other hand size, airflow and
deposition depend markedly on smoking pattern. A technique not
interferring with normal smoking has been recently proposed
(19). This allows the puff volume, inhaled amount and
respiratory deposition to be measured on volunteers with an
accuracy of $\pm$ 10%. The technique captures exhaled smoke with
an exhaust hood and establishes the amount of inhaled smoke by
monitoring puff volume, duration and timing and replaying the
exact smoking sequence with matched cigarettes. The mass of
captured smoke was measured and gave on 11 volunteers a
deposition range of 22 to 75% with an average value of 47%.

The relevant difference between the measured deposition and
the values expected on the basis of stable particle deposition
is due to the size increase for water captured by the particles
in the respiratory tract.

**Size increase of soluble particles**

The respiratory tract is a high humidity environment;
typically values of relative humidity of 99.5% are reported at
the wall of the third bifurcation for a flow rate of 431 cm$^3$
sec$^{-1}$ (TV = 750 cm$^3$) and inhalation time of 17.4 sec) and for
all the downstream regions. There may be instances, though,
where supersaturation occurs (20);  inhaling saturated air at

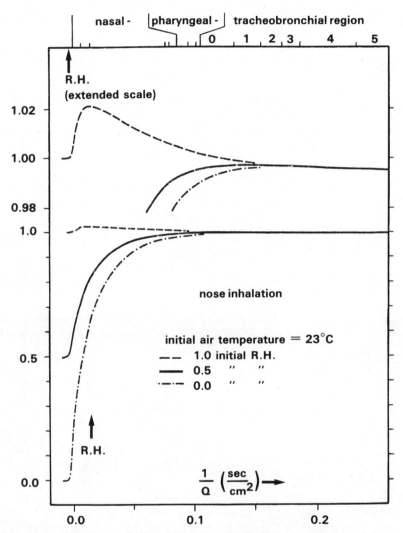

FIGURE 10:    Calculated relative humidity (R.H.) in the upper
              human airways.  The initial air temperature is
              23°C and the initial R.H.s are 1.0, 0.5 and 0.0.

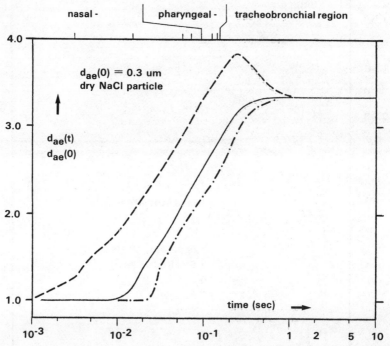

FIGURE 11:    Growth of a dry NaCl particle with an aerodynamic
              diameter $d_{ae}$ of 0.3 um according to the R.H.
              curves of Fig. 1.

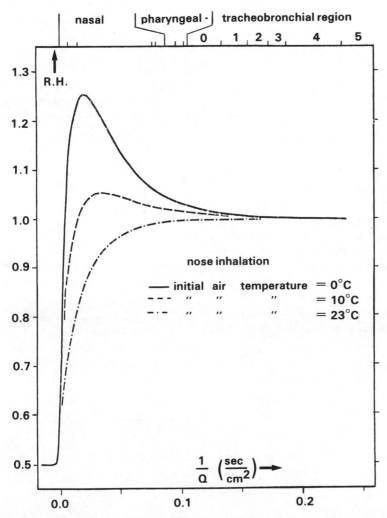

FIGURE 12: Calculated R.H. in the upper human airways. The initial air temperatures are 0°, 10° and 23°C, the initial R.H. is 0.5.

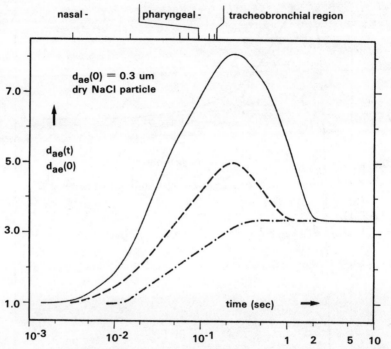

FIGURE 13:  Growth of a dry NaCl particle with an aerodynamic
diameter $d_{ae}$ of 0.3 um according to the R.H.
curves of Fig. 3.

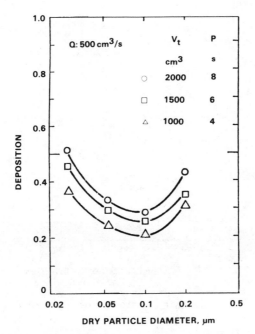

FIGURE 14:    Total deposition of sodium chloride aerosols as a
function of the dry particle size in the inhaled
air for different tidal volumes, $V^t$, and breath
periods, P, with the flowrate, Q, constant at 500
$cm^3/s$. The hygroscopic sodium chloride particles
absorb water in the lung to an unknown extent,
consequently the size of the particle when it
deposits is unknown.

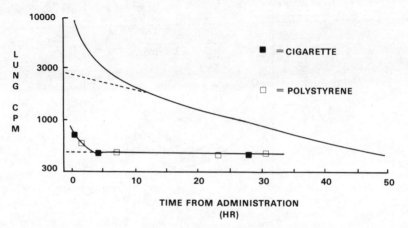

FIGURE 15:    Lung clearance curves for one subject.

$23^{\circ}C$ gives rise to 1.02 supersaturation ratio in the nasal passages down to the trachea due to the different thermal and vapour diffusivities. Higher supersaturation ratio is found when inhaling air at lower temperatures; in Fig. 10 are reported the calculated (20) relative humidities at initial temperature of $23^{\circ}C$ and for three different relative humidities and in the following Fig. 11 the growth factor.

For lower temperatures the relative humidity and growth factors are shown in Fig. 12 and Fig. 13 respectively (20). The growth factor in this instance is much higher. In the case of smoking, inhalation is not through the nose but the single bolus is undergoing cooling and considerable saturation can take place.

Total deposition of soluble monodisperse (NaCl) ultrafine particles has been recently measured by Blanchard and Willeke (21) who found the general trend of Figure 1 and 5, but shifted to lower particle sizes, as shown in Fig. 14. By assuming a fast growth and therefore similar deposition mechanisms for stable and soluble particles, the comparison of the sizes of minimum deposition gives a growth factor of at least 3.

**Regional deposition of cigarette smoke**

Although retention of tar can take place in the whole respiratory tract, there are regional differences that make it desirable to know deposition and clearance characteristics. Black and Pritchard (18) have recently presented a study on a comparison of the regional deposition and short term clearance of tar particulate material from cigarette smoke with that of 2.5 um polystyrene microspheres.

Polystyrene particles were labelled with Tc-99m and the cigarettes were labelled wtih I-123 cetyliodide. Both aerosols were inhaled via the mouth.

Deposition could be measured and retention followed for some days. In Fig. 15 a typical lung clearance curve is shown; it can be expressed as two-phase exponentials; the amount is found by extrapolation to zero time.

The fast phase half times are similar for polystyrene and tar particulate; it is then attributed to tracheobronchial clearance also in the case of cigarette smoke (18).

The slow phase half time for polystyrene, representing clearance from the pulmonary P region, is of the order of a hundred days and thus it shows no change during the period

Table I – Airways Structure According to Weibel's Model

| Airway | Gener-tion | Number per gene-ration | Diameter (mm) | Length (mm) | Total cross section (cm$^2$) | Cumulative volume (cm$^3$) | Velocity (cm/sec) + | Reynolds number + V R/ |
|---|---|---|---|---|---|---|---|---|
| Trachea | 0 | 1 | 18,0 | 120,0 | 2,5 | 30,5 | 196 | 2175 |
| Main bronchi | 1 | 2 | 12,2 | 47,6 | 2,3 | 41,8 | 213 | 1605 |
| Lobar bronchi | 2 | 4 | 8,3 | 19,0 | 2,1 | 45,8 | 231 | 1195 |
| Segmental bronchi | 4 | 16 | 4,5 | 12,7 | 2,5 | 50,7 | 196 | 555 |
| Terminal bronchi | 11 | $2,05.10^3$ | 1,09 | 3,9 | 19,6 | 84,8 | 26,1 | 17 |
| Terminal bronchioles | 16 | $6,55.10^4$ | 0,60 | 2,0 | 180 | 175 | 2,7 | 1,0 |
| Resp. bronchioles | 17–19 | $9,20.10^5$ | 0,50 | 1,2 | 994 | 371 | 1,1 | 0,3 |
| Alveolar ducts | 20–22 | $7,30.10^6$ | 0,43 | 0,6 | $5,9.10^3$ | 1085 | 0,09 | 0,04 |
| Alv. sacs | 23 | $8,40.10^6$ | 0,41 | 0,5 | $11,8.10^3$ | 1675 | 0,04 | |
| Alveoli | | $3.10^8$ | 0,3 | 0,2 | | 4875 | | |

+ For 15 resp./min, 100 cm$^3$ TV

Table II - Lung Clearance Half Times and Distribution

|  | FAST PHASE $T\frac{1}{2}$ h | SLOW PHASE $T\frac{1}{2}$ h | $\frac{P}{P + TB}$ % |
|---|---|---|---|
| POLYSTYRENE PARTICLES | Mean 1.98 SE $\pm.36$ | ND | 70.6 $\pm3.4$ |
| CIGARETTE TPM | 1.88 $\pm.07$ | 17.4 $\pm0.5$ | 36.2 $\pm1.1$ |

shown. For tar particles, however, a slow half time of 17 hours means that the label is soluble in the lung: since it is uniformly distributed with particle size and with particle material, this means that the tar particulate material must break up, if not dissolved in the lung, so that the label is available for clearance.

In Table 2 the clearance half times are shown together with the ratio of alveolar to intrathoracic deposition (P/P + TB); there is a large difference between the two values and with the lower alveolar deposition for tar particles. From the size distribution of mainstream smoke, as it was mentioned above, one should expect only alveolar deposition: this discrepancy can be resolved by assuming a growth factor of the order of 10. For this a considerable supersaturation has to take place within the airways.

## REFERENCES

1.  PRODI, V., Melandri, C. and Tarroni, G. (1982),
    The particles in the atmosphere and their intrapulmonary deposition.
    In: The Lung and its Environment. Ed. G. Bonsignore & G. Cumming, Plenum Press, New York.

2.  MERCER, T.T. (1973),
    Aerosol Technology in Hazard Evaluation.
    Academic Press, New York.

3.  FUCHS, N.A. (1964),
    The mechanics of aerosols.
    Pergamon Press, London.

4.  HEYDER, J., Gebhart, J. and Stahlhofen, W. (1980),
    Inhalation of Aerosols: Particle Deposition and Retention, Aerosol Generation and Exposure Facilities. K. W. Willeke, ed. Ann Arbor Science, Ann Arbor.

5.  WEIBEL, E.R. (1980),
    Morphometry of the human lung.
    Academic Press, New York, pp. 136-140.

6.  LIPPMANN, M. (1977),
    Regional deposition of particles in the human respiratory tract.
    In: Handbook of Physiology Sect. 9 : Reaction to Environmental Agents. K. H. K. Lee, H. L. Falk, S. D. Murphy and S.R. Geiger. Eds. Bethesda, M.D. : Am. Physiological Society, 213-232.

7.  TAULBEE, D.B. and Yu, C.P. (1976),
    Theory of particle deposition in the human lung.
    Ann. Meeting of the Gesellschaft fuer Aerosolforschung,
    Bad Soden/Ts.

8.  TARRONI, G., Melandri, C., Prodi, V., De Zaiacomo, T.,
    Formignani, M. and Bassi, P. (1980),
    An indication on the biological variability of aerosol
    total deposition in humans.
    Am. Ind. Hyg. Ass. Journal, 41, 826.

9.  YU, C.P. and Taulbee, D.B. (1977),
    A theory for predicting respiratory tract deposition of
    inhaled particles in man.
    Inhaled Particles IV, W.H. Walton, Ed.  Pergamon Press,
    Oxford, 35-46.

10. MELANDRI, C., Prodi, V., Tarroni, G., Formignani, M.,
    De Zaiacomo, T., Bompane, G.F. and Maestri, G. (1970),
    On the Deposition of Unipolarly Charged Particles in
    the Human Respiratory Tract, Inhaled Particles IV.
    W.H. Walton, Ed.  Pergamon Press, Oxford, 193-203.

11. HEYDER, J. and Rudolf, G. (1977),
    Deposition of Aerosol Particles in the Human Nose,
    Inhales Particles IV.  W. H. Walton, Ed., Pergamon
    Press, Oxford 107-125.

12. LIPPMANN, M., Albert, R.E. (1969),
    The effect of particle size on the regional deposition
    of inhaled aerosols in the human respiratory tract.
    Am. Ind. Hyg. Ass. J., 30, 257-275.

13. CHAN, T. L. and Lippmann, M. (1980),
    Experimental measurements and empirical modelling of
    the regional deposition of inhaled particles in humans.
    Am. Ind. Hyg. Ass. J., 41, 399.

14. STAHLHOFEN, W., Gebhart, J. and Heyder, J. (1980),
    Experimental determination of the regional deposition
    of aerosol particles in the human respiratory tract.
    Am. Ind. Hyg. Ass. J., 41, 385.

15. CHAN, T.L., Schreck, R.M. and Lippman,, M. (1978),
    Effect of turbulence on particle deposition in the
    human trachea and bronchial airways.
    71st Ann. Meeting. Am. Inst. Chem. Eng., Miami, Fl.
    Nov. 12-16.

16.  KNUTSON, K.O. and Whitby, K.T. (1975),
     Aerosol classification by electric mobility: apparatus
     theory and applications.
     J. Aerosol. Sci., 6, 443-451.

17.  HINDS, W. (1978),
     Size characteristics of cigarette smoke.
     Am. Ind. Hyg. Ass. J., 39, 48-54.

18.  BLACK, A. and Pritchard, J. (1983),
     A comparison of the regional deposition and short-term
     clearance of tar particulate material from cigarette
     smoke with that of 2.5 um polystyrene microspheres.
     XI GAeF Conference, Munchen, Sept. 14-16.

19.  HINDS, W., First, M.W., Huber, G.L. and Shea, J.W. (1983),
     A method for measuring respiratory deposition of
     cigarette smoke during smoking.
     Am. Ind. Hyg. Ass. J., 44, 113-118.

20.  FERRON, G.A., Haider, B., Kreyling, W.G. (1983),
     Conditions for measuring supersaturation in the human
     lung using aerosols.
     XI GAeF Conference, Munchen, Sept. 14-16.

21.  BLANCHARD, J.D., and Willeke, K. (1983),
     An inhalation system for characterising total lung
     deposition of ultrafine particles.
     To be published on Am. Ind. Hyg. Ass. J. Nov.

22.  PRODI, V., De Zaiacomo, T., Melandri, C., Tarroni, G.,
     Formignani, M., Olivieri, P., Barilli, L. and
     Oberdorster, G. (1982),
     "Description and applications of the inertial
     spectrometer". Aerosols in the Mining and Industrial
     Work Environment, V. A. Marple and B. Y. H. Liu, Eds.,
     Ann Arbor Science, Ann Arbor.

DISCUSSION

LECTURER: Prodi                                    CHAIRMAN: Cumming


JEFFREY:        What do you estimate the range of particle size
                to be, after growth, in tobacco smoke.

PRODI:          6.5 microns is the mean, with a geometric
                standard deviation of 1.5 to 2.0, giving a
                range of about 10 between the smallest and the
                largest.  There might also be a smaller
                component and we need to make further
                measurements under growth conditions and after
                hearing this discussion this is something which
                I might do.

JEFFREY:        We have heard how complex tobacco smoke is,
                could it be that particles of different sizes
                also have different compositions, or are the
                particles chemically homogeneous.

PRODI:          We could study this using our technique by
                fractionating the smoke into different particle
                sizes and then analyse each fraction.

RAWBONE:        I thought that there was evidence which
                suggests that the growth of particles in the
                respiratory tract was not very great, although
                this goes against most of the evidence that you
                have quoted.  I agree that where growth has
                been reported a geometric
                standard deviation of about 2 is correct, but
                have no information about the chemical
                homogeneity of different sized particles.

                We saw yesterday that smoking consists of
                taking a bolus of smoke into the mouth which
                then is mixed into the lung by a subsequent
                inspiration so that there is an initial high
                concentration which falls as inspiration
                proceeds.  This is different from the
                inspiration of a constant smoke concentration;
                would you predict differences in deposition in
                these two different circumstances?

PRODI:          Concentration does affect deposition.
                Calculations are based upon the assumption that

each particle behaves independantly of the others. With cigarette smoke the only interaction I would anticipate would be coagulation, otherwise the behaviour of particles as a whole reflects the behaviour of individual particles without interaction. The cloud may however behave as a unit because of the presence of carbon dioxide in the smoke, as an example behaviour of particles in a centrifuge may be atypical because the smoke particles behave collectively. A suitable precaution is to use the same gas for winnowing as the carrier gas. I see no other mechanism which would lead to a change of deposition with concentration.

HIGENBOTTAM:    There is some work on the chemical composition of particles of different sizes, using a spinning cone separator and there was a different concentration of polycyclic aromatic hydrocarbons in large and small particles. Smokers first inhale smoke and then exhale it, are there different considerations of deposition during exhalation?

PRODI:          There is deposition during exhalation, and in the nose the deposition efficiency is identical in the two phases of respiration. There is a lower particle number during exhalation and size distribution has been changed by residence in the lungs. The absolute amount of particles differs, but the deposition efficiency is the same.

GUYATT:         When a bolus of smoke is inhaled and the subject then exhales, presumably some of the particles will remain, diluted in the alveolar volume so there is a long time available for deposition. Have you any information on the recovery of particles in successive breaths?

PRODI:          Yes, about 92% of total deposition takes place during the first breath, but this concerns the mising of tidal air with residual air and this involves the level of ERV, turbulence and other factors.

CUMMING:        In England we would use the term functional reserve capacity (F.R.C.) rather than expiratory reserve volume (ERV) so that we may avoid semantic problems.

DENISON:        One of the most important features of the mammalian airway is its difference in geometry between inspiration and expiration. How can the deposition efficiency be the same if this is so?

PRODI:          I referred specifically to the nose.

CUMMING:        A simple experiment would be to inject a bolus of tobacco smoke into a chamber containing air saturated with water vapour at $37^{\circ}C$ and observe the change in particle size distribution.

PRODI:          I never work with tobacco smoke, but the instrumentation we have is applicable to the problem.

CUMMING:        The real problem is the understanding and measurement of particle retention by the smoker, and this could be done by measuring the size, distribution and mass of inspired particles, with similar measurements during expiration. Unfortunately this procedure is complicated by the fact that particle size changes of itself by residence in the lung. If therefore we can demonstrate the nature of this latter change we could make some sensible inference about particulate retention. Would your instrumentation help in looking at this problem?

PRODI:          This would involve measuring size distribution of particles in relation to enviromental parameters, as well as their chemical composition. One could make experiments by duplicating the environmental conditions in the airways and measuring the results.

CUMMING:        At that point I will close the discussion, with apologies to the many questions who have not been heard.

# ELASTIN-LYSINE DERIVED CROSS-LINKS STRUCTURE, BIOSYNTHESIS, AND RELATION TO LUNG EMPHYSEMA PATHOGENESIS

P. A. Laurent, A. Janoff and H. M. Kagan

Insem U 139 Hopital Henri Mondor, Creteil, France
State University of New York and Boston University

## ABSTRACT

A strong statistical correlation exists between cigarette smoking and the susceptibility to pulmonary emphysema, a disease characterized by excessive destruction of elastic fibres of the lung. The recruitment of elastases secreting cells or the decreased activity of lung antiproteases have been involved in elastin destruction of smoker's lungs. We have explored a third possibility, namely that cigarette smoking can also cause abnormalities in elastin cross-linking. The three-dimensional structure of elastin is highly stabilized by covalent cross-links between adjacent elastin peptide chains. Desmosine and isodesmosine are the major cross-linking amino acids in elastin and are synthesized by condensation of three residues of $\alpha$-aminoadipic-$\delta$-semialdehyde (allysine) with one lysine residue lying in close proximity to one another within and between adjacent polypeptide chains of tropoelastin, the soluble precursor molecule of elastin. Allysines, in turn, are formed extra-cellularly by the oxidative deamination of amino groups of lysine, a reaction catalyzed by the copper-dependent enzyme, lysyl oxidase.

We demonstrated that water-soluble components of the gas phase of filtered cigarette smoke inhibit the formation of covalent desmosine cross-links during the conversion of tropoelastin to elastin in vitro. These same smoke components also supress lysyl oxidase-catalyzed oxidation of lysine $\epsilon$-amino groups in tropoelastin in a dose-dependent fashion. However, gas phase cigarette smoke does not block the oxidation of diaminopentane by lysyl oxidase. Thus, gas phase cigarette smoke may possess substrate-directed (rather than enzyme-directed) inhibitory components capable of interferring with elastin cross-linking in vitro. Similar effects occuring in smokers' lungs could impede elastin repair and contribute to the development of pulmonary emphysema, it is noteworthy that other agents that inhibit elastin cross-link formation (eg. $\beta$ - aminoproprionitrile) greatly increase alveolar departitioning in experimental emphysema.

Alveolar tissue destruction and distension observed in emphysematous lung has been demonstrated to be related to elastic fibres destruction (1). In human beings emphysematous changes are particularly frequent among smokers. Only enzymes that hydrolyse fibrous elastin in vitro are able to induce emphysematous lesions when injected intratracheally to animals. The pathogeny of lung emphysema is thought to be related to an increase of elastolytic forces. This elastolytic activity appears when the equilibrium between proteases (Ic elastase) and antiprotease (Ie $\alpha_1$ antiprotease inhibitor) is unbalanced (1) (2).

There is an apparent discrepancy between this theory and the almost null turnover of elastin in adult and with the normal elastin content of the emphysematous lung. Yet it has been shown that elastin synthesis is stimulated in case of tissue aggression, giving evidence of the ability of the lung cells to resynthesize elastin and eventually to keep up with elastin destruction (3) (4) (5).

These data permit to hypothesize that lung emphysema could be the result of a lack of elastin resynthesis or a lack of synergy between elastin degradation and elastin reconstruction.

Numerous factors influencing the protease antiprotease balance have been found. Much less is known regarding factors which can modify the in vivo elastin resynthesis and concerning the possible role of tobacco smoke components as disturbing agents of synthesis.

## A - Biochemical nature of elastin

This topic has been extensively reviewed recently (6); we will summarise the most important features of elastin structure and of cross-links formation.

Elastin can be considered as a polymer (Fig. 1) of linear polypeptide chain (tropoelastin) (7) (8), stabilized by lysine derived cross-links such as desmosine, isodesmosine,

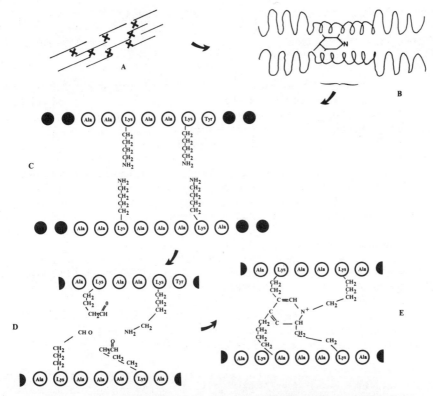

**FIGURE 1:  ELASTIN STRUCTURE AND DESMOSINE BIOSYNTHESIS**

A – Schematic aspect of cross linked elastin fiber.

B – Desmosine molecult links two molecules of tropo-elastin on a level of inelastic area (small loop) of the protein: those areas are alternating with elastic regions of the protein.

C – Tropp-elastin molecules in the coacervated state flock together lysyl side chains.

D – Three epsilon aminogroups of lysyl residues are enzymatically (lysyloxidase) transformed in aldehydes.

E – Chemical condensation of three aldehydes and of non modified lysylresidue give birth to a pyridinium ring named desmosine.

dehydrolysinorleucine or lysylnorleucine (9). Cross-links
(Fig. 2) are located in regions having helix structure. Those
regions are surrounded by large areas with $\beta$ spiral structure
which sustain the elastic properties of the molecule.

Elastin resistance to chemicals such as NaOH and
enzymatic degradation is related to the existence of those
cross-links (10). This resistance to degrading agents permits
its purification. If less cross-linked elastin occurs (11), it
would be more fragile and thus degraded during purification.

Tropoelastin is biosynthesized intracellularly like
other proteins in cells such as fibroblast or smooth muscle
cell (12). Newly synthesized proteins are released in the
extracellular space where tropoelastin molecules self aggregate
by means of hydrophobic forces (coacervated state). The cross-
linking process begins within coacervate and is initiated by a
specific enzyme: lysyloxidase (13). Thus cross-linking
synthesis is a post transcriptional modification (14) which
takes place in the interstitium.

Lysyloxidase is a copper dependent enzyme (16) and has
probably as cofactor Vit. B6. It is found closely associated
with elastin and collagen fibres (17).

Cross-links formation (18) is a two step reaction. The
first one is enzymic (Fig. 3): lysyloxidase deaminates
oxidatively the $\epsilon$ amino group of lysyl residue giving birth
to $\alpha$ aminoadipic $\delta$ semi aldehyde (allysine) (20). The second
step is a chemical reaction (Fig. 1) where allysine will react
with other allysine or lysine residues and condensate by Shiff
base and aldol condensation mechanisms to give the stable
desmosine or its isomer (19). The close proximity of lysine
residues allowed by the coacervation state is essential in
order to permit achievement of lysine residues chemical
condensation (14) (15).

Elastin turnover in tissue declines very rapidly from
birth to the age of twenty (21) after which, in normal tissues,
almost no synthesis occurs (22). The total body elastin
degradation per day can be estimated by measurement of the
desmosine content in urine to 2.5 mg/day (23). This estimation
is based on the number of 4 desmosines per 1000 amino-acids
found in purified elastin (6).

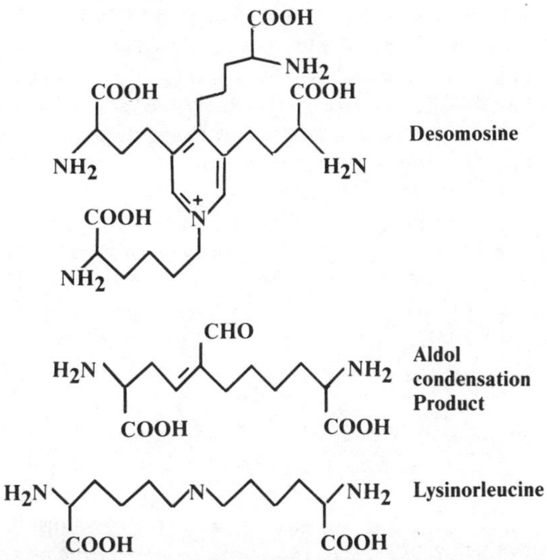

FIGURE 2: LYSYL DERIVED CROSS LINKS

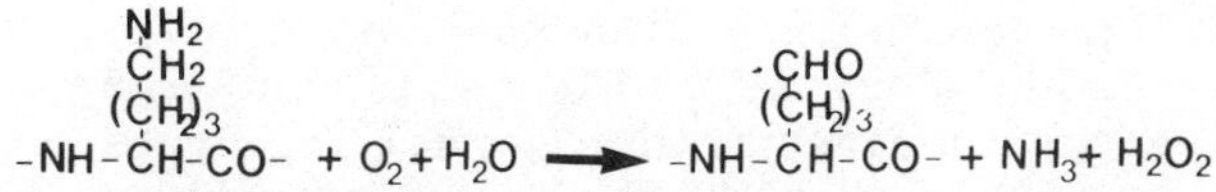

FIGURE 3: OXIDATIVE DEAMINATION BY LYSYL OXIDASE OF THE ε AMINO—GROUP OF LYSYL RESIDUE

## B – Role of elastin cross links in emphysema pathogenesis

Elastin cross links have been shown to play an important role in the formation and maintenance of a normal alveolar pattern (24) (25).

The first observations were made in growing rats fed with sweet peas (lathyrus odoratus) (26). These animals showed multiple abnormalities of connective tissues called lathyrism including pulmonary emphysema. The compound causing the alteration of connective tissues is closely related to β amino-propionitrile (BAPN) (27). This molecule is a very potent lysyloxidase inhibitor forming irreversible covalent bonds with the active site of the enzyme (28).

Administration of BAPN to growing animals has marked effects on their lung structure. Alveolar spaces and alveolar ducts are enlarged without alveolar wall rupture, both total number of alveoli and alveolar surface area are markedly reduced (29). Elastic fibres appeared equally reduced throughout the alveolar wall. The same morphological features have been observed in lung of blotchy mice, a strain with a congenitally acquired defect in lysyloxidase activity (30) and in the lungs of suckling animals with copper deficient diets (31). In BAPN treatment, animals lung function test showed increased compliance and a shift of the pressure volume curve upward and to the left. Cross-links of elastin and collagen as well as allysine contents have been shown to be decreased in those animals.

In contrast no anatomic or functional changes were found when BAPN was administrated to mature hamster (32). This apparent stability of elastin tissue in adults is no longer true in the case of tissue aggression. The great urinary desmosine enhancement observed in acute pulmonary infection giving evidence of elastin degradation in this very common disease (23) is contrasting with the absence of emphysema as a sequella. This provides evidence of the lung ability to resynthesise normal elastin with normal architecture.

In elastase-induced emphysema morphological evidence of rearrangement of lung structure has been observed up to two months after the instillation (33) (34) (35). In contrast, biochemical measurement shows almost normal lung elastin content two weeks after the enzyme instillation (36). Surprisingly, it has been shown that during the second and third weeks after the injection of elastase the synthesis of elastin and cross-links continues if followed with radiolabelled isotopes, assessing and accelerated turnover of elastin, the quantity of elastin resynthesized being equal to the degraded elastin (37).

In the same context a low dose of papain instilled in hamster lungs has been shown to stimulate the synthesis of connective tissue without production of emphysema, thus indicating that the repair mechanismsthe of connective tissue can keep up with the damage (38). A marked worsening of emphysematous lesions is observed in hamsters intratracheally injected with low doses of porcine pancreatic elastase and fed with BAPN, compared to controls without BAPN diet (39). Cross-links measurements eight weeks after elastase injection in BAPN-treated animals disclosed a normal desmosine level, but there was indirect evidence that other lysine derived cross-links were lowered. In that study there was no indication of a possible modification of the turnover of elastin and of elastin cross-links. The enhancement of lesions observed could result in a slow down of a fully cross-linked elastin synthesis with as possible result a less resistant elastin to aggressions. An alternative hypothesis is an accelerated turnover of poorly cross-linked elastin due to non-specific proteases thus facilitating connective tissue remodeling.

Finally we can assume that fully cross-linked elastin resynthesis is an important process in reducing the intensity of emphysematous lesions after elastin degradation.

## C – Relation between tobacco smoke and elastin cross-links processing.

Knowing the strong statistical correlation between cigarette smoking and susceptibility to pulmonary emphysema (40), we wanted to examine the effects of cigarette smoke on desmosine cross-links formation in vitro (41).

We incubated pure lysyloxidase with pure tropoelastin and measured desmosine formation by radioimmunoassay (23). Aqueous cigarette smoke extract prepared by bubbling smoke through buffer (41) when added into the lysyloxidase-tropoelastin mixture, was shown to inhibit almost completely desmosine synthesis (table 1). We partially characterized the inhibiting agent contained in that extract. It has come to light that this or those water soluble compounds were contained in the gas phase (obtained after smoke filtration on a Cambridge pad) and were negatively charged between pH 5.5 and 7.6 as they were retained on an anionic exchange resin (table 1). Such an experiment explored both the enzymic and the chemical steps of desmosine formation.

**TABLE 1:     INHIBITION OF DESMOSINE CROSS—LINK FORMATION BY DIFFERENT FRACTIONS OF CIGARETTE SMOKE AQUEOUS SOLUTION.**

| Smoke Fraction | Smoke Volume (ul) | Experiments (n) | Desmosine Recovery * (% control) |
|---|---|---|---|
| None | 0 | 10 | 100 |
| Whole | 10 | 7 | 18 ± 11 – |
| Gas phase | 10 | 1 | 49 |
| Gas phase | 15 | 1 | 23 |
| Acidic | 15 | 2 | 17 ± 9 – |
| Basic | 15 | 3 | 86 ± 6 |

Values are mean ± SD. Recovery of desmosine/24 h. in the absence of smoke (control) ranged between 4.0 and 17.5 ng. depending on the individual batch of lysyl oxidase used in any given experiment.   In each experiment, percent desmosine recovered in the presence of smoke was calculated by comparison with the control desmosine recovery (no smoke) for that same experiment (arbitrarily 100%), and all data were then pooled for statistical analysis.   In two of the control experiments, reproducibility of duplicate incubations using the same batch of enzyme without smoke (estimate of error of method) was 100 ± 8% and 100 ± 17% (± 1 SDM).
†p versus no smoke < 0.001.

From ref: (38) with permission.

In another experiment using insoluble substrate (15) we were able to demonstrate that lysyloxidase deaminating capacity per se was inhibited by cigarette smoke extract (fig. 4). In that experiment we could not go beyond 50% inhibition. However the fact that three out of four lysine residues have to be deaminated to allow desmosine synthesis suggests that 25% inhibition of lysyloxidase activity is sufficient to give 100% inhibition of desmosine synthesis.

In a third experiment we used as substrate for lysyloxidase a small molecule (diamino-pentane); cigarette smoke extract, when added to the enzymic substrate mixture, did not inhibit the reaction (table 2). This demonstrates that the catalytic site of the enzyme is not involved in the inhibition mechanism. The exact mechanism of cross-links synthesis inhibition by smoke extract has not been demonstrated.

Lysyloxidase and tropoelastin form an heterogeneous phase; in such a system it is extremely important that enzyme binds to its substrate for the enzymic reaction to be carried out. So we suggested that inhibition could result from a charge modification on tropoelastin or on the enzyme, impairing their binding. Such enzyme substrate binding modification has been well demonstrated with elastin. Enzymic activity of lysyloxidase or pancreatic elastase on elastin can be greatly modified by charged substances added to the medium (42).

The concentration of smoke used in our experiments seems to be relevant to in vivo conditions since such a concentration could be reached in the alveolar spaces after inhalation of 20 puffs of smoke. Moreover, decreased protease inhibitor activity reported under the influence of smoke cigarette (43) can only be obtained at a 40 fold greater concentration than those use in our experiments.

These in vitro experiments do not shed any light on the possibility of elastin cross links synthesis inhibition in vivo by cigarette smoke. Recently (44) smoke exposure of rodents with elastase induced emphysema have been shown to worsen the intensity of lesions. This effect is obtainable only if the animal has been exposed to smoke several days before elastase administration. As we have no data on lung antiprotease screen we cannot conclude on the actual mechanism of this phenomena. More conclusive are data showing a reduction in cross-linking formation in elastase treated animals exposed to cigarette smoke (45). However the demonstration of in vivo inhibition of lysine derived cross-links in elastin by tobacco smoke remains to be studied.

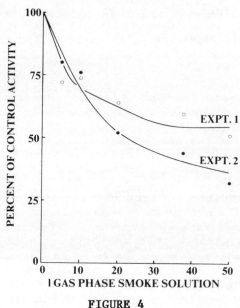

**FIGURE 4**

Gas-phase-smoke-induced inhibition of $^3$H release from $^3$H-lysine-labelled insoluble elastin (chick embryo aorta) during oxidation catalyzed by purified lysyl oxidase (bovine aorta). The insoluble aortic elastin substrate was prepared as described elaswhere (15) from 16-day-old chick embryo aortas that had been pulsed with L-4,5-$^3$H-lysine prior to extraction to isolate the insoluble elastin pellet. Each assay for elastin oxidation contained 3 ug of lysyl oxidase (specific activity of 500,000 cpm $^3$H released per milligram enzyme per 2h at 37°C), and 125,000 cpm of elastin substrate in 0.1 M sodium borate, 0.15 M NaCl (pH, 8.0) in a final volume of 750 ul. Reactions were stopped at 2 h by freezing, and $^3$HHO was isolated by distillation in vacuo and counted by liquid scintillation spectrometry. Aliquots of smoke fractions were added at zero degrees just prior to setting the assay tubes at 37°C. Ordinate shows $^3$H release as percent of control value (in the absence of smoke solution). Experiments 1 and 2 were run under identical conditions. Each point is the average of duplicate determinations;  standard errors equal $\pm$5%.

TABLE 2: **EFFECT OF GAS PHASE SMOKE SOLUTION ON OXIDATION OF 1,5-DIAMINOPENTANE BY LYSYL OXIDASE FROM BOVINE AORTA**

| Smoke Solution Added per 2 ml Reaction Volume (ul) | $H_2O_2$ Production – (nmol/min) | Activity (%) |
|---|---|---|
| 0 | 0.65 ± 0.05 | 100 ± 8 |
| 100 = | 0.75 ± 0.05 | 115 ± 8 |
| 200 = | 0.65 ± 0.05 | 100 ± 8 |
| 200 = | 0.65 ± 0.05 | 100 ± 8 |

Enzyme activity against diaminopentane as substrate was determined using the peroxidase-coupled fluorometric assay described elsewhere (49). Assays were conducted in 0.05 M sodium borate, 1.2 M urea (ph, 8.2) at 37°C in a final volume of 2.0 ml. Only initial rates taken from the initial linear phase of the reaction are reported. It was established that the smoke fraction had negligible effects on the enzyme-coupled assay system by testing for the peroxidase-dependent utilization of $H_2O_2$ production.

† Values are mean ± 1 SD.

‡ Smoke and enzyme added at time zero of assay

§ Enzyme was preincubated with smoke in borate buffer for 1 h at 37°C in the absence of the other assay components  Other assay components were added at time zero to initiate the reaction.

From ref (38) with permission.

The normal elastin content human emphysematous lung prove that quantitative biochemical measurements are inadequate to fully understand the emphysema pathogenesis. Minor modifications in enzyme kinetic may permit other factors to exert their offensive effects. This statement is supported by the varying degrees of emphysematous changes observed after the same elastase aggression in lungs submitted to different mechanical stresses. Lesions were worst in the lungs of animals doing physical exercise (46). Conversely no lesions were seen in lungs treated in vitro by elastase (47).

Finally we can suppose that most people with unbalanced protease-antiprotease lung screen (48) are probably able to resynthesize new elastin rapidly, thus permitting connective tissue to keep a normal configuration.

If elastin resynthesis could be slowed down among smokers the connective tissue repair should be delayed allowing stress due to breathing to gradually distort alveolar wall architecture until sufficient amount of elastin is synthesized.

# REFERENCES

1.  KUHN, C., Slodkowska, J., Smith, J., Starcher, B.C. (1980)
    The tissue response to exogenous elastase.
    Clin. Respir. Physiol;, 168 (suppl). 127-139.

2.  SNIDER, G.L. (1981).
    Pathogenesis of emphysema and chronic bronchitis.
    Med. Clin. North. Am., 65, 647-665.

3.  MITMAN, C. (Editor) (1972),
    Pulmonary emphysema and proteolysis.
    New York Academic, 1-537.

4.  PIERCE, J.A., Mocott, J.B., Ebert, R.V. (1961),
    The collagen and elastin content of the lung in
    emphysema.
    Ann. Intern. Med., 55, 210.

5.  FITZPATRICK, M. (1967),
    Studies of human pulmonary connective tissue. Chemical
    changes in structural protein with emphysema.
    Am. Rev. resp. Dis., 96, 257.

6.  SANDBERG, L.B., Soskel, N.T., Leslie, J.G. (1981),
    Elastin structure, biosynthesis and relation to disease
    state.
    New Engl. J. of Med., 566–579.

7.  SANDBERG, L.B., Weissman, N., Smith, D.W. (1969),
    The purification and partial characterization of a
    soluble elastin like protein from copper deficient
    procine aorta.
    Biochemistry, 8, 2940–2945.

8.  FOSTER, J.A., Bruenger, E., Gray, W.R., Sandberg, L.B.
    (1973),
    Isolation and amino-acid sequence of tropoelastin
    peptides.
    J. Biol. Chem., 248, 2876–2879.

9.  PARTRIDGE, S.M., Elsden, D.F., Thomas, J., Dorfman, A.,
    Telser, A., (1966),
    Incorporation of labelled lysine into desmosine cross-
    bridges in elastin.
    Nature, 209, 399.

10.  MECHAM, R., Foster, J.A., Franzblau, C. (1977),
     Proteolysis of tropoelastin.
     Adv. Exp. Med., Biol, 79, 209–220.

11.  PAZ, M.A., Keith, D.A., Traverso, H.P., Gallop, P. (1976),
     Isolation, purification and cross-linking profiles of
     elastin from lung and aorta.
     Biochem., 15, 4912.

12.  HINEK, A., Thyberg, J. (1977),
     Electron microscopic observation of the formation of
     elastic fibres in primary culture of aortic smooth
     muscle cells.
     J. Ultrastrut. Res., 60, 12–20.

13.  SIEGEL, R.C. (1979),
     Lysyl-oxidase,
     Int. Rev. Connect. Tiss. Research, 8, 73–118.

14.  NARAYANAN, S.N., Page, F., Kuzan, C., Gaylor-Cooper, G.G.
     (1978),
     Studies on factors influencing the formation of
     desmosine by lysyl-oxidase action on tropoelastin.
     Biochem. J., 173, 855.

15. KAGAN, H.M., Hewitt, N.A., Saceldo, L.L., Franzblau, C.
    (1974),
    Catalytic activity of aortic lysyl-oxidase in an
    insoluble enzyme-substrate complex.
    Biochem. Biophys. Acta., 365, 223-234.

16. SNIDER, R., Faris, B., Verbitzki, V., Moscaritolor, J.F.,
    Franzblau, C., (1981),
    Elastin biosynthesis and cross-links formation in
    rabbit aortic smooth muscle cell culture.
    Bioch., 20, 2614-2618.

17. RAYTON, J.K., Harris, E.D. (1979),
    Induction of lysyl-oxidase with copper.
    J. Biol. Chem., 3, 621-626.

18. NARAYANAN, S.N., Siegel, R.C., Martin, G.R. (1974),
    Stability and purification of lysyl oxidase.
    Arch. Biochem. Biophys., 162, 231-237.

19. GUAY, M., Lamy, F. (1979),
    The troublesome cross-links of elastin.
    T.I.B.S., pp.160.

20. PINNEL, S.R., Martin, G.R. (1968),
    The cross-linking of collagen on elastin enzymatic
    conversion of lysine in peptide linkage to alpha-amino-
    adipic- S. semialehyde (allysine) by an extract from
    bone.
    Proc. Nat. Acad. Sci. USA., 61, 708-716.

21. GUNJA-SMITH, Z., Boucek, R.J. (1981),
    Desmosine in human urine.
    Biochem. J., 193, 915-918.

22. HANCE, A.J., Crustal, R.G. (1976),
    Collagen: a biochemical basis of pulmonary function.
    Ed: Crysta, R.G., Series: Lung biology in Health and
    Disease, Vol. 2., N.Y. Marcel DEKKER, pp.215-271.

23. HAREL, S., Janoff, A., Yu, S.Y., Hurewitz, A.,
    Bergofsky, E.H. (1980),
    Desmosine radioimmunoassay for measuring elastin
    degradation in vivo.
    Am. Rev. Resp. Dis., 122, 769-779.

24. EMERY, J.L. (1970),
    The postnatal development of the human lung and its
    implications for lung pathology.
    Respiration, 27 (suppl). 41-50.

25. LOOSLI, C.G., Potter, E.L. (1959),
    Pre and postnatal development of the respiratory
    portion of the human lung with special reference to the
    elastic fibres.
    Am. Rev. Resp. Dis, 80, part 2, 5-23.

26. GEIGER, B.J., Steenbock, H., Parsow, H.T. (1933),
    J. Nutr. 427-442.

27. PAGE, R.C., Benditt, E.P. (1967),
    Molecular diseases of connective and vascular tissues:
    II - amine oxidase inhibition by the lathyrogen β-
    aminopropionitrile.
    Biochemistry, 6, 1142-1148.

28. TANG, S.S., Trackman, P.C., Kagan, H.M. (1983),
    Reaction of aortic lysyloxidase with β-
    aminopropionitrile.
    J. Biol. Chem., 258 (7), 4331-4338.

29. KIDA, K., Thurlbeck, W.M. (1980),
    The effects of β-aminopropionitrile on the growing rat
    lung.
    Am. J. Pathol., 101, 693 -708.

30. FISK, D.E., Kuhn, C. (1976),
    Emphysema like changes in the lung Blotchy mouse.
    Am. Rev. Respir. Dis., 113, 787-797.

31. O'DELL, B.L., Kilburn, K.H., McKenzie, W.N.,
    Thurston, R.J. (1978),
    The lung of the copper deficient rat: a model for
    developmental pulmonary emphysema.
    Am. J. Pathol., 91, 413-432.

32. KUHN, C., Starcher, B. (1976),
    The significance of connective tissue repair in
    elastase induced emphysema.
    Am. Rev. of Resp. Dis., 113 (suppl) 209.

33. KAPLAN, P.D., Kuhn, C., Pierce, J.A. (1973),
    The induction of emphysema with elastase, the evolution
    of the lesions and the influence of serum.
    J. Lab. Clin. Med., 82, 349-356.

34.  KUHN, C., Tassavoli, F. (1976),
     The scanning electron microscopy of elastase induced
     emphysema, a comparison with emphysema in man.
     Lab. Invest, 34, 2-9.

35.  SNIDER, G.L., Sherter, C.B. (1977),
     A one-year study of the evolution of elastase induced
     emphysema in hamsters.
     J. Appl. Physiol., : Respir. Environ. Exercise Physiol,
     43, 721-729.

36.  KUHN, C., Yu, S.Y., Chraplyvy, W., Linder, H.E.,
     Senior R.M., (1976),
     The induction of emphysema with elastase: II changes in
     connective tissue.
     Lab. Inves., 34, 372.

37.  YU, S.Y., Keller, N.R., Yoshida, A. (1978),
     Biosynthesis of insoluble elastin in hamster lung
     during elastase emphysema.
     Proc. Soc. Exp. Biol. Med., 157, 369-373.

38.  COLLINS, J.F., Durnin, L.S., Johanson, N.G. (1978),
     Papain induced lung injury: alteration in connective
     tissue metabolism without emphysema.
     Exp. Mol. Pathol, 29, 29-36.

39.  KUHN, C., III, Starcher, B.C. (1980),
     The effects of lathyrogen on the evolution of elastase
     induced emphysema.
     Am,. Rev. Resp. Dis., 123, 453-460.

40.  AUERBACH, O., Hammond, E.C., Garfinkel, L., Benante, C.
     (1972),
     Relation of smoking and age to emphysema: whole-lung
     section study.
     N. Engl. J. Med., 286, 853-857.

41.  LAURENT, P., Janoff, A., Kagan, H.M. (1983),
     Cigarette smoke blocks cross-linking of elastin in
     vitro.
     Am. Rev. Resp. Dis., 127 (2), 189-192.

42.  KAGAN, H.M., Tseng, L., Simpson, D.E. (1981),
     Control of elastin metabolism by elastin ligands.
     Reciprocal effects on lysyl oxidase activity.
     J. Biol. Chem., 256, 5417-5421.

43.  JANOFF, A., Carp. H., Laurent, P., Raju, L. (1983),
     The role of oxidative processes in emphysema.
     Am. Rev. Respir. Dis., 127, 531-538.

44.  HOIDAL, J.R., Starcher, B.C. (1983),
     Cigarette smoke inhalation of lathyrogens on the
     evolution of elastase-induced emphysema.
     Am. Rev. Resp. Dis., 127, 478-481.

45.  OSMAN, M., Cantor, J., Roffman, S., Turino, G.M.,
     Mandl, J. (1982),
     Tobacco smoke exposure retards elastin repair in
     experimental emphysema.
     Abstract: Am. Rev. Resp. Dis., 123 (2), 213.

46.  SAHEBJAMI, H., Vassalo, C.L. (1976),
     Exercise stress and enzym-induced emphysema.
     J. Appl. Physiol., 41, 332-335.

47.  KARLINSKY, J.B., Snider, G.L., Franzblau, C., Stone, P.J.
     Hoppin, F.G.Jr. (1976),
     In vitro effects of elastase and collagenase on
     mechanical properties of hamster lungs.
     Am. Rev. Respir. Dis., 113, 769-777.

48.  CARP, H., Miller, F., Hoidal, J.R., Janoff, A. (1982),
     Potential mechamism of emphysema: alpha 1 proteinase
     inhibitor recovered from lungs of cigarette smokers
     contains oxidized methionine and has decreased elastase
     inhibitory capacity.
     Proc. Natl. Acad. Sci.,USA, 79, 2041-2045.

DISCUSSION


LECTURER: Laurent                                    CHAIRMAN: Cumming

CUMMING:            Do I understand your hypothesis correctly if I
                    say that the destruction of lung tissue is not
                    direct, but due to an inhibition of synthesis
                    of new tissue brought about by the presence of
                    smoke particles?

LAURENT:            There is certainly parenchymal destruction by
                    elastase but this is not sufficient and re-
                    synthesis inhibition must also be involved.

# EFFECTS OF CIGARETTE SMOKE ON THE METABOLISM

# OF ARACHIDONIC ACID AND PROSTAGLANDINS IN LUNG

Y.S. Bakhle

Department of Pharmacology
Institute of Basic Medical Sciences
Royal College of Surgeons
Lincoln's Inn Fields, London WC2

## INTRODUCTION

The lung, in common with many other organs, is capable of synthesizing and inactivating prostaglandins (PG) and thromboxanes (TX). However, there are at least two features unique to the lung, which make PG matabolism in this tissue of particular importance.

Firstly, the lung and its pulmonary circulation are placed at the point of division between venous and arterial circulations and thus receive the entire cardiac output. For this reason if PGs are removed from venous blood or added to the blood perfusing the pulmonary circulation, then those PGs or their absence is distributed immediately and ubiquitously to the systemic arterial circulation. This "exporting" metabolic function represents a pseudo-endocrine function of lung.

Secondly, and more relevant to this paper, the lung is designed to give the closest possible contact between inhaled gases and the blood. It is thus a place in which the external environment is brought continuously into very close apposition to the internal milieu of the body. The acute administration of gaseous anaesthetics has been shown to depress the metabolism of noradrenaline in the pulmonary circulation (Bakhle & Block 1976) and exposure to high concentrations of oxygen decreased metabolism of angiotensin I, 5-hydroxytryptamine and PGE$_2$ in the pulmonary circulation (Bakhle 1983; Block & Fisher, 1977).

Thus it is not surprising that considerable effort has

305

been made to study the effects of a more chronic inhaled insult
to the lungs - tobacco smoke - on the metabolism of a variety
of substrates in the lung, particularly as many of those
substrates have activity on the cardiovascular system.   The
correlation between smoking and cardiovascular disease is now
well established (U.S. Surgeon General, 1979) but the
mechanisms responsible still remain to be elucidated.   I want
here to deal chiefly with the synthesis and subsequent
metabolism of PGs and Txs, compounds with important
cardiovascular effects (Whittle & Moncada, 1983), by lung and
the effects thereon of exposure to tobacco smoke.

## Prostaglandin synthesis and inactivation

The metabolic pathways leading to the synthesis and
inactivation of PGs and Txs are outlined in Fig. 1.   The
initial step, the synthesis of the endoperoxides ($PGG_2$ and
$PGH_2$), is catalysed by cyclo-oxygenase-an enzyme that requires
free arachidonic acid (AA) as substrate.   Subsequent
transformations of the endoperoxides to a mixture of PGs and
$TxA_2$ are controlled by the activity of a range of different
enzymes all demonstrably present in lung tissue.   Thus after
infusion of AA through the pulmonary circulation of perfused
isolated lung from several species including man, all the
products shown may be identified in the pulmonary venous
effluent.

Further metabolism of the PGs and Txs in perfused lung
has also been demonstrated.   Although $PGI_2$ is not metabolized,
all the other PGs and $TxB_2$ are further oxidised at the 15-
hydroxy group to give 15-oxo-derivatives with much decreased
biological activity (Bakhle, 1983).   Both $TxA_2$ and $PGI_2$ undergo
spontaneous chemical hydrolysis to yield virtually inactive
derivatives, $TxB_2$ and $6-oxo-PGF_1$ respectively. Although this
chemical breakdown is fast, ca. 30 sec in absolute terms, it is
probably not an important pathway of inactivation in vivo where
the whole body circulation time is 15 - 17 sec and the transit
time from pulmonary circulation to peripheral vascular beds is
probably in the order of 5 sec.

All the enzymic pathways, synthetic or catabolic, are,
in principle, susceptible to alteration by external factors,
since both synthesis and inactivation take place in lung, the
net effect of an external factor can only be assessed by
measuring both processes.   In our experiments, therefore, we
measured the synthesis of PGs and Tx from infused AA and the
inactivation of $PGE_2$ in lungs from rats exposed to tobacco
smoke in vivo or in rat isolated lungs ventilated with tobacco
smoke.

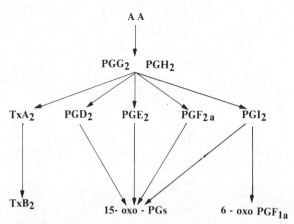

**Pathways of COP Metabolism**

FIGURE 1: The pattern of AA metabolism via the cyclo-oxygenase pathway. Both the endoperoxides ($PGG_2$ and $PGH_2$) appear to serve equally well as substrates for the subsequent transformations to PGs and $TxA_2$. Note that $TxB_2$ and 6-oxo-$PGF_1$ are <u>chemical</u> degradation products of $TxA_2$ and $PGI_2$ respectively.

**METHODS**

The details of the experimental procedures are given elsewhere (Bakhle et al, 1979; Mannisto et al., 1981; Mannisto & Uotila, 1982) but, in general, two main procedures for smoke exposure were adopted. "Chronic" exposure denotes the exposure in vivo of male rats to the smoke from 4 cigarettes during one hour per day of exposure. "Sham exposed" rats were placed in an equivalent chamber with pure air flow for an equivalent time each day. "Acute" exposure was achieved by ventilating the isolated lungs from rats with an atmosphere of tobacco smoke in which cigarette smoke was diluted with room air.

All experiments utilized isolated lungs perfused via the pulmonary circulation with Krebs solution warmed ($37^O$) and oxygenated (95% $O_2$; 5% $CO_2$) at a rate of 8 ml/min. Test substances, either radiolabelled with $^{14}C$ or unlabelled, were added to the perfusing solution immediately before the perfusate entered the cannula in the pulmonary artery. For radiochemical assays, the effluent from the lung was collected in timed fractions and analysed by thin layer chromatography (t.l.c.) and subsequent measurement of radioactivity in areas corresponding to marker compounds by liquid scintillation. For bioassay, the effluent from lung superfused isolated smooth muscle strips (rabbit aorta and coronary artery; rat or hamster stomach) which were calibrated with known amounts of $PGE_2$

**RESULTS**

Effects on prostaglandin inactivation

In our first experiments (Bakhle et al., 1979) we used lungs from rats with "chronic" exposure for 1 day and 10 days only, testing for effects on $PGE_2$ inactivation. Using bioassay, we were able to show an increased survival (i.e. decreased inactivation) after 1 day's exposure but no significant effect after 10 days exposure. The latter result was due to an increased survival in the sham-exposed group. In this series we also studied bradykinin, 5-hydroxytryptamine and angiotensin I metabolism and only angiotensin I metabolism was also changed after 1 day's exposure.

Subsequent investigations using $^{14}C$-$PGE_2$ and radiochemical analysis have shown a decrease in $PGE_2$ inactivation after acute exposure. Thus in rat and hamster lungs ventilated with air mixed with mainstream cigarette smoke, $PGE_2$ metabolites in the effluent were decreased by 25 - 30%. This acute effect was not altered by chronic pre-exposure of the animals in vivo for 3 weeks (Mannisto & Uotila, 1982). In a further series, the inhibition of $PGE_2$ metabolism was correlated with the tar content of the cigarette used to provide the ventilating smoke (Matintalo et al 1983). Cigarettes with a tar content of 18 mg or over (medium and high tar) inhibited $PGE_2$ metabolism whereas the low tar (up to 5 mg) cigarettes caused no significant change.

## Effects on PG and Tx synthesis

The synthesis of PGs and Tx for exogenous AA was also investigated radiochemically. Using bolus injections of $^{14}C$-AA in hamster lungs, chronic exposure to cigarette smoke either for 30 min or for 10 days (one hour per day) did not change the pattern of AA metabolites in the lung effluent. However, ventilation of normal lungs with cigarette smoke, caused a 54% increase in $PGF_2$ and almost 100% increase in $PGE_2$ synthesized from AA. No significant changes were seen in the amounts of 6-oxo-PGFl or $TxB_2$, representing $PGI_2$ and $TxA_2$ respectively (Mannisto et al., 1981). Correlation of this effect with tar content (Matintalo & Uotila, 1983) again showed that the high and medium tar cigarettes had the most influence on $PGE_2$ levels. A more striking change was in the labelling of lung lipids by the $^{14}C$-AA infused. During smoke ventilation, label in triglycerides specifically was increased from 25% to 250% depending on tar content.

## DISCUSSION

It is reasonably clear from our results that PG inactivation was decreased but that $PGI_2$ and $TxA_2$ synthesis were less obviously changed by ventilation of isolated lungs with cigarette smoke (acute exposure). Chronic exposure had no effect on either synthesis or metabolism. How does this compare with the results of others?

The greater effect of acute vs. chronic exposure has been recorded for lipolysis in rat lungs (Hartiala et al 1980) for 5-HT metabolism in rat lungs (Karhi et al 1982) for $TxB_2$ and 6-oxo-$PGF_1\alpha$ levels in smokers and non-smokers (Mehta & Mehta, 1982). In this last report, plasma levels of $TxB_2$ and 6-oxo-$PGF_1\alpha$ in smokers and non-smokers had doubled the previous concentration of $TxB_2$ in plasma whereas the smokers showed no change.

Masotti et al (1981) demonstrated, by bioassay, that

the generation of a $PGI_2$-like substance in blood was less in smokers than non-smokers during a non-smoking period. However, non-smokers were more affected by cigarette smoking acutely than were the habitual smokers. In this study the concentrations of $PGI_2$ reported were considerably higher than usual so the identification of the bioactive substance as $PGI_2$ might be questioned. Nevertheless, smoking undoubtedly brought about marked changes.

More recently, decreased $PG_2$ production during cigarette smoking by smokers but not by non-smokers was reported by Nadler et al (1983). In smokers also the increase in $PGI_2$ induced by i.v. noradrenaline was reversed by smoking. These authors estimated $PGI_2$ production by measuring 6-oxo-$PGF_1$ in urine. It is difficult to assess the contribution of pulmonary $PGI_2$ to either plasma or urinary $PGI_2$. In any case, since the $PGI_2$ is thought to derive from endothelium, a change in endothelial cell biochemistry in the renal vessels might be shared by all endothelial cells.

There seems thus to be general agreement that exposure to tobacco smoke acutely alters PG synthesis such that a net pro-platelet-aggregatory effect is brought about, most probably through decreased $PGI_2$ production. There is less agreement on the component of cigarette smoke that is responsible for the effect.

Carbon monoxide did not affect $PGI_2$ synthesis by rat aortic rings (Hartiala et al., 1982). Nicotine is favoured by Sonnenfield & Wennmalm (1980) and by Nadler et al (1983), but it too has its detractors (Hartiala et al, 1982). Another possible mechanism involves the rise in non-esterified fatty acids stimulated by smoking and suggests that it is the raised fatty acid levels that are the proximal cause of decreased $PGI_2$ production (Mikhailides et al, 1983). The increase in lipolytic activity of rat isolated lungs during ventilation with cigarette smoke (Hartiala et al., 1980) would be compatible with this proposal.

We face therefore a familiar situation - agreement about the end but not about the means. Although under resting conditions, circulating $PGI_2$ is too low to have any effect on platelets, the local concentration immediately adjacent to the endothelium could be much higher than in the bulk phase and the defects described above would then be important deficiencies in the total of endogenous anti-aggregatory factors. The analyses so far have concentrated in $PGI_2$ and $Tx_2$ as the prime mediators of the cardiovascular effects of smoking, but more attention ought to be paid to the other PGs both as indicators of endoperoxide formation – the endoperoxides are potent pro-aggregatory agents in their own right – and as anti-aggregatory agents synthesized in lung ($PGD_2$ or $PGE_1$).

There are now enough indications that smoking alters PG synthesis in lung and in other tissues. What remains is to establish a causal relationship between disturbed PG metabolism and the cardiovascular defects associated with exposure to tobacco smoke.

## ACKNOWLEDGEMENTS

I would like to thank all my colleagues in the Department of Physiology, Turku University for their enthusiasm in collaboration, and the Royal Society for their support.

## REFERENCES

BAKHLE, Y.S. (1983).
Synthesis and catabolism of cyclo-oxygenase products.
In: Prostacyclin, Thromboxane and Leukotrienes,
eg. Moncada, S.
Br. med. Bull, 39, pp.214–218.

BAKHLE, Y.S. & Block, A.J. (1976).
Effects of halothane on pulmonary inactivation of noradrenaline and prostaglandin $E_2$ in anaesthetized dogs.
Clin. Sci. mol. Med., 50, 87–90.

BAKHLE, Y.S., Hartiala, J., Toivonen, H. & Uotila, P. (1979).
Effects of cigarette smoke on the metabolism of vasoactive hormones in rat isolated lungs.
Br. J. Pharmac., 65, 495–500.

BLOCK, E.R. & Fisher, A.B. (1977).
Depression of serotonin clearance by rat lungs during oxygen exposure.
J. appl. Physiol., 42, 33–38.

HARTIALA, J., Simberg, N. & Uotila, P. (1982).
Exposure to carbon monoxide or to nicotine does not inhibit $PGI_2$ formation by rat arterial rings incubated with human platelet-rich plasma.
Artery, 10, 412–419.

HARTIAL, J., Viikari, J., Hietanen, E. & Toivonen, H. (1980).
    Cigarette smoke affects lipolytic activity in isolated rat
    lungs.
    Lipids, 15, 539–543.

KARHI, T., Rantala, A. & Toivonen, H. (1982)
    Pulmonary inactivation of 5-hydroxytryptamine is decreased
    druing cigarette smoke ventilation of rat isolated lung.
    Br. J. Pharmac., 77, 245–248.

MANNISTO, J., Toivonen, H., Hartiala, J., Bakhle, Y.S. &
    Uotila, P. (1981).
    The effect of cigarette smoke on the metabolism of
    arachiodonic acid in hamster isolated lungs.
    Prostaglandins, 22, 195–204.

MANNISTO, J. & Uotila, P. (1982).
    Cigarette smoke ventilation decreases prostaglandin
    inactivation in rat and hamster lungs.
    Prostaglandins, 23, 833–839.

MASOTTI, G., Roggesi, L., Galanti, G. Trotta, F. &
    Serieni, G.G.N. (1981).
    Prostacyclin production in man.
    In: Clinical Pharmacology of Prostacyclin, eds. Lewis, P.J.
    &
    O'Grady, J., pp.9–20, New York: Raven Press.

MATINTALO, M., Kuusisto, T., Mannisto, J. & Uotila, P. (1983).
    The metabolism of $PGE_2$ is decreased by high and medium-tar
    but not by low-tar cigarette smoke in isolated rat lungs.
    Acta Pharmacol. et Toxicol., 52, 230–233.

MATINTALO, M. & Uotila, P. (1983).
    The effects of low-, medium- and high-tar cigarette smoke
    on the fate of arachidonic acid in isolated hamster lungs.
    Acta Pharmacol. et Toxicol. (in press).

MEHTA, P. & Mehta, J. (1982).
    Effects of smoking on platelets and on plasma thromboxane-
    prostacyclin balance in man.
    Prostaglandins, Leukotrienes and Medicine, 9, 141–150.

MIKHAILIDES, D.P., Barrads, M.A., Jeremy, J.Y. & Dandona, P.
    (1983).
    Cigarette smoking inhibits prostacyclin formation.
    Lancet, ii, 627–628.

NADLER, J.L., Velasco, J.S. & Horton, R. (1983).
    Cigarette smoking inhibits prostacyclin formation.
    Lancet, i,  1248-1250.

SONNENFELD, T. & Wennmalm, A. (1980).
    Inhibition by nicotine of the formation of prostacyclin-
    like activity in rabbit and human vascular tissue.
    Br. J. Pharmac., 71, 609-613.

U. S. Surgeon General (1979).
    Smoking and Health.   pp. 1:13 - 1:15.
    U.S. Department of Health Education and Welfare.

WHITTLE, B.J.R. & Moncada, S. (1983).
    Pharmacology of prostacyclin and thromboxanes.
    In: Prostacyclin, Thromboxane and Leukotrienes, et.
    Moncada, S.
    Br. med. Bull., 39, pp.232-238.

DISCUSSION

LECTURER: Bahkle                                    CHAIRMAN: Cumming

CUMMING:              May I comment on how pleased I was that you had
                      separated the two major cigarette related
                      diseases from the two target organs – those in
                      which the smoke is deposited in the lungs with
                      the appropriate consequences, and the
                      cardiovascular group which must come from a
                      substance which had left the lung and entered
                      the circulation. I am only sorry that you were
                      unable to tell us what that substance was.

DENISON:              In your perfusion experiments, you put your
                      marker into a system with holes in it, larger
                      at the arterial than the venous end. Any
                      molecule entering that system depends on the
                      geometry and the size of the holes, and you
                      pointed out the difficulty in distinguishing
                      the mechanical fate of a marker from its
                      biochemical fate in the lung. There is a
                      technique of putting many passive markers of
                      different sizes into the circulation and
                      comparing their fates by a dimensionless
                      number, the Peclet number which is a statement
                      about the geometry. It might be useful to
                      combine this procedure with your biochemical
                      experiments. Another point; when a biochemical
                      reaction takes place in the lung it could occur
                      there as a means of influencing the whole body,
                      because the lung has the highest oxygen partial
                      pressure and some of the reactions you
                      mentioned are served by oxidases and I wondered
                      what is known about the Michaelis constants of
                      these reactions – do they require a high $PO_2$
                      for optimum function.

BAHKLE:               I accept completely your first comment, and we
                      are always at the mercy of changes in the flow
                      pattern. My hope is that all the tubes are
                      lined with endotheluim where the reaction takes
                      place although the time course will be
                      different. We have tried to change the efflux
                      pattern of prostaglandin $E_2$ and it is extremely
                      difficult to do except in cases where the lung
                      is severely oedematous or if the lungs are
                      poisoned with oxygen. Prostaglandin

dehydrogenase is an enzyme which is destroyed by hyperoxic environments and is much reduced by passage through the lung. Other enzymes are not crucially dependant on high levels of oxygen but this idea has not been tested formally. The thing which distinguishes lung from peripheral tissues is what the lung will not do biochemically – the lung will not metabolise prostacycline or histamine, or adrenaline or augiotensin 2, peripheral tissue metabolise everything.

CUMMING:          You showed the transit time distribution with and without smoking, and since the radioactive dose was the same the area under the curves should have been identical which it was not, but in the table they added up to 94+1. An alternative hypothesis to a change in a transit time is that veno-arterial communications are opened by smoking, transit time being shortened because the capillary meshwork is avoided by some blood.

BAHKLE:          I was not aware that one could pass from the pulmonary artery to the pulmonary vein without passing through a capillary network. Is that untrue?

DENISON:          It is not necessary to create new fast channels, but merely change the volume of the system. For the same flow halving the volume of the bed would halve the transit time. The bed would therefore need to construct rather than dilate.

BAHKLE:          If one gives constrictor drugs, the transit time of these markers is unchanged.

CUMMING:          Then how can you explain that?

BAHKLE:          You physiologists are always dealing with non-metabolisable markers and your rules do not apply when the marker is metabolised. Most of the radioactivity which comes out of the lung from prostaglandin is not prostaglandin itself, but its metabolites so that what vasoconstruction must do is to exceed the saturation for the uptake system. That is difficult to do in view of the large capacity and the small dose.

CUMMING:       Whilst there is hidden a half truth in what you have just said, whatever metabolic process takes place the involved molecules must first obey the rules for the transit of molecules so I think you still have a problem with the observation that vasoconstriction produces no change in transit time. There may be an offsetting change operating, that is a possibility.

LAURENT:       During the transformation of arachidonic acid into $PG_2$ there is possible co-oxidation of substances like benzopyrine metabolites to co-carcinogens. What do you think is the possibility of producing metabolites in your system which are co-carcinogens.

BAHKLE:       This may be an important aspect of arachidonic acid metabolism in lung associated with cigarette smoke. Most of our experiments have indicated an increased turnover of arachidonic acid, though metabolites may be different. The initial enzyme is the same - cyto-oxygenase which generates the co-oxygenation radical which seems to be involved in the benzopyrine experiment. Increased arachidonic acid turnover implies increased oxidation with consequent increase in co-oxidation and metabolites which may act as co-carcinogens. This may be the link between cigarette smoking and carcinoma of the lung. The cardiovascular effects of arachidonic acid may be totally irrelevant.

HEATH:       We know that endothelial cells in the pulmonary artery are not fixed structures, they are labile in form and change with functional circumstances, responding to haemodynamic stresses so that they may come to resemble an aortic endothelial cell. The cells also change in structure with age and there is a very delicate ultrastructure. In the cells you describe you relate dysfunction to inability to take up substrate, is there any evidence of structual or ultrastructural changes which may have accompanied the results which you found.

BAHKLE:            The short answer is no. A paper by Wolf suggests that in animals exposed to tobacco smoke vascular endothelial cells are damaged and he describes a pre-atheromatous form of lesion, whatever that means, but he does describe this damage. There is controversial evidence relating to hypoxia – endothelial cells convert angiotension I to angiotension II and this is rapidly and reversibly changed by exposure to a hypoxic environment when the enzyme appears no longer to function. This can only apply when the enzyme is attached to the cell since the isolated enzyme will function happily in 100% nitrogen, so that it is a structural change. The enzyme is on the outside of the cell and constitutes an intruiging puzzle as to how this can lose up to 50% of its function on being rendered hypoxic.

RAWBONE:           Can you tell us about the method of smoke generation and the degree of smoke dilution used in your studies? In the acute ventilation studies have you been able to study the time course of the results you showed and in particular was there any recovery?

BAHKLE:            The early experiments have been published and I cannot now remember the figures you have asked for. The acute ventilation was done with 40 puffs per minute of 2 ml diluted into one litre of air. As to recovery, we did not do this, but I would guess that they would recover rather easily.

RAWBONE:           People smoke by taking in intermittent boli and you showed no chronic changes, so maybe there is recovery between each smoking puff.

BAHKLE:            You could be right. Blood levels in man between exposure show no difference but they respond differently to the acute exposure. It may be that we cannot determine the chronic effect with the tools at our disposal.

CUMMING:           We will draw the discussion to a close at this point.

# THE CONSEQUENCES OF SMOKING AND SMOKERS AT RISK

Fiona M. Langley

The Midhurst Medical Research Institute
Midhurst, England

The efficiency with which inspired gas is mixed with resident gas appears to be a useful indicator of ventilatory function and may provide a mechanism for detecting those smokers at risk of developing chronic lung disease.

This paper is concerned mainly with the methods for the derivation of alveolar mixing efficiency, and attempts to show its reproducibility, and some of the results we have obtained by studying a rural sample of smokers and never smokers. Alveolar mixing efficiency is derived from the multi-breath nitrogen washout technique of Cumming (1967), its advantages being that it is non-invasive, requires minimal patient co-operation and is not effort dependent.

The variables in the multi-breath nitrogen washout are listed below.

FeN2 = Fractional end expired nitrogen concentration.

Ve = Expired volume.

Te = Expiratory time.

Vi = Inspired volume.

Ti = Inspiratory time.

VN2 = Volume of nitrogen.

normalised VN2 = Volume of nitrogen per litre of lung volume.

remaining VN2 = Nitrogen remaining in one litre of lung volume
                following a breath.

ideal residual VN2 = Ideal residual nitrogen in one litre
                     of lung volume.

TO = Turnover number

These are all calculated breath by breath, and are necessary to obtain the nitrogen washout curve. The lung volume (VL) obtained at Functional Residual Capacity is also computed at the end of the study, from the total volume of recovered nitrogen.

All measurements are made in the laboratory, where the subject is seated with a noseclip on and attached by a mouthpiece to a valve-box and bag-in-box system. The valve-box has two inspiratory ports and one expiratory port. The inspiratory valves are operated electromagnetically and the expiratory valve is a gravity flap (Lee, K.D. and Crisp, A. H. 1974). The inspiratory gas can be switched with the subject being unaware of any change. The subject breathes from one of two inspiratory bags and out into the expiratory bag in the box. Inspired and expired concentrations of gas are measured by mass spectrometer. The probe is sited in the valve-box in mid-axial stream, close to the mouth. Flow is measured by a pneumotachograph and micromanometer fitted to the bag-in-box system. The continuous signals of flow and gas concentration are recorded on an analogue tape recorder.

The subject breathes air for two minutes, or until he is sufficiently relaxed, and to aid relaxation the subject listens to music of his own choice through headphones. The subject is free to select his own tidal volume, and at the end of an inspirate and unknown to the subject, the nitrogen washout is commenced by the inspired gas being changed from air to a nitrogen free mixture of 79% Argon and 21% Oxygen.

Breathing is continued until the end expired nitrogen concentration is less than 2%.

The recorded data are digitised every 20 ms and the digital information fed into a computer for analysis. The signals from the mass spectrometer and flow device are handled by the computer such that for each sampling

interval the product of concentration and flow yield a
volume of gas expired in that interval (figure 1).
These volumes are summed over the breath and plotted on the
ordinate of a graph, whilst expired volume is plotted on the
abcissa. The resulting curve for each individual breath is
handled as a second order polynomial regression as
described by Cumming and Guyatt (1982), the intercept on
the abcissa being designated the series dead space volume or
'anatomical dead space' (VdS). Thus from the signals of
flow and gas concentration we obtain the following
variables breath by breath of the nitrogen washout; Vi, Ti,
Ve, Te, VN2, and VdSN2, and are now in a position to
calculate the lung volume measured at Functional Residual
Capacity (FRC).

The Cummimg nitrogen washout technique assumes that the
lung volume, of which 80 per cent is nitrogen, is made up of
two parts; the volume of gas removed (Sum $VN_2$ (breath by
breath) * 100/80), and the volume of gas remaining in the lungs
followimg the last breath of the washout, the end tidal
nitrogen concentration of the last breath being indicative of
the volume of nitrogen still in the lungs.

The volume of nitrogen leaving the lung in the first
breath divided by the lung volume, gives the normalised
nitrogen volume, that is, the volume of nitrogen per litre of
lung which is lost in the breath. Since the lung contains
80% nitrogen, then at the start of the washout one
litre contains 800 millilitres of nitrogen, and following
the first breath of the washout (800 - normalised
nitrogen expired) ml of nitrogen will remain. A similar
procedure for each breath yields a plot of the volume of
nitrogen remaining in each litre of lung volume as the
washout proceeds. This procedure corrects for differences in
lung volumes between subjects. In the ideal lung gas mixing
is perfect and there is no dead space. However in each breath
it is possible to calculate a series dead space, therefore
the volume of nitrogen remaining in the ideal lung in which
there is no dead space must be calculated, breath by breath as
follows:-

Ideal residual nitrogen = Ideal residual =   $\dfrac{VL}{VL+Vi}$
nitrogen (n)                nitrogen (n-1)

where

n = breath number;  VL = lung volume at FRC;
Vi = inspired breath volume.

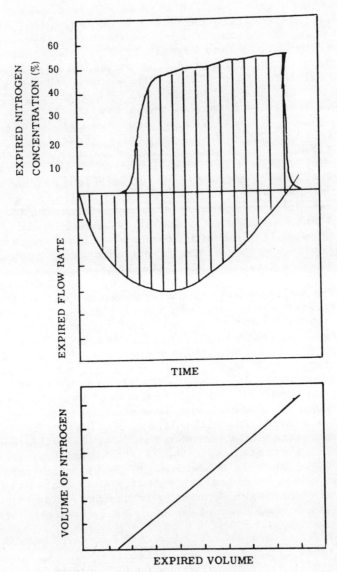

**FIGURE 1**    Expired concentration of nitrogen is shown in the
upper panel, expired flow rate in the centre panel,
the product of which yields a volume of nitrogen as
shown in the bottom panel.

To correct for differences in tidal volume the abcissa plots turnover number, such that when the summed tidal volume is equal to the lung volume, this is 'one turnover' (TO). It is now possible to plot the actual nitrogen washout curve and the ideal washout curve assuming no dead space, as shown in figure 2. The process of plotting nitrogen remaining in one litre of lung against turnover number normalises for all variables.

When only 80 millilitres of nitrogen remain in one litre of lung, and the ideal number of turnovers taken to reach that point are expressed as a percentage of the actual number of turnovers, the 'ventilatory efficiency' to 90% clearance can be determined. The ventilatory dead space (VdV) at 90% is determined by first subtracting the ventilatory efficiency from 100% which is an expression of the inefficiency, and if the mean tidal volume for the washout is multiplied by the inefficiency, this gives the ventilatory dead space.

Ventilatory dead space is made up of a series or anatomical dead space (VdS), and alveolar dead space (VdA). A tidal volume constitutes the VdS and the volume available for gas mixing beyond it, the tidal mixing volume (ViA). This in turn is made up of two parts, the alveolar dead space (VdA) which takes no part in gas mixing, and the effective alveolar mixing volume (VmA). Hence: $Vt = VdS + ViA.$, $VdV = VdS + VdA.$, and $ViA = VmA + VdA.$ multi-breath alveolar mixing efficiency is $VmA/ViA*100\%.$ or alternatively $(Vt - VdV)/(Vt - VdS)*100\%$   Since this derivation is designed to exclude the series dead space, it is assumed that it is an index of function distal to the stationary interface.

In order to test the repeatability of the measurement, (Buckman et al, 1983) 16 asymptomatic, naive subjects were tested at fixed times, daily on seven occasions, over the course of 8 days. Forced expiratory manoeuvres were made and served as a check on the daily lung function of each subject. No significant trend was observed with time (Page's 1 trend test, 1963) over the course of the study in FEV1.0, PEF, $\dot{V}50$, $\dot{V}75$ and FEV1.0/FVC. This suggested that the condition of the subjects had remained stable over the 8 days. Similarly no significant trend was observed with time in the lung volume (VL) measured at FRC, frequency of breathing and multi-breath alveolar mixing efficiency. The mean alveolar mixing efficiency for all subjects over the 7 visits was 64.5%, with a standard deviation between subjects of ±9.6%, and within subjects of ±4.1%. Figure 3 shows the means and standard deviations for the results of each

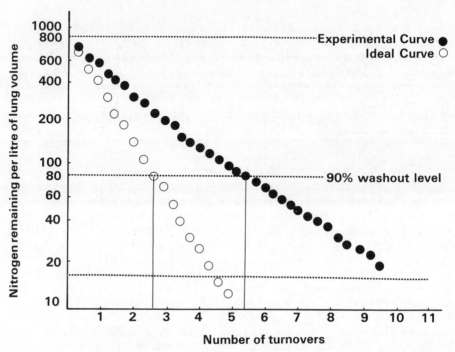

**FIGURE 2**     A nitrogen decay curve showing the volume of
nitrogen remaining in one litre of lung volume
plotted against turnover number. (One turnover is
when the summed tidal volume is equal to the lung
volume at end expiration). The ideal decay curve
is that which would be obtained if there were no
dead space and gas mixing was perfect;  the
experimental decay curve is that obtained from the
subject breathing normally until the washout is
terminated.

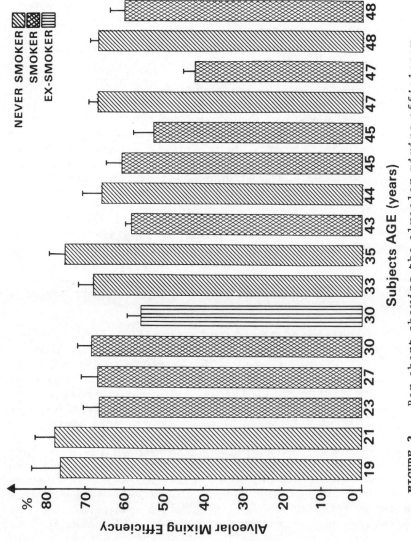

**FIGURE 3**    Bar chart showing the alveolar mixing efficiency within subjects and according to age.

individual in the form of a bar chart according to age. Since the subjects included smokers, ex-smokers and never smokers with ages ranging from 19 years to 48 years, the between subject variability, was as expected, more than within subjects. It was concluded from the low variability within subjects that multi-breath alveolar mixing efficiency for the 16 subjects on 7 occasions was a repeatable index.

In a cross-sectional study 134 smokers were compared with 162 never-smokers using this index. The subjects were all volunteers who were recruited by distributing leaflets, by advertising in local newspapers and on commercial radio. Some subjects lived as far away as 80 kilometers. It was requested that smokers should be fit and healthy men and women who had habitually smoked 15 cigarettes or more each day for at least 5 years. Similarly it was requested that the never-smokers should be fit and healthy and never have smoked as much as one cigarette a day or its equivalent in other tobaccos.

Volunteers attended the laboratory on 2 occasions one week apart. On the first visit they completed gas mixing and lung mechanics tests, and the Medical Research Council questionnaire on respiratory symptoms (1966). A chest X-ray, an electrocardiogram and a medical examination were also carried out. On the second visit they repeated only the tests of lung function.

Results showed there was a significant difference (p<0.01) using an unrelated 't' test between the alveolar mixing efficiency of smokers 58.1% ±10.1SD and that of never smokers 67.8% ±7.2SD. Figure 4 shows a scattergram of the alveolar mixing efficiencies of both smokers and never smokers against age. The regression coefficients of alveolar mixing efficiency (AME) on age (a) were steeper in smokers (AME=75.55-0.42a), than in never smokers (AME=76.78-0.22a). The slope of these curves indicate that smokers have a decline in efficiency which is more rapid than that of never smokers. Smokers decline at 0.42% per annum and never smokers 0.22% per annum. Since these subjects were all recruited on the basis that they were fit and healthy both by self assessment, and by medical examination, we would like to suggest that particular attention should be paid to monitoring those subjects whose alveolar mixing efficiency falls below the expected efficiency obtained from the sample of never smokers, since these may include most of those subjects at risk of becoming the victims of chronic lung disease.

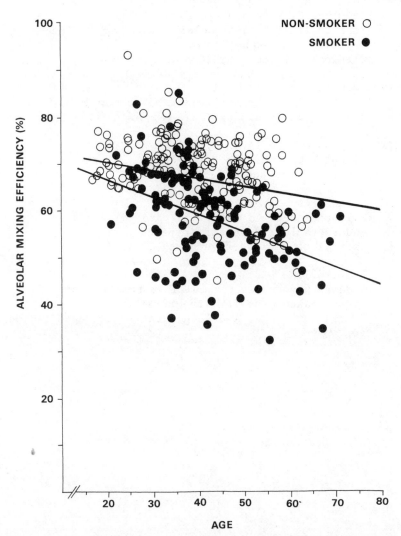

**FIGURE 4**   A scattergram showing the individual values of alveolar mixing efficiency and the regression lines obtained in never smokers and smokers plotted against age.

**REFERENCES**


BUCKMAN, M.P., Parker, S.S., Guyatt, A.A.R.,  O'Bree, C.P.,
        Cumming, G. and Langley, F.M. (1983).
        Reproducibility of Alveolar  Mixing  Efficiency
        measured  in  man.
        Clinical Science, 65, 2P-3P.

CUMMING, G.  (1967).
        Gas Mixing  Efficiency  in  the  human lung.
        Respiration Physiology, 2,  213-224.

CUMMING, G. and Guyatt, A.R.  (1982).
        Alveolar Gas  Mixing Efficiency   in  the  human  lung.
        Clinical   Science,   62, 541-547.

LEE, K.D. and Crisp, A.H. (1974).
        Solenoid-operated valve  box providing sudden
        undetectable  changes  of  inspired  gas.
        Journal of Applied Physiology, 36,  765-766.

PAGE, B.E. (1963).
        Ordered hypotheses for multiple treatments: A
        significance test for linear ranks.
        Journal  of  the  American  Statistical  Association,
        58,  216-230.

DISCUSSION

LECTURER: Langley                          CHAIRMAN: Cumming

HIGENBOTTAM:    My first question relates to the dose of smoke
                which the subjects received. Did you
                standardise for the number of cigarettes smoked
                in measuring the efficiency of mixing.
                Secondly, did you observe what happened on
                giving up smoking?

LANGLEY:        Not exactly. We have information on duration of
                smoking and the number of cigarettes. Duration
                correlates well, but the number does not
                correlate well, with alveolar mixing
                efficiency. We have looked at the effect of
                stopping smoking and there was no detectable
                change in alveolar mixing efficiency after one
                year. The numbers who managed to give up were
                only about 25 and the time of study was short,
                so the significance of the result is not high.

BAKE:           Thank you for an unusually clear presentation.
                Why do you think this index has anything to do
                with small airways?

LANGLEY:        The establishment of the stationary interface
                of concentration between inspired and resident
                gas was established by several authors about
                ten years ago, and denotes the point in the
                airways where convective flow downwards and
                diffusive flow upwards becomes equal so that
                events beyond the stationary interface are
                determined by gaseous diffusion only, events
                upstream being determined by interaction
                between gaseous diffusion and convective flow.
                Mixing efficiency measures events distal to the
                stationary interface, which is measured in each
                breath. Mixing is produced by gaseous
                diffusion which has two variables - the
                diffusion distance, from the stationary
                interface to the alveolar wall and the area of
                cross section of contact between the gases. A
                deficiency of mixing must therefore be
                attributable to one of these variables, and
                what produces a deficiency is a change in the
                nature of the interface which we know to be
                produced in the small airways and not connected

directly with the distributive function of the large airways.

BAKE:  I put it to you that mixing efficiency is determined by large airway function which is not reflected in the $FEV_{1.0}$. What evidence do you have for your hypothesis?

LANGLEY:  Since it is impossible to make a direct measurement, and it is only possible currently to analyse expired gas, evidence can only be by inference as with any other hypothesis in this area.

BAKE:  I entirely agree.

FLETCHER:  Was there any difference in health history between any of the subjects?

LANGLEY:  We rejected all subjects with clinical evidence of lung disease or heart disease, but smokers with cough and phlegm determined by the M.R.C. questionnaire were included whilst asthmatics were excluded.

FLETCHER:  Was the nitrogen slope more or less discriminatory than the more complex alveolar mixing efficiency?

LANGLEY:  Although we have the information about the nitrogen slope we have not yet studied it in the way you suggest.

CUMMING:  The nitrogen slope is expressed between 750 ml and 1250 ml of expirate and all our measurements were made with tidal breathing so that the slope we measure will not be comparable with the standard test. The advantage of alveolar mixing efficiency as an index is that it can be measured during quiet breathing and calls for no patient co-operation.

FLETCHER:  But you are interested in people who do not have lung disease so that problem would not arise?

STANESCU:  How does this test compare in regard to sensitivity and specificity with other tests of distribution of ventilation?   Secondly, I was surprised that you found no change after giving up smoking for one year, which is at variance with other results.

LANGLEY:    We did note an improvement in tests of airflow limitation but not in alveolar mixing efficiency.

CUMMING:    You have raised an important question.   On quitting smoking all lost their cough, they felt better and their mechanical ventilatory efficiency was improved.   However their alveolar mixing efficiency was unchanged and this illustrates that changes in one part of the bronchial tree may be demonstrable by one test but changes in another part requires a different test.   If it is the small airways that are attacked by tobacco smoke this may be irreversible whilst large airway function is in part reversible as has been shown by many people.   You used the term ´distribution of ventilation´ and we now believe that analysis of Phase 3 gives almost no information about the distribution of ventilation but that such information is to be found in analysis of Phase 2.   If one analyses Phase 2 and Phase 3 together, the information obtained contradicts the hypothesis that evidence for distribution is found in Phase 3.

BAKE:       Can you answer the question on sensitivity and specificity?

CUMMING:    The question on sensitivity has been answered in part by the standard deviations, but as to specificity since we are unable to define what we are supposed to be measuring we have some difficulty in answering that question.

LEE:        What percentage of smokers would be below the normal range of never smokers, and how would this relate to all the other measures of respiratory function?

LANGLEY:    It looks like 25 or 30%, but the difficulty is the need to set arbitrary levels.

LEE:        How does it compare with other tests in its

ability to discriminate smokers from non-smokers?

LANGLEY:    This is the only test we have found which separates smokers into two groups, but there appears to be no test which will discriminate smokers from non-smokers, but some smokers have a higher efficiency than some non-smokers. What has impressed us is not that cigarette smoking causes chest and heart disease but that it does so only in a minority of cases, and what is interesting is that characteristics of the majority which renders cigarette smoking relatively innocuous. This test offers the possibility of identifying the polar members of these two groups so that a study might identify those characteristics which produce susceptibility or immunity.

BAKE:       My talk is very similar to this one and I would suggest that the single breath nitrogen test is equally valid in identifying the two groups.

FLETCHER:   I would expect that since a minority is affected that the population distribution would be skewed, whilst the whole population would be normally distributed. I suppose your numbers are too small to establish this?

LANGLEY:    I have the histograms which I can show you and the shapes are indeed different, but largely because of a difference in their standard deviations.

FLETCHER:   But are the normals Gaussian and the others not?

LANGLEY:    You should remember that the nature of the test implies a skew distribution since values above 100% cannot occur, whereas low values are found commonly.

CUMMING:    At that point I will draw the discussion to a close.

# CILIARY ACTIVITY IN THE RESPIRATORY TRACT AND

# THE EFFECTS OF TOBACCO SMOKE

Michael A. Sleigh

Department of Biology
University of Southampton
United Kingdom

## INTRODUCTION

Efficient mucociliary clearance from the respiratory tract depends upon successful interaction between a normal propulsive mechanism (the cilia) and a normal transport medium (the mucus). Proper functioning of the ciliary component requires (1) an extensive (but not necessarily complete) coverage of cilia, (2) a normal orientation of the cilia on the epithelial cells, such that all cells propel mucus in the same direction, and (3) unimpaired ciliary beating, coordination within cells and between contiguous cells being an automatic consequence of normal beating. Proper coupling of ciliary activity to mucus transport depends both on mucus rheology and on the presence of a correct depth of periciliary sol fluid. These various aspects of the mucociliary mechanism will be described in sufficient detail to permit understanding of disturbances resulting from contact with tobacco smoke.

During the smoking of a cigarette there is a reduction and sometimes total cessation of the beating activity of tracheal cilia, and a consequent reduction in the rate of transport of mucus by the cilia. Both beating activity and mucociliary transport recover in the absence of smoke. Ciliated epithelia repeatedly exposed to cigarette smoke have an increased number of goblet cells, a higher rate of mucus production, a thicker epithelium and a higher mitotic cell count than control epithelia; the ciliated cells may have shorter cilia and a higher proportion of the cells are atypical in structure and/or function in epithelia subjected to long-term exposure to tobacco smoke than in normal epithelia. Local

areas of ciliated epithelia, particularly in central airways,
may show severe and long-lasting effects of tobacco smoke, but
many areas quickly recover so that the overall mucociliary
clearance of habitual smokers may remain unimpaired during
periods between cigarettes. The evidence concerning these
various effects of cigarette smoke on the mucociliary mechanism
will be outlined.

## THE CILIA AND THEIR ACTIVITY

### The cilia and their distribution

Each ciliated cell of the respiratory tract commonly
carry about 200 cilia, packed at a density of about 8 to the
square micrometer at the cell apex, and interspersed with some
short microvilli. The cilia themselves are usually between 5
and 7 um long, those on cells lining the smaller bronchioles
being at the lower end of the range; they have the usual
diameter of about 0.25 um. The cilia have a normal internal
structure of 9 doublets and two single central microtubules
(Sleigh, 1977).

Interspersed among these ciliated cells are various
secretory cells, principally the mucus-secreting goblet cells,
and the Clara cells, the nature of whose secretion remains
uncertain (Jeffrey and Reid, 1977). Brush cells with a border
of microvilli up to 2 um long and undifferentiated intermediate
cells also occupy part of the epithelial surface. Jeffrey and
Reid found the proportion of ciliated cells in the rat to be
lowest in the upper trachea and highest in the bronchioles;
both ciliated cells and goblet cells become less frequent in
the smaller bronchioles, disappearing altogether in the
terminal bronchioles. The ciliated cells often form groups or
patches separated by non-ciliated cells; in some scanning
micrographs isolated ciliated cells are seen, but more commonly
most of the surface appears ciliated because the projecting
cilia hide the smaller surfaces of non-ciliated cells.

### Ciliary Movement

Cilia of mammalian respiratory tracts commonly beat at
frequencies of between 10 and 25 Hz at body temperatures, with
the higher frequencies in the larger airways. The patterns of
beating of cilia from the trachea of the rabbit have been
described by Sanderson and Sleigh (1981) and those of cilia
from the human bronchus have been described by Marino and
Aiello (1982); these two descriptions are similar in all main
features.

Individual cilia show a rest phase at the end of the effective stroke, when they lie bent over with their tips pointing in the direction of mucus transport;  this rest period is perhaps more marked on patches of rabbit tracheal epithelium than in preparations from human bronchi.  From this rest position a cilium enters its recovery stroke as the basal region of the ciliary shaft bends to the right and backwards (relative to the direction of mucus propulsion), drawing the ciliary tip low across the cell surface in a clockwise arc (as seen from above the surface).  During this movement the bend of the cilium enlarges and extends up the cilium, so that by the time the cilium has swung through an arc of $180-^{o}$ the ciliary shaft is nearly straight and has completed its recovery stroke. From this position the cilium moves directly into the effective stroke in which the almost straight cilium swings around its base in an almost vertical plane, to reach the bent position of the rest phase.  The beat cycle is thus markedly three-dimensional (Fig. 1).

During their effective stroke the cilia start from an inclination of $30 - 40^{o}$ to the cell surface and move through an arc of $110 - 120^{o}$ before coming to rest at an angle of about $30^{o}$ to the cells surface.  At a common frequency of 20 Hz the effective stroke takes about 15 ms and its tip moves at a speed of about 0.75 mm $s^{-1}$;  the recovery stroke commonly takes about 25 ms and the rest period about 10 ms at this frequency.

The rate of ciliary beating is variable, and is widely believed to be under the control of the animal.  Examples of ciliary control by nerves in invertebrate animals are well-established, and the rate of ciliary beating on the frog palate is under cholinergic nervous control, but doubts remain about the extent and nature of control of the activity of cilia on respiratory epithelia of mammals (see Sleigh, 1977).

The cilia of respiratory epithelia all beat in the general direction of the oropharynx.  Some small deivations of beat direction are seen on normal respiratory epithelia, but a general uniformity of the beat direction of adjacent groups of ciliated cells can be seen in both live preparations and scanning electron micrographs, eg. in Fig. 2 where it is seen that the inactive cilia lying in their rest positions all point in approximately the same direction.  The effective stroke of cilia is found to take place in the direction towards particular doublet microtubules within the ciliary shaft, and these in turn are at that side of the cilium from which a basal foot of the basal body projects into the ciliated cell.  It is found that the basal feet of all basal bodies of a normal cell of respiratory epithelium normally project in the same

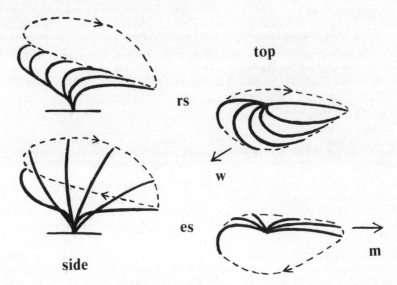

**FIGURE 1:** A rabbit tracheal cilium 6 um long performs a beat cycle that can be divided into a recovery (or preparatory) stroke, r.s., upper figures, and an effective stroke, e.s., lower figures. The figures at the left show these two strokes seen in side view, and those at the right show the two strokes as seen by an observer looking down on the ciliated surface. The dashed line shows the path taken by the ciliary tip throughout the cycle. Waves of coordination (w) travel backwards and to the right relative to the direction in which effective strokes propel the mucus (m).

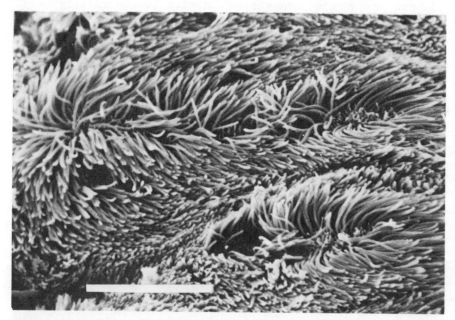

**FIGURE 2:** A scanning electron micrograph of an area of tracheal ciliated epithelium from a rabbit shows the majority of cilia lying in their rest position, with tips pointing towards the right. Amongst these are several groups of cilia caught by fixation during different stages of the beat cycle shown in Fig. 1. In recovery strokes the more strongly curved cilia move downwards and to the left, while the more erect cilia move to the right in the effective strokes. The magnification bar represents 10 um. (Electron micrograph by M. J. Sanderson).

direction - towards the oropharynx. The uniform direction of
beating of respiratory cilia is hence a feature of
morphogenetic origin; the controls acting during development
of ciliated cells determine the orientation of the cilia that
develop upon these cells and hence their direction of beat. In
some lower organisms the direction of beat of the fully formed
cilia can be varied under the control of the organism, but
there is no evidence that the direction of beat of mammalian
cilia can be changed once the orientation of the ciliary basal
bodies has been determined within the ciliated cell.

## Ciliary coordination

Since cilia are densely packed upon the cell surfaces
of respiratory epithelia, it is impossible for them to perform
such three-dimensional beat cycles without strongly interacting
with neighbouring cilia. These interactions lead to the
formation and propagation of the metchronal waves of coordinated ciliary movement described from rabbit trachea by
Sanderson and Sleigh (1981). When a cilium moves from rest at
the start of its recovery stroke, it pushes against the cilia
lying behind it and to its right and excites these cilia to
move in the same way. They in their turn communicate
excitation to cilia lying behind them, and so on. Such
excitation involves a small delay and so a wave of activity
spreads backwards and to the right with respect to the
direction of the effective stroke (which is also the direction
of mucus propulsion), and results in small areas of coordinated
activity within patches of resting cilia, as shown in Fig. 2.

These small metachronal waves may pass unhindered
across one or more cell boundaries before they die away,
propagation presumably being interrupted because there is some
physical discontinuity, such as a non-ciliated cell, that
prevents the spread of interciliary excitation. The ciliated
epithelium is thus made up of many small areas of only a few
cells (i.e. 20 um or so across) within each of which repeated
waves of activity develop, are propagated and die away. If the
frequency of ciliary beating, and hence the strength of
interciliary interaction, increases, adjacent coordinated areas
may join together by propagation of metachronal waves across
boundaries that previously provided barriers. The waves on
human bronchial epithelia appear to be longer than those on
rabbit tracheal epithelium and pass across a larger number of
cells (Marino and Aiello, 1982); presumably there are fewer
barriers to communication of excitation between the cilia of
adjacent human cells.

## THE PROPULSION OF MUCUS BY CILIA

Mucus is secreted from submucosal glands and to a lesser extent from epithelial goblet cells of the respiratory tract. The characteristic glycoproteins of mucus are released in this secretion and aggregate to form patches or .a distinct layer of viscoelastic material that extends outwards for a variable distance from a line at about the level of the ciliary tips. The cilia themselves are surrounded by a layer of periciliary fluid which is virtually free of glycoproteins. The depth of the periciliary layer is such that the highly viscous layer that overlies it is only penetrated by the tips of the more erect cilia in their effective strokes, while the remainder of the ciliary movement, including all of the recovery stroke, takes place in the low-viscosity periciliary fluid beneath the mucus (Fig. 3). Evidence for the existence of these two layers has been discussed for many years and is strongly supported by recent research (see Sanderson and Sleigh, 1981).

Propulsion of the mucus is enhanced if the ciliary tips penetrate the mucus layer for as great a part of the effective stroke as possible and also move as quickly as possible during this stroke, and if the ciliary shaft is kept as close as possible to the cell surface and well within the periciliary layer during the recovery stroke. Because only the ciliary tips penetrate the mucus layer, most of the propulsive force is exerted by the cilia at their extreme tips. If the cilia were longer they could develop higher ciliary tip speeds in the effective stroke, but would require greater stiffness than is provided by the ciliary structure if they were not to bend backwards and lose most of their propulsive efficiency. It appears for these reasons that the optimal length for mucus propelling cilia is about 6 - 7 um (Sleigh, 1982). Such cilia normally propel mucus at speeds of 100 - 300 um $s^{-1}$.

The viscoelastic nature of mucus means that the mucus of one region that has been propelled forward by the effective strokes of a group of cilia will merely recoil elastically when the propulsion ceases unless the cilia of surrounding regions co-operate by continually contributing their own propulsive effort in the same direction. The propulsion of a patch of mucus is therefore achieved by the repeated activity of numerous groups of cilia that provide many sites of application of propulsive force randomly scattered in both time and space (Sleigh, 1981). This can maintain the flow of a layer of mucus that might be crudely compared to an exceedingly loosely-woven layer of sticky elastic fibres. The cohesion of the mucus sheet allows the maintenance of a steady flow of mucus in spite

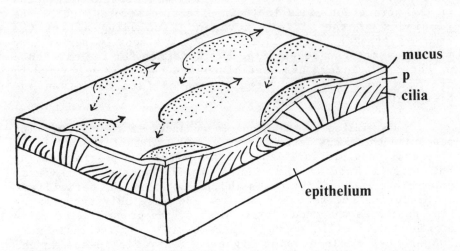

**FIGURE 3:** A diagram on a small area of mucus-transporting epithelium. A thin sheet of mucus is carried at the tips of cilia, above a layer of periciliary fluid (p). Cilia performing their effective strokes penetrate the lower surface of the mucus, but the recovery strokes move below the mucus. Areas of propulsion push the mucus in the direction of the solid arrows, but these waves of activity are propagated in the direction of the dotted arrows. (After Sleigh, 1981).

of the scattered sites of application of force, and
irregularities of both the mucus sheet and the rate of ciliary
beating.

The layer of mucus provides a transport medium for any
foreign materials in this system; if the mucus of a
mucociliary epithelium is severely depleted, particles placed
on it are no longer transported until either a natural mucus or
synthetic mucus-like covering material is provided. Propulsion
of the mucus will only occur if the depth of the periciliary
fluid falls within fairly narrow limits (Sleigh, 1982/1983).
It is essential if the mucus is to be propelled that the tips
of the cilia penetrate the mucus during their effective
strokes, so the periciliary fluid must not be deeper than about
90% of the ciliary length. If the periciliary fluid layer is
too shallow, the cilia remain in contact with the mucus
throughout the recovery stroke and therefore merely carry back
the mucus that had been carried forward in the effective
stroke, so that no net propulsion can take place. It is
believed that if the periciliary fluid layer is so deep that
cilia cannot reach the mucus, then the periciliary fluid is
carried away by ciliary action until its depth is reduced to a
little less than the ciliary length. When there is too little
periciliary fluid for the mucus to be propelled, it is believed
that fluid is produced by secretion from the epithelial cells
to maintain an adequate depth of the periciliary layer.
Possibly some specialised ionic transport mechanisms are
involved in regulation of this depth. Under normal conditions
there is thought to be little net transport of periciliary
fluid, which is merely moved to and fro beneath the mucus by
the effective and recovery strokes of the cilia.

## EFFECTS OF CIGARETTE SMOKE

### Effects on mucociliary transport

More than forty years ago it was observed by Mendenhall
and Shreeve (1937) that the transport of carmine particles
along isolated tracheae from calves was reduced by exposure to
cigarette smoke or dissolved components of this smoke. In more
refined experiments Kensler and Battista (1963) found that the
transport of tracer particles on excised tracheae of rabbits
was totally inhibited by the smoke from between 3 and 6 puffs
of any of 7 types of cigarette. The time-course of the
response was described by Falk, Tremer and Kotin (1959) who
studied particle transport on frog oesophageal epithelium and
tracheobronchial epithelia of rats and rabbits; in typical
cases a 30-second exposure to cigarette smoke was followed
initially by an acceleration of particle transport, but by 15

minutes particle transport was reduced to a minimum rate which
was a small fraction of the control level and by 45 minutes the
particle transport rate had shown a substantial recovery back
towards the level of the control. Human subjects showed a
reduction of mucus transport rate to half the original level
during smoking (Pavia, Thomson and Pocock, 1971), and a
reduction of mucociliary transport rate to one-third or less of
control levels has been reported from anaesthetised dogs
following exposure to the smoke of 5 - 8 cigarettes (Isawa et
al. 1980).

        The recovery of mucus transport after exposure to smoke
noted by Falk et al. can be seen as an explanation of the
numerous observations that repeated exposures to tobacco smoke
do not necessarily result in permanent reductions in the rate
of transport of mucus. For example, Iravani (1972) found no
significant change in the mucus transport rate of rats that had
been exposed to smoke from 230 - 460 cigarettes, and although
after exposure to smoke for 300 days hamsters showed a reduced
rate similarly exposed rats showed a faster rate (Iravani and
Melville, 1974). Human subjects who are habitual smokers have
been found to have mucociliary rates that are not significantly
different from those of non-smokers (eg. Thomson and Pavia,
1973). However, the rats exposed to smoke by Iravani (1972)
and Iravani and Melville (1974) showed several ciliary features
characteristic of bronchitis to a much greater extent than
controls; thus the smokers had many areas of inactive or
abnormal cilia, with some disturbances of coordination and
reversed beating associated with local perturbations of mucus
transport, for example in whirlpools of mucus movement up to
200 um across; these abnormalities were mostly in the trachea
and main bronchi, decreasing in number and size peripherally.

## Effects on ciliary beating

        The acute reduction in mucociliary transport described
above can be attributed to the direct effects of components of
tobacco smoke on the beating of cilia. Dalhamn and Rylander
(1964) reported that the beating of cilia on rat tracheal
preparations stopped after exposure to the smoke of 5 - 6
cigarettes. When Kikta, Holser and Lessler (1976) exposed frog
lung cilia to Ringers solution (10 ml) through which 10 puffs
of cigarette smoke have been bubbles, the ciliary beat
frequency fell to about half its original value. At least some
of the ciliostatic effects of cigarette smoke are due to
components in the gas phase, and when Walker and Kieffer (1966)
exposed cilia of the mollusc Anodonta to fractions of cigarette
smoke separated by gas chromatography, it was found that the
strongest inhibition of ciliary beating was produced by

hydrogen cyanide and acrolein fractions, while the acetaldehyde fraction produced a more temporary reduction of ciliary activity.  Iravani (1972) noted his unpublished observations that when tobacco smoke is blown over the respiratory mucosal surface of rats the amplitude and frequency of ciliary beating diminish progressively until the cilia stop.

Following cessation of exposure to cigarette smoke the cilia recover their activity.  Evidence for this is provided not only by observations that the mucociliary transport of experimental animals and huma subjects shows recovery, but also by direct observations on ciliary beat rates.  In the experiments of Iravani (1972) and Iravani and Melville (1974) measurements of ciliary beat frequency mirrored those of mucociliary transport;  thus rats that had been exposed to smoke from several hundred cigarettes showed a normal beat frequency, and after exposure to smoke for 300 days hamsters showed a reduced rate of beat and rats an elevated rate of beat compared with controls;  the raised rate of beat of rat cilia could be due to an increased stimulation provided by a higher rate of mucus production.

## Other effects on respiratory mucociliary epithelia

Increases in the production of mucus by tracheal epithelia of experimental animals exposed to cigarette smoke (Iravani, 1972; Iravani and Melville, 1974) indicate one of the more frequently reported changes within the epithelium – an increase in the number of goblet cells.  Thus Jones, Bolduc and Reid (1972) found an almost three-fold increase in the number of tracheal goblet cells of rats after six weeks exposure to tobacco smoke, and Chang (1957) found that the goblet cells of human bronchial epithelia were distended in smokers compared with those of non-smokers.  Chang also noted that the bronchial epithelium of these smokers had a thicker epithelium, with more atypical cells and more frequent metaplasia than in non-smokers;  the smokers' bronchial epithelia also showed more basal cell activity and appeared to have shorter cilia.  Rats exposed to smoke from 25 cigarettes a day for six weeks by Jones, Bolduc and Reid (1972) showed a marked increase ( >75%) in the tracheal epithelial thickness and an increase in mitotic activity of tracheal cells by more than six times.  The increase in epithelial thickness is accompanied by an increase in the number of ciliated cells, which became taller and narrower (Jeffrey and Reid, 1977).  Even short-term exposures to smoke can cause an increase in the rate of cell division; for example, Wells and Lamerton (1975) found that after a two-hour exposure of rats to tobacco smoke there was an increase in the mitotic rate of tracheal cells which reached a maximum

after 24 hours, when the rate was five or more times the normal
level.  In the bronchi of human smokers the areas of metaplasia
resulting from such accelerated division rates were often
squamous or of cells with very short or no cilia (Chang, 1957).
Such accelerated division also appears to lead to more errors
of morphogenetic control of the polarity of developing cilia,
leading either to whole cells with abnormal polarity or to a
lack of uniformity of orientation of cilia within the ciliated
cells;  this last effect could result from a regeneration of
cilia in cells infected by viruses.

## CONCLUSIONS

Cigarette smoke has an inhibitory effect on mucociliary
transport through the direct action of several of its
constituents, notably hydrogen cyanide and acrolein, on the
cilia that propel the mucus.  This effect dies away after the
exposure to smoke has ceased.  Cigarette smoke also causes an
increase in the cell division rate of basal cells in
respiratory epithelia, resulting in longer-lasting changes,
namely the production of more goblet cells that release more
mucus (which some reports suggest is more sticky than normal),
and also the production of more ciliated cells.  These ciliated
cells may be abnormal in one or more ways;  the cilia may be
abnormally short or even absent (producing a squamous surface),
or the ciliated cells may develop with an abnormal polarity,
resulting in areas of reversed mucus propulsion and local
whirlpools of mucus flow.

The general level of ciliary activity and mucociliary
transport may be normal in habitual smokers, suggesting a
normal relationship between the beating cilia and the
transported mucus.  However, smokers may have local areas with
little or no net flow of mucus, either because cilia are
missing or because the beat direction has been disturbed, and
during each period of smoking the ciliary activity and hence
mucus transport may temporarily slow or stop.

## REFERENCES

CHANG, S.C. (1957),
   Microscopic properties of whole mounts and sections of human
   bronchial epithelium of smokers and non-smokers.
   Cancer, 10, 1246-62.

DALHAMN, T. and Rylander, R. (1964).
Ciliostatic action of smoke from filter-tipped and non-tipped cigarettes.
Nature (Lond), 210, 401-2.

FALK, H.L., Tremer, H.M. and Kotin, P. (1959).
Effect of cigarette smoke and its constituents on ciliated mucus-secreting epithelium.
J. Nat. Cancer Inst., 23, 999-1012.

IRAVANI, J. (1972).
Effects of cigarette smoke on the ciliated respiratory epithelium of rats.
Respiration, 29, 480-487.

IRAVANI, J. and Melville, G. N. (1974).
Long-term effect of cigarette smoke on mucociliary function in animals.
Respiration, 31, 358-366.

ISAWA, T., Hirano, T., Teshima, T. and Konno, K. (1980),
Effect of non-filtered and filtered cigarette smoke on mucociliary clearance mechanism.
Tohoku J. Exp. Med., 130, 189-197.

JEFFERY, P.K., and Reid, L. M. (1977).
The respiratory mucous membrane.
In: Brain, J. D., Proctor, D.F. and Reid, L.M. (eds):
"Respiratory Defense Mechanisms".
New York, Marcel Dekker, Pt. 1, pp.181-245.

JONES, R., Bolduc, P. and Reid, L. (1972).
Protection of rat bronchial epithelium against tobacco smoke.
Brit. Med. J., 2, 142-144.

KENSLER, C.J. and Battista, S.P. (1963).
Components of cigarette smoke with ciliary-depressant activity. Their selective removal by filters containing activated charcoal granules.
New Eng. J. Med., 169, 1161-1166.

KIKTA, M.C., Holser, J.M. and Lessler, M.A. (1976).
Effect of tobacco smoke on ciliated lung epithelium of Rana pipiens
Ohio J. Sci., 76, 27-31.

MARINO, M.R. and Aiello, E. (1982).
   Cinematographic analysis of beat dynamics of human
   respiratory cilia.
   Cell Motility, Suppl. 1, 35-39.

PAVIA, D., Thomson, M.L. and Pocock, S.J. (1971).
   Evidence for temporary slowing of mucociliary clearance in
   the lung caused by tobacco smoking.
   Nature (Lond), 231, 325-326.

SANDERSON, M.J. and Sleigh, M.A. (1981).
   Ciliary activity of cultured rabbit tracheal epithelium:
   beat pattern and metachrony.
   J. Cell Sci., 47, 331-347.

SLEIGH, M.A. (1977).
   The nature and action of respiratory tract cilia.
   In: Brain, J. D., Proctor, D.F. and Reid, L.M. (eds):
   "Respiratory Defense Mechanisms".  New York, Marcel Dekker,
   Pt. 1, pp. 247-288.

SLEIGH, M.A. (1981).
   Ciliary function in mucus transport.
   Chest, 80. Supply., 791s-5s.

SLEIGH, M.A. (1982).
   Movement and coordination of tracheal cilia and the relation
   of these to mucus transport.
   Cell Motility, Suppl. 1, 19-24.

SLEIGH, M.A. (1983).
   Ciliary function in transport of mucus.
   Eur. J. Respir. Dis., 64, (Suppl. 128), 287-292.

THOMSON, M.L. and Pavia, D. (1973).
   Long-term tobacco smoking and mucociliary clearance from the
   human lung in health and respiratory impairment.
   Arch. Environ. Health, 26, 86-89.

WALKER, T.R. and Kieffer, J.E. (1966).
   Ciliostatic components of the gas phase of cigarette smoke.
   Science, 153, 1248-1250.

WELLS, A.B. and Lamerton, L.F. (1975).
   Regenerative response of the rat tracheal epithelium after
   acute exposure to tobacco smoke: a quantitative study.
   J. Nat. Cancer Inst., 55, 887-891.

## DISCUSSION

LECTURER: Sleigh                                    CHAIRMAN: Bonsignore

DENISON:        The surface area of the walls of the bronchial
                tree diminishes from the periphery to the
                trachea by a factor of several thousand, and if
                mucus covered the whole surface it would either
                pile up in great quantities as it approached
                the mouth or it would increase its velocity
                greatly.  An alternative would be that mucus
                does not cover the whole surface, but only in
                patches, and if this were so large parts of the
                distal tree would be unprotected by mucus.
                Which of these two possibilities is true?

SLEIGH:         The first is true, but according to the
                circumstances, in time of infection the blanket
                may be more continuous, or during particulate
                stimulation.  The normal situation is that
                smaller airways contain only small flakes of
                mucus, perhaps connected together with strings
                of glycoprotein.

RICHARDSON:     Dust particles can indeed hit bronchial wall
                which is devoid of mucus, but they then
                stimulate mucus production and so help with
                clearance of the particles.

DENISON:        Does that mean the particles are buoyed up the
                sol phase?

RICHARDSON:     We do not know, but the particles might be
                lifted by the ciliary beat.

DENISON:        You spoke as if each cilium was aware of its
                neighbour hitting it and that each excited the
                next to beat which is why the beat was in
                phase.  What evidence is there of this
                awareness, or is an alternative explanation the
                mechanical load since if it tries to beat out
                of phase it must do more work which would then
                slow it until it came back into phase.

SLEIGH:         I think this is exactly the same phenomenon.
                Each cilium has a layer of fluid around it
                which interacts with its neighbours, then cilia
                do not have to touch.  Therefore a cilium does

know that something else is there provided that
it is moving.

DENISON:      If some distal part of the bronchial tree is
              stimulated to produce more mucus, how do the
              upstream elements of the mucus escalator know
              that they must do more work?  If it does not
              know do the cilia beat continuously, which
              seems rather wasteful?

SLEIGH:       The evidence is that cilia do beat
              continuously, but they are stimulated to
              greater mechanical activity by greater
              mechanical load.  The cilia can cope with a
              greater load by a simple mechanism - if a
              single cilium cannot move the load it is joined
              by others continuously because of the synchrony
              until sufficient propulsive force is generated,
              and even with a large depth of mucus there is
              sufficient energy.  With an increased
              mechanical load there is also an increased rate
              of activity, as shown experimentally in frogs.

FLETCHER:     From epidemiological studies there is
              overwhelming evidence that an increase in mucus
              secretion is associated with an increase in
              bronchial infection, and this answers the
              question posed in your talk.  However it seems
              to have no effect on the sort of lung damage
              which makes people breathless, measuring mucus
              hypersecretion by that which is expectorated.
              What we can discover nothing about is that
              mucus which lies in the periphery of the lung
              in the small airways which may play a part in
              infection at that site.

BAHKLE:       May I suggest in response to Denison's
              geometrical argument that if the mucus from
              peripheral airways were in full production it
              would choke up the trachea and therefore we
              cough.

HEATH:        The beautiful pictures of Clara cells which lie
              between the ciliated cells suggest that their
              functions might be inter-related.  Do you think
              there is such an association?  Secondly where
              in animal evolution does this close proximity
              of Clara cells and ciliated cells being, and is
              it confined to the mammalian bronchial tree.

SLEIGH:     I am rather ignorant of this, and I do not
            believe that the function of Clara cells has
            yet been established, but perhaps someone in
            the audience might help.  I do not know whether
            Clara cells exist outside mammals, they may be
            in birds but I do not know.

JEFFREY:    The situation in the diseased lung is very
            different to the normal.  In the normal the
            mucus producing apparatus is not found in
            airways smaller than 2 mm in diameter and
            distal to that site there is a large ciliary
            field but no mucous glands and no goblet cells.
            Thus the general question is what are the
            ciliary doing in the absence of mucus to remove
            particles.  We must be careful in analysing the
            experiments of Irivani and Melville, who did
            not use specific-pathogen-free rats and the
            effect might have been due either to infection
            or to tobacco smoke.  Other work suggests that
            the cilia are normal in animals exposed to
            tobacco smoke, a number of studies have shown a
            direct effect of infection on ciliary function.
            May I ask what would be the effect of ciliary
            shortening as the bronchi get smaller, and if a
            deepening peri-ciliary layer renders mucus
            removal difficult why is it that in Mytilus
            ciliary activity is normal whilst the depth of
            the peri-ciliary layer in infinite — sea water.

SLEIGH:     If a cilium becomes too short it produces two
            problems.  One is the difference between the
            heights of the effective stroke and the
            recovery stroke, and if there is not a distinct
            difference there is no propulsion.  The other
            problem is the minimum radius of curvative to
            which a cilium can be bent by its internal
            mechanism and this limits the minimum height to
            about four micrometres.  In Mytilus the mucus
            travels at the level of the ciliary tips and it
            is transported in contact with the cilia even
            if it is upside down.

STANESCU:   There is some evidence that beta-
            sympathomimetic agents can impair mucus
            clearance, and you produced evidence of an
            acute effect of cigarette smoke.  What is the
            effect of chronic smoking?

SLEIGH:            Some people who smoke for many years may still
                   have normal ciliary function, so that there is
                   probably recovery after each exposure to a
                   normal rate of activity.

RICHARDSON:        Barnes has recently reported the presence of
                   adreno-receptors in the mammaliam respiratory
                   tracts.  The cells could not be identified but
                   do you think that these might regulate ciliary
                   activity?

SLEIGH:            They might, but the evidence for regulation by
                   mechanical means is direct, and can take place
                   in an isolated clump of cells.

RICHARDSON:        Adrenergic activity would increase the
                   concentration of ATP within the cells, do you
                   think this could accelerate the rate or the
                   force of ciliary action?

SLEIGH:            Both may occur but a more important effect
                   might be ionic since ATP does not normally
                   limit the rate of ciliary beat, this is
                   normally an ionic phenomenon associated with
                   enzyme breakdown of ATP.

CUMMING:           David Denison referred to the paradox that a
                   thousand-fold increase in lateral wall surface
                   does not produce clumping in the large airways
                   and he made two suggestions.  A third
                   possibility might be that the thickness of the
                   mucus layer increases - is there any evidence
                   that this varies in bronchi of different sizes?

SLEIGH:            I know of no consistent evidence.

CUMMING:           A fourth possibility is that whilst the mucus
                   velocity is constant through the tree, there is
                   fluid resorption as the mucus travels up the
                   tree so that it becomes more concentrated as it
                   approaches the trachea.  Is there any evidence
                   that the composition of mucus varies through
                   the bronchial tree?

SLEIGH:            I am not aware of any.

SADOUIL:           Some authors describe small claws on the tip of
                   the cilia, do you think these play a role in
                   the adherence and transportation of mucus?

SLEIGH:          I beleive that these claws permit a better
                 engagement in the fibrous matrix of the mucus,
                 since the claws may be retracted very easily
                 from the mucus at the end of the stroke and
                 enhances the adherence and transportation.

BONSIGNORE:      I will close the discussion at that point.

# THE GAS PHASE OF TOBACCO SMOKE AND THE

# DEVELOPMENT OF LUNG DISEASE

Tim Higenbottam and Colin Borland

Department of Respiratory Physiology
Addenbrookes Hospital, Cambridge

## Introduction

In this paper we will present a number of observations which suggest that the gas phase of tobacco smoke contributes to the development of lung disease. We acknowledge with great thanks the contributions made by our collaborators to the studies to be described. They include Professor Geoffrey Rose of the London School of Hygiene and Tropical Medicine, Professor Tim Clark of Guys Hospital and Professor Brian Thrush at the Department of Chemistry, Cambridge University. Also we thank our long suffering Research colleague, Andrew Chamberlain, from the Addenbrookes Respiratory Physiology Laboratory.

Our thesis is that the gas phase of tobacco smoke can itself cause lung damage. This lung damage can be divided into 2 parts, an acute stage which rapidly reverses when the smoker abstains from cigarettes and a more chronic stage which is probably not reversible and leads to a persistent functional impairment of ventilation. The acute stage, probably located in the peripheral parts of the lung takes the form of an alveolitis and bronchiolitis. Complex chemical reactions continue in the tobacco smoke particularly in the gas phase and these reactions are capable of generating highly reactive chemical species - the free radicals.

## 'Tar' or Particulate Matter

Tobacco smoke is arbitarily divided into 2 parts. The 'tar' or particulate matter retained when smoke is passed

through a Cambridge filter, which extracts particles greater
than 0.1 micron in size, and the gas phase which is not so
retained (1).  This arbitary division has limitations, as
certain elements, for example nicotine whilst mainly in the
'tar' is also found in the vapour phase.  Furthermore water
vapour associated with particulate matter is also retained by
the filter making it possible for water soluble gases to be
retained.

Before describing the effects of gas  phase, it is
important to keep in mind that 'tar' does constitute a hazard
to the health of the smoker.  This is now well established from
the work of Hammond (2), the overall mortality ratios for
cigarette smokers, that is deaths from all causes, increased
for both men and women with the tar/nicotine yields of the
cigaretts they usually smoke.

## Lung Function And Tar Yields Of Cigarette

We investigated the influence of cigarette tar yield on
lung function and respiratory symptoms by re-examining the data
from the Whitehall survey (3).  This included over 18,000 male
civil servants ranging from 40 to 64 years (Table 1).  Each
subject completed a questionnaire on respiratory symptoms and
performed spirometry as well as recording past and present
smoking habit.

Of respiratory symptoms only phlegm appeared related to
cigarette smoking, increasing with daily consumption (Figure
1).  Increased prevalence of phlegm in cigarette smokers was
similar in the 40-49 and 60-64 years old.  This taken with the
fact that prevalence of phlegm was comparable in ex-smokers and
non-smokers, further supports Professor Fletchers' observation
that the sputum production in smokers reflects current
cigarette smoking and diminishes on smoke cessation (4).
Phlegm production also increased with tar yield of cigarette
(Figure 2).  The effect of tar yield was less marked in smokers
using more than 20 cigarettes per day.  It is important to note
that the lowest tar groupings in this study was 18.23 mgm as a
result of the marked reduction in tar yields over the last 10
years most cigarettes sold today being below this level.

Forced expired volume in one second (FEV1) in smokers
was lower than non-smokers (Figure 3).  The reduction in FEV1
increased with daily cigarette consumption, such that smokers
of more than 20 cigarettes per day have 400 mls less FEV1 than
non-smokers.  By contrast with the effect of tar yield on
phlegm production, it has only a limited effect on FEV1 (Figure
4).

**TABLE 1**

The Whitehall study

Profile of age and smoking habits

| Age | Non-smokers | Ex-smokers | Pipe and cigar smokers | Cigarette smokers | Total |
|---|---|---|---|---|---|
| 40-49 | 1,845 | 2,658 | 258 | 2,962 | 7,723 |
| 50-59 | 1,318 | 3,129 | 297 | 3,532 | 8,276 |
| 60-64 | 289 | 906 | 83 | 1,106 | 2,384 |
| Total | 3,452 | 6,693 | 638 | 7,600 | 18,383 |

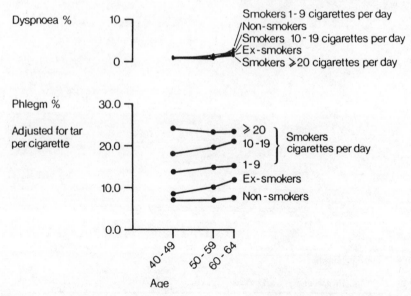

**FIGURE 1:** The prevalence of the symptoms of phlegm and dyspnoea according to age and daily cigarette consumption. Prevalence of phlegm is adjusted for tar yield.

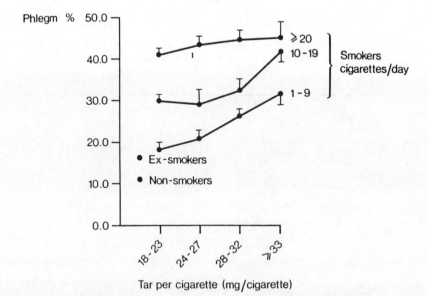

**FIGURE 2:** Prevalence of the symptom of phlegm, mean and standard errors, values standardized for age.

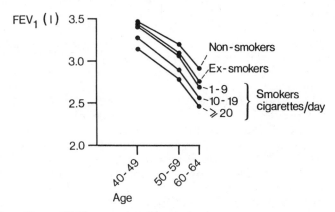

**FIGURE 3:**   Mean FEV1 standardized for employment grade, tar yield
and height, standard errors range from 15–45 nls.

FEV$_1$ for tar yield of cigarettes smoked

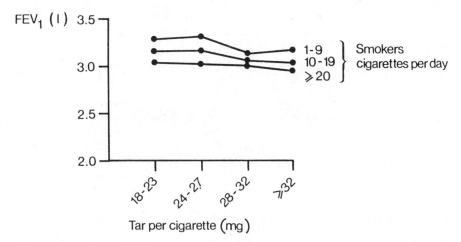

**FIGURE 4:**   Mean FEV1 standardized for age, employment grade and
height, standard errors range from 12–60 nls.

Whilst tar yield has a major effect on phlegm production its influence on airflow obstruction is less, raising the possibility that the gas phase may also responsible for obstruction.

A limitation of the Whitehall study was that lung function was recorded once only. Ideally the influence of tar yield on lung function of cigarette smokers would be better assessed prospectively. Recently Sparrow and colleagues reported the results of a 5 year prospective study of the effect of cigarette tar yield on the rate of decline of FEV1 (5). They showed that the rate of decline of FEV1 was just as great in smokers of low tar cigarettes as in those with higher tar cigarettes. This would further support the view that tar yield is a poor predictor of disturbance of lung function in smokers and that the gas phase may make a significant contribution towards its development.

## The Persistence of Abnormalities of Lung Function in Ex-Smokers

A method to discover which constituent or constituents of tobacco smoke are most important as a cause of lung dysfunction would be to alter the composition of the smoke yielded by cigarettes and then attempt to detect any change in function. In most smokers (Figure 3) the reduction in function with most recognised tests is small. Furthermore even after total abstinence from tobacco smoking FEV1 for example does not quite return to non-smokers values. This is shown in Table 2 where ex-smokers of the Whitehall survey are classified according to calculated past cigarette consumption and duration of abstinence from smoking. Some improvement in FEV1 up to 6 years of abstinence can be detected but then no further differences are apparent, the FEV1 never fully returning to non-smoking levels (Figure 3).

This suggests that there are two phases of smoke induced lung dysfunction, a reversible phase and a non-reversible phase. How these two phases are related remains unknown. The measurement of FEV1 or similar spirometric measurements appear unsuitable as a means of assessing either the acute reversible effects or the persistent effects of smoke as the magnitude of the change is small.

## Lung Epithelial Permeability and Disturbance of Lung Function

An ideal measurement of smoke induced change in lung function would be one which reverses shortly after abstaining from cigarettes and where the magnitude of the difference

**TABLE 2**

Lung function in relation to total previous cigarette consumption

| Total cigarette consumption | Ex - smokers | | | Current smokers |
|---|---|---|---|---|
| | Number of years of abstinence | | | |
| | ⩾ 13 n = 3113 | 7 - 12 n = 1440 | ⩽ 6 n = 1984 | n = 7564 |
| ⩽ 90 | 3. 29 | 3. 30 | 3. 26 | 3. 22 |
| > 90 ⩽ 180 | 3. 21 | 3. 25 | 3. 22 | 3. 06 |
| > 180 | 3. 08 | 3. 10 | 3. 18 | 2. 87 |

between smokers and non-smokers is large. We were therefore particularly excited by the reported observations of Gareth Jones and colleagues (6) that the permeability of the alveolar-capillary barrier is increased in smokers. Permeability was measured by inhaling an aerosol of 99m Technetium chelated to diethylene triamine penta acetate (Tc 99m DTPA) in saline (492 Daltons). Clearance of the isotope from the alveolus is determined by the permeability of the alveolar-capillary membrane and can be followed using a scintillation counter over the right upper chest. An index of permeability, the time to clear half the isotope was calculated, DTPA T1/2. The DTPA T1/2 appears related to current cigarette smoke exposure as measured by carboxyhaemoglobin level (7) and improves (lengthens) within 3 days of stopping smoking.

We have been studying some 100 smokers whilst they gave up their smoking habit. To satisfy their craving for nicotine each had been given 3 months supply of nicotine containing chewing gum. In one third we also measured DTPA T1/2 at intervals during abstinence. Twenty individuals have been followed up for 6 months.

Their ages ranged from 22 to 35 years and all had normal lung function as measured by FEV1 (mean value 106% of predicted). Each was studied initally then at 1 week, 4 weeks, 12 weeks and 24 weeks. Using mixed expired carbon monoxide levels (8) as a guide to truthfulness 10 subjects completely abstained whilst 10 reverted back to their smoking habit (Figure 5). Amongst abstainers the DTPA T1/2 returned towards normal after a week, followed by a further rise at 1 month, whilst no significant change occurred amongst non-abstainers (Figure 6).

On each occassion DTPA T1/2 was measured, we also measured FEV1, vital capacity, total lung capacity, maximal expiratory flow at 50% of VC, specific airway conductance and closing volume. A single breath nitrogen wash out (9) was used to measure closing volume, and to obtain a measure of inhomogeniety of ventilation we also recorded the slope of phase III, % $N_2$/L. Sonia Buist and colleagues have reported that this is a sensitive measure of smoking induced lung dysfunction (10). In abstainers or non-abstainers the only measurement to change was slope of phase III, there was no alteration in any of the other measures of lung function. Interestingly the slope of phase III fell within a week of abstaining (Figure 7) continuing to improve over the 6 months in abstainers, measuring the change observed in DTPTA T1/2. No such change occurred in non-abstainers.

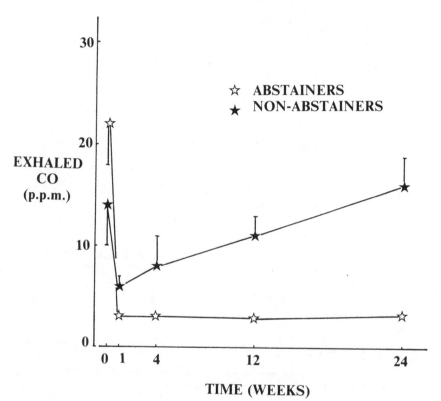

FIGURE 5: Mixed expired carbon monoxide levels of 10 abstainers and 10 smokers who relapsed and began to smoke again. Mean values and standard errors are shown.

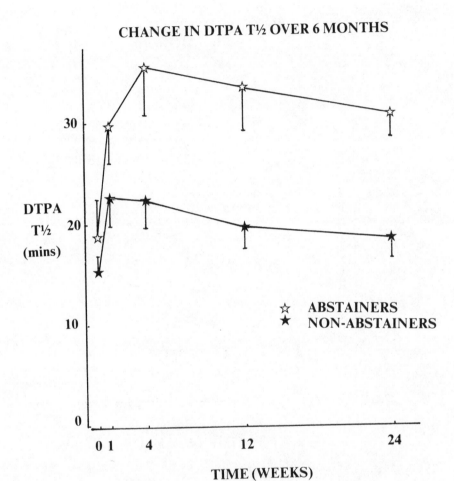

FIGURE 6: The mean values and standard errors for DTPA T1/2 in 10 abstainers and 10 failed abstainers.

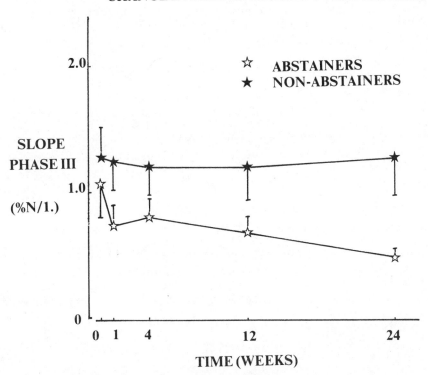

FIGURE 7: The mean values and standard errors for the slope of phase III of a single breath nitrogen washout for 10 abstainers and 10 non-abstainers.

We suggest that the reduction of DTPT1/2 and the increased slope of phase III seen in cigarette smokers reflects the macrophage alveolitis (11) and inflammatory bronchiolitis seen in all smokers (12). Tobacco smoke not only alters the integrity of the alveolar capillary barrier but also the elastic properties of the alveoli and bronchioli. Both changes may be reversible when the smoker quits the habit.

**The Gas Phase and the Acute Smoke Induced Lung Dysfunction**

The prompt reversal of the increased lung epithelial permeability when smokers quit their habit offers a means of studying how alteration to tobacco smoke will effect lung function.

We measured lung epithelial permeability in 5 smokers, again with normal lung function. Then for one week each smoked cigarettes through a cigarette holder containing a Cambridge filter to remove 'tar'. The DTPA T1/2 was again measured, after which they then abstained from cigarettes for a further week and permeability measured. Each received chewing gum containing nicotine to reduce 'nicotine craving' and mixed expired carbon monoxide levels were recorded at each laboratory visit.

Figure 8 shows that after a week of smoking only gas phase there was no increase in DTPA T1/2 whilst on abstinence it increased. This suggests that gas phase alone can cause the acute peripheral lung damage. Whilst smoking cigarettes through Cambridge filters cigarette consumption fell, as shown by the fall in exhaled CO, which might suggest that the gas phase is more toxic to the lung than whole smoke. Indeed Kilburn has noted a similar enhanced cytotoxicity of gas phase to hamsters lungs (13).

**Gas Phase Constituents**

A number of ciliotoxic and carcinogenic chemical species travel in the gas phase of tobacco smoke (Table 3). We will discuss only 2 of these, carbon monoxide, quantitatively the most important and nitrogen oxides which have been so extensively linked with the development of chronic lung disease.

To study the effects of carbon monoxide on lung function we returned to the Whitehall survey. As a result of a study of old cigarettes by Dr. Wald and colleagues (13) we were able to obtain the carbon monoxide yields of two thirds (4,278) of the original smokers of the Whitehall survey. We were able

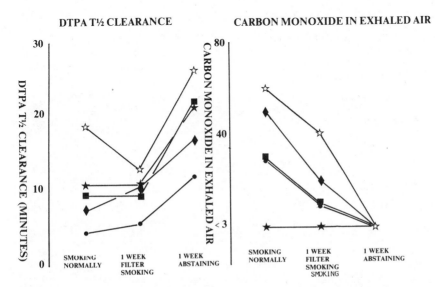

**FIGURE 8:** The individual values for DTPA T1/2 and exhaled CO level in 5 subjects (each with a separate symbol) initially then after smoking through a Cambridge filter and after abstinence.

**TABLE 3**

MAJOR TOXIC AGENTS IN THE GAS PHASE OF
CIGARETTE SMOKE (AMERICAN BRANDS)

| | |
|---|---|
| Carbon monoxide | 17 mg |
| Acetaldehyde | 800 ugm |
| Nitrogen oxides | 350 ugm |
| Hydrogen cyanide | 110 ugm |
| Acrolein | 70 ugm |
| Ammonia | 60 ugm |
| Formaldehyde | 30 ugm |
| Pyridine | 10 ugm |
| Acrylonitide | 10 ugm |
| 2-Nitropropane | 0.92 ugm |
| Urethine | 30 ngm |
| Hydrozine | 32 ngm |
| Vinyl chloride | 12 ngm |
| Dimethylnitrosomine | 13 ngm |
| Nitrosopyrolidine | 11 ngm |

to compare the FEV1 levels of smokers using cigarettes of differing carbon monoxide yields (Table 4). As CO yield rose so also did FEV1 and the effect was significant as a p value <0.05. In other words the higher CO yield cigarette smokers had better lung function so reducing the probability that carbon monoxide contributes to the pulmonary dysfunction of smokers.

Oxides of nitrogen in tobacco smoke have been implicated in the development of lung disease. Nitrogen dioxide both experimentally (14) and in man (15) causes a toxic alveolitis. However the principal oxide of nitrogen in fresh tobacco smoke is nitric oxide (16), which appears not to cause lung damage. Therefore for oxides of nitrogen in tobacco smoke to cause lung damage it is necessary for the nitric oxide to be converted to nitrogen dioxide.

The rate of oxidation of nitric oxide depends upon its concentration (Table 5). Each puff of tobacco smoke may contain up to 1000ppm. This however is diluted with room air (17) giving a concentration between 20-100 ppm which would be expected to react very slowly with oxygen (Table 5).

The question remains as to the fate of inhaled nitric oxide and to answer this we have experimentally exposed humans to the low concentrations of nitric oxide expected to be found in the lungs when tobacco smoke is inhaled.

In the first study 3 normal subjects inhaled a series of volumes of a gas mixture of 20 ppm NO and 11% Helium in air, ranging from 100 to 2000 mls. Each volume was inhaled from FRC, the breath was held for 10 seconds and then volume exhaled. The exhaled gas was analysed for He and NO using respectively a Katharometer and chemiluminescent analyser. The retention of nitric oxide was calculated by the following:-

$$\frac{NO\ exhaled}{NO\ inhaled} \quad x \quad \frac{He\ inhaled}{He\ exhaled} \quad x \quad 100\%$$

There was no significant retention of NO until the inhaled volume exceeded 500 mls (Figure 9), only when the gas mixture was ´exposed´ to alveolar capillary membrane. We concluded that within a 10 second period NO is not oxidized to $NO_2$ even when in contact with epithelium of the conducting airways. Furthermore uptake of NO appears confined to the alveolar epithelium not airway epithelium.

Nitric oxide combines with both reduced (18) and oxyhaemoglobin (19), exhibiting an affinity for haemoglobin in

## TABLE 4

Mean Values of $FEV_1$ in Relation to CO yields

(adjusted for (a) age only (b) age, number of cigarettes, employment grade and tar)

Mean $FEV_1$ (1) (standard error)

| CO yield | < 18 mg | 18 to 20 mgm | > 20 mgm |
|---|---|---|---|
| a) | 2.96 (0.015) | 3.04 (0.012) | 310 (0.024) |
| b) | 2.97 (0.015) | 3.01 (0.011) | 3.04 (0.023) |

## TABLE 5

$$2NO + O_2 \longrightarrow 2NO_2$$

$$\text{RATE} = K\,(NO)^2\,O_2$$

$$K = 1.52 \times 10^4.\ L.^2\ \text{MOLE}^{-2}\ \text{SEC}^{-1}\ \text{AT}\ 300^\circ K$$

| NO COM (PPM) | TIME OF PRODUCTION OF (5 PPM $NO_2$) | $1/2$ TIME | |
|---|---|---|---|
| 1000 | 1 SEC | 100 SEC | |
| 500 | 4 SEC | 3 MINS | PUFF |
| 100 | $1\tfrac{1}{2}$ MINS | 16 MINS | CONCENTRATION |
| 20 | 41 MINS | 1.4 HOURS | LUNG |
| 1 | – | 27 HOURS | CONCENTRATION |

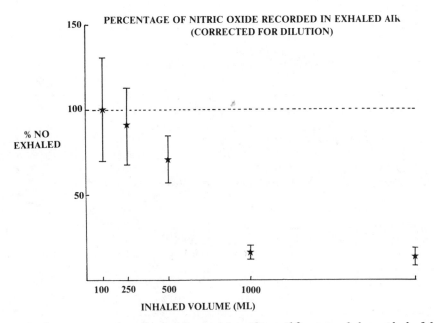

**PERCENTAGE OF NITRIC OXIDE RECORDED IN EXHALED AIR (CORRECTED FOR DILUTION)**

% NO EXHALED

INHALED VOLUME (ML)

**FIGURE 9:** Recovery of nitric oxide after 10 second breath hold of different values, mean values and standard deviations are shown.

excess of that seen for carbon monoxide. We therefore decided to study whether the factors which influence transfer of carbon monoxide also affect transfer of NO. To achieve this we modified the standard method (20) for measuring gas transfer for carbon monoxide ($TL_{CO}$) to simultaneously measure gas transfer for nitric oxide ($TL_{NO}$).

Fifteen subjects inhaled a gas mixture (11% He, 20 ppm NO and 0.2% CO) from residual volume to total lung capacity. Breath was held for 7.5 seconds then after discarding 500 mls of the expirate the gas was sampled for CO, NO and Helium. The standard equations were used to calculate:-

$$TL_{CO} = \frac{23.33}{BHT} \times Alv\ Vol \times Log_e \frac{(CO\ in \times HE\ ex)}{(CO\ ex \quad He\ in}$$

| | | |
|---|---|---|
| BHT | = | Breath hold time |
| Alv Vol | = | Alveolar volume |
| CO | = | Inspired CO |
| CO ex | = | Expired CO |
| He ex | = | Expired He |
| He in | = | Inspired He |

To calculate $TL_{NO}$ the inspired and expired concentrations of NO were substituted.

There was a high degree of correspondence between $TL_{NO}$ and $TL_{CO}$ (r = 0.69) (Figure 10). The average $TL_{NO}$ was approximately 4 times the average value of $TL_{CO}$ perhaps reflecting the greater affinity of NO for haemoglobin.

Next we studied whether exercise similarly perturbs $TL_{CO}$ and $TL_{NO}$. Five subjects were studied at rest and immediately after 3 levels of exercise on a treadmill giving pulse rates of 100, 130 and 160 per minute. The same simultaneous measurement of $TL_{NO}$ and $TL_{CO}$ was made. As the pulse rate increased (Figure 11) both $TL_{NO}$ and $TL_{CO}$ increased, plateauing out at high rates of work. These observations suggested to us that both NO and CO are taken up in the alveolar capillaries by combining with haemoglobin.

Sceptics could argue that NO is simply dissolved in

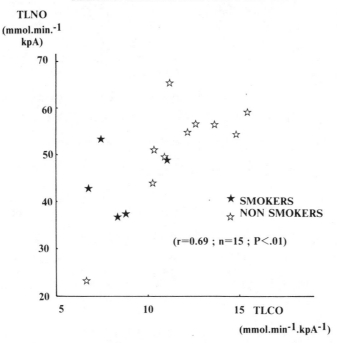

**FIGURE 10:** The correspondence between $TLN_O$ and $TLC_O$, mean values of 3 replicates for each of 15 subjects.

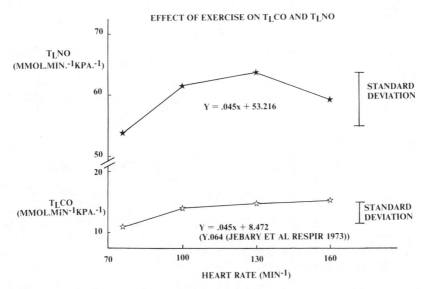

**FIGURE 11:** The effect of exercise on $TLN_O$ and $TLC_O$ with the overall variance shown.

plasma. For such a gas the transfer would depend upon the pulmonary blood flow (cardiac output) and its solubility eg:-

$$TL_{NO} = B \times Q$$

$$Q = \text{Cardiac output}$$

$$B = \text{Solubility (capacitance coefficient at } 37^{\circ}C \text{ for water} = (0.016)$$

$$TL_{NO} = 48.5 \text{ (average result for our 15 subjects)}$$

This would require a cardiac output:-

$$Q = \frac{48.5}{0.016} = 3031 \text{ litres/min, clearly impossible even when exercising.}$$

Whether NO reacts with reduced or oxyhaemoglobin remains unclear. Studies with radiolabelled $N_{15}O$ suggest that the reaction is with oxyhaemoglobin to form methaemoglobin and nitrate (21). The $N_{15}$ was found in red cells and urine as nitrate.

Whichever reaction with haemoglobin is most important NO shows in vitro a greater affinity than CO has from haemoglobin. This raises the interesting possibility that measurement of $TL_{NO}$ may provide a closer measurement of alveolar capillary membrane diffusion than $TL_{CO}$.

$$\frac{1}{TL} = \frac{1}{DM} + \frac{1}{BQ + OVC}$$

$$DM = \text{Diffusing capacity of the alveolar capillary membrane}$$

$$B = \text{Solubility}$$

$$Q = \text{Cardiac output}$$

$$O = \text{Rate of combination of NO with haemoglobin}$$

$$V = \text{Capillary blood volume}$$

In case of NO O is increased so the second part of the equation lessens in size, and TL approaches the values of DM. It is therefore possible that the $TL_{NO}$ is a better guide to diffusing capacity of the alveolar capillary membrane than $TL_{CO}$.

## Reactions of Nitric Oxide in Tobacco Smoke

It could be argued that all we have studied with respect to nitric oxide is its behaviour in air and not in tobacco smoke. To show that the rates of reaction of NO are similar in smoke we measured the rate of loss of nitric oxide from tobacco smoke as it ages in a glass syringe.

We used a standard single port smoking machine to provide a bell shaped puff of 35 mls every minute. The 3rd puff was collected in a glass syringe. The NO concentration was then measured after different intervals of time, two duplicates were performed at each time. In one study whole smoke was analysed and in the second, the smoke was first passed through a Cambridge filter to remove 'tar'.

Figure 12 shows that NO disappears more slowly from whole smoke than from the gas phase. In whole smoke the rate of oxidation is similar to that observed for NO in air. However in the gas phase NO is more rapidly oxidized, and this continues for up to 10 minutes. We suggest that more reactive chemical species (for example oxygen based free radicals), are contributing to this reaction. Such free radicals have been described by Pryor and colleagues in the gas phase but only last for seconds (22) not for 10 minutes. We therefore suggest that a chain reaction is occurring which not only oxidizes NO but regenerates the oxygen free radicals:-

$$RO_2. \; + \; NO \; = \; RO. \; + \; NO_2$$

$$RO. \; + \; RH \; = \; ROH \; + \; R.$$

$$R. \; + \; O_2 \; = \; RO_2$$

Where $R = CH_3$, $C_2H_5$, $C_3H_7$, $C_4H_9$ etc.

Such a chain reaction could explain the reactions of the oxidation of NO observed in the gas phase. It raisesthe interesting idea that the presence of tar, in whole smoke, quenches the free radicals and may help to explain Killburns observation that the gas phase of tobacco smoke is more cytotoxic than whole smoke (13).

We cannot from this experiment predict accurately what happens in the lungs, the glass syringe shares few similarities with bronchial epithelial. However the fact that complex chemical reactions continue in tobacco smoke after it has been generated and that these are modified by the presence of tar

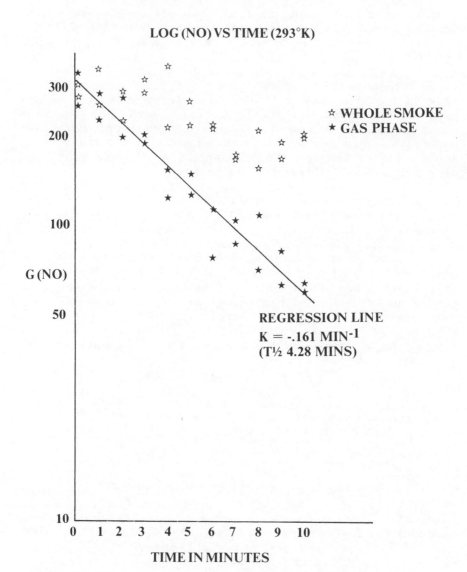

**FIGURE   12:**   The amount of NO remaining in the glass syringe after various periods of aging (expressed as the logarithm) whole smoke and gas phase are shown.

illustrates further the complexities of tobacco smoke toxicology. The determination to whether NO undergoes conversion to $NO_2$ in inhaled tobacco smoke must await further experimentation, although we believe that it is unlikely in the presence of tar.

## Conclusion

The gas phase of tobacco smoke may damage the lung, and this damage takes the form of an acute reversible change in the structural integrity of the alveolar capillary membrane associated with a disturbance of alveolar ventilation, perhaps the functional manifestations of macrophage alveolitis and bronchiolitis as seen in smokers. All smokers in varying degrees also aquire permanent loss of their lung function, presumably the result of irreversible structural changes of the alveolar interstitium and bronchioli. Both forms of damage appear related to cigarette consumption.

The gas phase of tobacco smoke alone is able to cause the acute changes in alveoli.

Of the gas phase constituents we do not believe that carbon monoxide or nitric oxide is important in the lung damage seen in cigarette smokers.

Finally it appears that free radicals can be self generating in tobacco smoke in the absence of tar and these reactive chemical species may have greater importance in lung damage than other more familiar chemical species.

## REFERENCES

1. ROTHWELL, K., Grant, G.A.
   Standard methods for analysis of tobacco smoke.
   Research. Paper II. Second Edition (London).

2. HAMMOND, E.C., Garfinkel, L., Seidman, H., Low, E.A.
   (1976).
   'Tar' and nicotine content of cigarette smoke in
   relation to death rates.
   Environmental Research 12: 263-274.

3. REID, D.D., Brett, G.Z., Hamilton, P.J.S., Jarrett, R.J.,
   Keen, H., Rose, G. (1974).
   Cardio-respiratory disease and diabetes among middle
   aged, male, civil servants. Lancet i: 469-473.

4.   FLETCHER, C., Peto, R., Tinker, C., Speiger, F.E. (1976).
     The natural history of chronic bronchitis and
     emphysema.
     Oxford University Press, Oxford.

5.   SPARROW, D., Stefos, T., Bosse, R., Weiss, S.T. (1983).
     The relationship of tar content to decline in pulmonary
     function in cigarette smokers.
     Am. Rev. Resp. Dis. 127: 56-58.

6.   JONES, J.G., Minty, B.D., Royston, J.P. (1982).
     The physiology of leaky lungs.
     Br. J. Anaesth. 52: 705-721.

7.   JONES, J.G., Minty, B.D., Royston, D., Royston, J.P.
     (1983).
     Carboxyhaemoglobin and pulmonary epithelial
     permeability in man.
     Thorax 38: 129-133.

8.   JAVIS, J.M., Russell, M.A.H., Salongee, Y. (1980).
     Expired air carbon monoxide level; a simple breath test
     of tobacco smoke intake.
     Br. Med. J. ii: 484-485.

9.   TRAVIS, D.M., Green, M., Don, H. (1973).
     Simultaneous comparison of helium and nitrogen
     expiratory 'closing volumes'.
     J. Appl. Physiol. 34: 304-308.

10.  BUIST, A.S., Sexton, G.J., Nagg, J.M., Ross, B.B. (1976).
     The effect of smoking cessation and modification on
     lung function.
     Am. Rev. Resp. Dis.  114: 115-123.

11.  CRYSTAL, R. (1982).
     Alveolitus: the key to the interstitial lung disorders.
     Thorax, 37: 1-10.

12.  NIEWOEHNER, D.E., Klaneuman, J., Rice, D.B. (1974).
     Pathological changes in the peripheral airways of young
     cigarette smokers.
     New Eng. J. Med. 165: 755-758.

13.  KILBURN, K.H., McKenzie, W. (1975).
     Leucocyte recruitment to airways by cigarette smoke and
     particulate phase in contrast to cytotoxicity of
     vapour.
     Science 189: 634-637.

14. HAYDON, G.B., Davidson, J.T., Lillington, G.A.,
        Wasserman, T. (1967).
        Nitrogen oxide induced emphysema in rabbits.
        Am. Rev. Resp. Dis. 95: 797-805.

15. SCOTT, E.G., Hunt, W.B. Jr. (1973).
        Silo fillers disease.
        Chest, 63: 701-706.

16. STEDMAN, R.L. (1968).
        The chemical composition of tobacco smoke.
        Chemical Reviews, 68: 153-207.

17. RAWBONE, R.G., Murphy, K., Tate, M.E., Kone, S.J.
        The analysis of smoking parameters, inhalation and
        absorption of tobacco smoke in studies of human smoking
        behaviour.
        In: Smoking Behaviour, Thornton, R.E. (Ed), Churchill
        Livingstone, p 171.

18. GIBSON, Q.H. (1959).
        The kinetics of reactions between haemoglobin and
        gases.
        Prog. Biophys. Chem. 9: 1-54.

19. DOYLE, M.P., Hoekstra, J.W. (1981).
        Oxidation of nitrogen oxides by bound dioxygen in
        haemoproteins.
        J. Inorg. Biochem. 14: 351-358.

20. COTES, J.E. (1963).
        Effect of variability in gas analysis on the
        reproducibility of pulmonary diffusing capacity by the
        single breath method.
        Thorax, 18: 151-154.

21. YOSHIDA, K., Kasama, K., Kitabatake, M., Kuda, M.,
        Imai, M. (1980).
        Metabolic fate of nitric oxide.
        Int. Arch. Occup. Environ. Health, 46: 71-77.

22. PRYOR, W.A., Prier, D.G., Church, D.F. (1983).
        Electron spin resonance study of mainstream and
        stidestream cigarette smoke: nature of the free
        radicals in gas phase smoke and in cigarette tar.
        Environ. Health Persp. 47: 345-355.

DISCUSSION

LECTURER: Higenbottam                    CHAIRMAN: Bonsignore

CUMMING:    You are presenting two hypotheses – one is that nitric oxide leaves the lungs due to its rapid combination with haemoglobin and this would be acceptable were it not for the fact that the rate of transfer is three or four times greater than for carbon monoxide, whereas the affinity is much greater for carbon monoxide so my first point is that the figures seem to be incompatible. As to the fascinating experiment in the syringe where the NO disappears as the result of a chain reaction, this is more attractive and renders the first hypothesis unnecessary. If the NO disappeared into the blood, the amount of $NO_2$ would be identical, whereas in the syringe the quantity of $NO_2$ should increase. My question is therefore have you analysed the $NO_2$ in the expirate and is it less, and have you analysed the $NO_2$ in the syringe and is it more?

HIGENBOTTAM:    The NO combines more rapidly with haemoglobin than does CO, the problem is that the experiments were done under very non-physiological condition. Historically the difference in transfer rates of CO and NO has never been appreciated, perhaps because of the difficulties in measuring NO. In terms of the syringe experiments, we have planned further work to clarify the point you raised. I suspect that NO is used because it happens to be in the syringe, whilst it may not be present in the human lung inhaling smoke.

GUYATT:    If the nitric oxide is taken up rapidly by the lung, perhaps there will not be time for the first order reaction to take place.

HIGENBOTTAM:    I am not able to answer the question as to what takes place in the lung, only in the syringe.

LEE:    You quoted results from the Whitehall study relating F.E.V. to number of cigarettes smoked and tar level, I was surprised to hear that you thought there was a stronger relationship between number of cigarettes than with tar level. My analysis showed the same improvement in FEV if the tar level was halved or the number of cigarettes smoked was halved.

HIGENBOTTAM: Multiple regression analysis does yield the relationship shown.

LEE: Where were the hand-rolled smokers grouped?

HIGENBOTTAM: In the first group, and this is an important observation since 10% of the population smoke hand-rolled.

DENISON: What is the nature of link between haemoglobin and nitric oxide? Is it the same kind of loose link as between oxygen and carbon dioxide?

HIGENBOTTAM: The reverse reaction of nitrosyl-haemoglobin has never been demonstrated experimentally so the bond is certainly very strong, but it remains a possiblity that it could enter the cell and form nitrogen dioxide. Once it enters the red cell any damage it causes is probably outside the lungs in the systemic circulation.

FLETCHER: How long had the smokers been smoking before they stopped? Related to the improvement in nitrogen slope.

HIGENBOTTAM: There was significant improvement at one week.

FLETCHER: How many years had they been smoking, or how old were they?

HIGENBOTTAM: Their ages were between 22 and 35.

FLETCHER: They were relatively young smokers who might not yet have lung damage.

HIGENBOTTAM: The damage to the lungs appears to occur in two phases - an acute one which can be detected with the B.T.P.A. half life and the nitrogen slope. Its relationship to the permanent damage is not clearly defined but may be by the route of white cells and proteases as suggested by Dr. Spadofora.

SPADOFORA: Regarding the susceptibility of the different

lung cells to oxygen. Since endothelial cells are more susceptible to injury by oxidants it has been suggested that they might be deficient in superoxide dysmutase activity. Secondly some interesting data from Canada indicates that surfactant can act as an anti-oxidant and this might also explain why endothelial cells are more susceptible than type 2 pneumocytes.

HIGENBOTTAM:      The role of surfactant in this regard is interesting, and its principal role in man has been suggested as detoxifying inhaled materials, and there is probably a quantitative link between the quantity of surfactant in the lungs and the quantity of inhaled smoke.

CUMMING:          Two questions regarding nitrosyl-haemoglobin. Do we know anything about the amount of this substance in the circulation of the smoker, and do you think this is a candidate for the damage sustained in the systemic circulation?

HIGENBOTTAM:      I have not looked for nitrosyl-haemoglobin, the estimation of which calls for electron-spin resonance techniques. One study has measured this in relation to cigarette smoking and there are many times less than there is carbon monoxide, but whether it causes damage is unknown.

BONSIGNORE:       We must stop now, and I thank the speakers and the discussants.

# CHANGES IN LUNG FUNCTION AFTER SMOKING CESSATION INTER-
# RELATIONSHIP BETWEEN SMOKING, LUNG FUNCTION AND BODYWEIGHT

Dan C. Stanescu

Cardiopulmonary Laboratory
Cliniques Universitaires St. Luc
1200 Brussels, Belgium

## A.  CHANGES IN LUNG FUNCTION AFTER SMOKING CESSATION

Smoking, and especially cigarette smoking, appears as the main culprit in the genesis of chronic obstructive lung disease (COLD).  Indeed, epidemiological data have shown that COLD, defined by a decrease in forced expiratory volume in 1 second/vital capacity ratio ($FEV_{1.0}/VC$) does not occur in non-smokers (1) and that not a single one of the lifelong non-smokers, in a large industrial population, was certified as dying of COLD (2).

In the absence of efficient therapy, smoking cessation appears to be the main element which may improve prognosis in this disease.  The influence of smoking upon lung function has been investigated in either cross-over and prospective studies in smokers, non-smokers and former smokers or in follow-up studies in subjects who gave-up smoking.  Selection of subjects was variable, being either members of staff, or friends and relatives of the investigator chosen for convenience, people from smoking cessation clinics or hospitalized patients, usually with well preserved lung function and little apparent disability.

The first table presents several lung function tests in a sample of 54 lifelong non-smokers (free of respiratory symptons), 51 ex-smokers and 105 current smokers, blue collar workers, 45 to 55 years old which we have studied in a steel plant (3).  One may note that almost all lung function values in ex-smokers are in between the values of smokers and non-smokers.  These results are quite typical of several crossover studies comparing ex-smokers with never and current smokers. From our data it cannot be concluded whether the better pulmonary function of ex-smokers, as compared with smokers, simply resulted from a shorter smoking history, or whether there was any actual improvement after smoking cessation. We therefore calculated in ex-smokers expected values for several lung function tests, i.e. values reached if lung function had evolved at non-smoking rates after smoking cessation (Fig. 1). We assumed a linear evolution of lung function indices with age and similar initial values in both non-smokers (as they started to smoke). We also took into account both smoking duration and time since smoking cessation. We compared actual values with expected ones with a rank test, and found that except specific airway conductance and maximal expiratory flow rates at 50% and 75% of expired vital capacity, all other differences were statistically significant, i.e. closer to the mean values of non-smokers than to the calculated expected values. In other words, cessation of smoking not only stopped the smoking induced faster decline in lung function, but by slowing the progression rate below that of non-smokers, it lead to an improvement in lung function. In order to see whether this improvement begins early after smoking cessation, we looked separately at 22 subjects who stopped smoking 5 years in average before the study. In these "late quiters" we found significant differences between expected and observed values for residual volume, $FEV_{1.0}$, vital capacity ratio, slope of $N_2$-plateau, closing volume and closing capacity, meaning that improvement, after smoking cessation, starts relatively early. A somewhat related approach was used by Beck et al (4) who used a comparison between expected and observed values of lung function tests which they called residual values, in non-smokers, ex-smokers and continuous smokers. Their conclusions were similar to ours: giving-up smoking is followed by an improvement of lung function, which starts early after stopping smoking. However, cross-over studies have evident limitations, since the course of the events is supposed but not observed. They can give indications which have to be verified by longitudinal studies.

The prospective studies, I will comment on were done by Higgins et al (5), over a 9 year peroid, by Fletcher et al (6), over 8 years and by Bosse et al (7) over 5 years. Higgins et al., studied a random sample completed with an age stratified one, from an industrial town in England. Four occupational groups were investigated: miners, ex-miners, foundry and ex-foundry workers. They found that the mean annual decline of $FEV_{0.75}$, over a 9 year period, in the age group 55 - 64 years, was 32 ml for non-smokers, 44 ml for light smokers, 54 ml for heavy smokers and 37 ml for ex-smokers. The decline did not

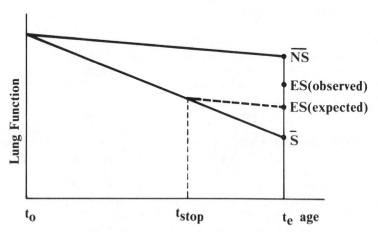

**FIGURE 1:**   Schematic presentation of how "expected" values
for ex-smokers (ES) were calculated : ES (expected)

$$= S + \frac{t_e - t_{stop}}{t_e - t_o} \quad (NS - S, \text{ where S is mean smoker}$$

value, NS mean non-smokers' value; $t_e$ is time at
examination, $t_o$ at onset of smoking, $t_{stop}$ at
smoking cessation.
Observed values were compared to "expected" value.

(Reproduced with permission from Nemery et al.,
Am. Rev. Resp. Dis., 125, 122-124, 1982).

appear to be closely related to occupation, suggesting, once again, that smoking is a more important factor in the development of respiratory disability than occupation. Fletcher et al (6) studied a non-random sample of working men, still fit enough to remain in full time employment. "Data were extrapolated over 20 years from observations made over 8 years, in 3 groups of men who had smoked about the same amount until the age of 40, developing definite but mild airflow obstruction. One group then gives-up smoking, one group smokes less than 15 cigarettes/day and the third group smokes over 15 cigarettes/day. By the age of 60 only about 5 per cent of those who had stopped smoking at 40 would have an $FEV_{1.0}$ below 1 litre (value which is usually associated with disability), while a third of those who had continued to smoke less than 15 cigarettes per day, plus more than half of those who smoked more than this, would have an FEV of less than a litre and would probably be severely disabled" (6). These conclusions came as a blessing at a time when the prevailing view, emphasized by P. Macklem (8), was that once FEV is decreased, the course of the COLD cannot be abated. The study of Bosse et al. (7), was of 850 male volunteers, participating in a longitudinal study of normal aging, free of chronic lung disease (both before and during the follow-up period). Of 452 who smoked at entry to the study, 98 quit during a 5 year period. The decrease in forced vital capacity (FVC) and $FEV_{1.0}$ for men who quit smoking was significantly less than that for current smokers (p = 0.02 for FVC and p < 0.001 for $FEV_{1.0}$). The authors also suggested that the effect of smoking cessation is relatively immediate, confirming thus the results we obtained from a cross-over study (3). Although there are differences in the absolute rate of decline of $FEV_{1.0}$ in ex-smokers, current smokers and non-smokers among different authors, there is nevertheless a consensus, showing that the decline in lung function is less important in former smokers than in smokers. This is true both in smokers without any ventilatory impairment, as in the study of Bosse et al (7), and in people with a decrease in $FEV_{1.0}$, studied by Fletcher et al (6).

Of particular interest are studies of smokers before and after giving-up the habit (9 - 18). These studies include asymptomatic subjects; others with chronic cough, expectoration and shortness of breath. In some papers no details are given concerning the clinical status. Only a limited number of smokers have been followed for up to 30 months, most studies being shorter. This is explained by the relapse of the habit. One of the first studies was done by Krumholz et al (9) on 10 smokers studied 3 to 6 weeks after stopping smoking. Functional residual capacity decreased, while specific airway conductance, peak expiratory flow rate, maximal voluntary ventilation and single breath diffusing capacity increased.   Ingram and O'Cain (12) studied 6 asymptomatic smokers who displayed frequency dependence of compliance.   One to eight weeks after smoking cessation compliance became frequency independent.   Slope of nitrogen alveolar plateau, closing capacity and residual volume were observed to decrease in 12 smokers who gave-up the habit for 25 to 48 weeks (16).   These results were confirmed by Buist et al. in two successive studies (15, 17) in 15 smokers, followed for 30 months after giving-up smoking: closing capacity and slope of $N_2$ alveolar plateau decreased, while $FEV_{1.0}$ and FVC increased.   All these changes were statistically significant. Functional improvement was greater in people who started with impaired function than in those whose function was initially normal.   Respiratory symptoms, such as cough, sputum production, wheezing and shortness of breath usually disappeared in 3 to 4 months.   Decrease in sputum production, probably reflected a decrease in the hypersecretion of the mucous glands in the large airways.   Among different studies, there was a trend for indices reflecting intrapulmonary distribution of ventilation (including $N_2$ slope of phase III, and closing volume) to improve earlier than tests of ventilatory capacity.

Previous studies were dealing with asymptomatic smokers or subjects with chronic cough and production of phlegm, but usually a well preserved ventilatory capacity.   However, it is pertinent to ask whether functional changes are still reversible in the overt, advanced stages of COLD.   Earlier studies of Jones et al (19), Burrows et al (20) and Johnston et al (21) were rather disappointing.   One hundred patients were studied by the former authors over 32 months.   They concluded that smoking cessation induced no change, though they conceded that the small numbers of ex-smokers made it impossible to draw any general conclusion.   Burrows et al (20) suggested (from 200 COLD patients followed for 4 to 8 years) that reduced cigarette smoking was correlated with a favourable course of the vital capacity.   In the study of Johnston et al (21), who followed 111 COLD patients up to 10 years, the decrease in the lung diffusing capacity was significantly less ($p < 0.025$) in patients who gave-up smoking.   More comforting results came from the survey of Hughes et al (22).   Fifty-six male patients with pulmonary emphysema were followed for a minimum period of 3 years (range 3 to 13).   There were 19 ex-smokers and 37 current smokers.   Their initial $FEV_{1.0}$ was 1.17 L and 1.44 L respectively.   $FEV_{1.0}$ and VC declined at a significantly faster rate in those who continued to smoke that in ex-smokers.   No statistical difference between the two groups was observed for

the CO permeablity index and PaO$_2$.  Body weight increased in
ex- but not in current smokers, and this difference was
statistically significant.  These results suggest that even in
patients with overt COLD, giving-up smoking may prevent a
further decline in ventilatory capacity.

We conclude that smoking cessation appears to protect
against accelerated loss of pulmonary function.  This effect is
of relatively rapid onset.  Intrapulmonary distribution of air
and related tests (such as closing volume and closing capacity)
improve early, followed by tests of ventilatory capacity
(FEV$_{1.0}$).  Abandoning the smoking habit seems to be of benefit
not only in the early, but also in the advanced stages of
chronic obstructive lung disease.

## B.  INTERRELATIONSHIP BETWEEN SMOKING, LUNG FUNCTION AND BODY WEIGHT

There is evidence that patients with advanced stages of
COLD, especially pulmonary emphysema, do lose weight (23).  On
the other hand smokers who give-up the habit gain weight (10,
21).  Non-smoking men are on the average slightly heavier than
smokers, although it has been suggested that this is the case
only in manual workers (24).  Smoking results in susceptible
people in a more rapid decline in lung function than in non-
smokers and in most other smokers.  We asked ourselves if there
is not an interrelationship between cigarette smoking, lung
function and body weight.

The physical data, body mass index (an index closely
related to body fatness) and FEV$_{1.0}$/VC ratio in the steel plant
workers, have been presented earlier (3). Age and weight were
comparable in smokers, non-smokers and former smokers.
However, smokers weighed significantly less than both non-
smokers and ex-smokers (p < 0.001) (25).  Smokers were further
classified according to whether or not they had an FEV$_{1.0}$/VC
ratio less than 66.6%, that is lower than the one-sided 95%
limit derived from the data of the asymptomatic non-smoker

(26). The lower weight and body mass index of the smokers was apparently accounted for by the lower body mass index (p < 0.005) of smokers with airflow obstruction.

We therefore found among middle-aged steelworkers that smokers weighed less than non-smokers and former smokers. More important, the lower weight and body mass index of smokers, as a group, was in fact due to a subset of smokers with airflow obstruction (25). Body mass index and weight were comparable in asymptomatic non-smokers and smokers without airflow obstruction. Both body mass index and $FEV_{1.0}/VC$ ratio were highly significantly (p < 0.001) correlated in smokers, but not in non-smokers (Fig. 2) The correlation coefficient was only 0.34, and therefore only part of the variance of these indices could be explained by their correlation. This is not surprising taking into account the many different determinants of body weight and lung function. In a letter to the editor, Kaufmann (27) reported similar results in 1049 workers, when data were analyzed in a similar fashion. This author divided her workers into two groups: with a higher or lower than 66.6% $FEV_{1.0}/VC$ ratio. In both moderate (less than 15g tobacco per day) and heavy smokers (more than 15g per day), body mass index was significantly less (p < 0.01 and p < 0.001 respectively) in those with $FEV_{1.0}/VC$ ratio less than 66.6%. In non-smokers and former smokers body mass index was similar in those with a higher or lower $FEV_{1.0}/VC$ ratio. The reason for these differences are not evident, but 3 hypotheses can be envisaged.

The first is that body weight influenced the lung function. Indeed, it is known that increasing weight is accompanied to some extent by an increase in lung function indices (the so-called "muscularity effect"), but further increase in weight results in a decrease in lung function ("obesity effect"). This results in a virtual plateau of the $FEV_{1.0}VC$ vs body weight relationship over the usual weights (28). This was not so in our smokers who showed a significant relationship between body mass index and $FEV_{1.0}/VC$ ratio. Therefore this hypothesis is not supported.

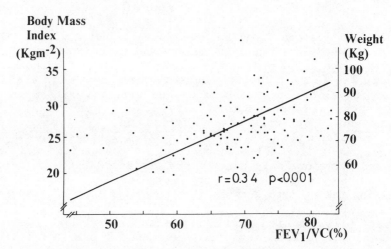

**FIGURE  2:**    The  relationship  of  body  mass  index  (weight,  height $^{-2}$)  vs  $FEV_{1.0}/VC$  ratio  in  105  current  smokers  (the  slope  of  the  regression  line  is  given  by  the  orthogonal  regression  coefficient).    Weight  for  a  standardized  height  of  169  cm  is  presented  in  the  right  side  of  the  ordinate.

(Reproduced  with  permission  from  Nemery  et  al.,  Br.  Med.  J.,  286,  249–251 , 1983).

The second hypothesis is that airflow obstruction was in some way responsible for the weight loss in the affected smokers. Progressive weight loss occurs in patients with overt COLD, especially emphysema. This has been attributed to loss of appetite and reduced food intake (23), increased energy requirements (29) or reduced production of anabolic steroid hormones (30). However, the subjects we studied are different from those hospital patients. The former were active workers, without (except in 2) shortness of breath when walking on a level floor and with $FEV_{1.0}/VC$ ratio well above those of the latter. If further data would support this explanation, then, weight loss, which is known to be an unfavourable prognostic factor in patients with overt COLD, may well start earlier in the course of the disease, than was hitherto considered.

The third possibility is that body weight and loss of lung function are produced by some common factors. For example, cigarette smoking may affect both lung function and energy metabolism. Decrease in basal $O_2$ consumption, protein-bound iodine concentration and 30-minute postprandial blood glucose have been observed shortly after giving-up smoking (31). On the other hand, smoking women have a smaller weight increase during pregnancy than non-smoking women, food intake being apparently similar in both groups (32).

Whatever the reasons for the link between body weight and lung function in smokers, from a practical point of view, loss of excess weight in a middle-aged smoker, should not always be regarded as favourable, since it could point to deteriorating lung function and susceptibility to chronic obstructive lung disease (25).

**REFERENCES**

1.  HUHTI, E.
    Chronic bronchitis in non-smokers - does it exist?
    Europ. J. Respir. Dis., 63, suppl. 118, 35-42, 1982.

2.  PETO, R., Speizer, F.E., Cochrane, D.L., Moore, F. et al.
    The relevance in adults of air-flow obstruction, but
    not of mucus hypersecretion, to mortality from chronic
    lung disease.
    Am. Rev. Respir. Dis., 128, 491-500, 1983.

3.  NEMERY, B., Moavero, N.E., Brasseur, L., Stanescu, D.C.
    Changes in lung function after smoking cessation : an
    assessment from a cross-sectional survey.
    Am. Rev. respir. Dis., 122-124, 1982.

4.  BECK, G.J., Doyle, C.A., Schachter, E.N.
    Smoking and lung function.
    Am. Rev. Respir. Dis., 123, 149-155, 1981.

5.  HIGGINS, I.T.T., Gilson, J.C., Ferris, B.G., Waters, M.E.
    et al.
    Chronic respiratory disease in an industrial town : a
    nine-year follow-up study.
    Preliminary report. Amer. J. Publ. Hlth, 58, 1667-1676,
    1968.

6.  FLETCHER, C., Peto, R., Tinker, C., Speizer, F.E.
    The natural history of chronic bronchitis and
    emphysema.
    Oxford University Press, Oxford 1976.

7.  BOSSE, R., Sparrow, D., Rose, C.L., Weiss, S.T.
    Longitudinal effect of age and smoking cessation on
    pulmonary function.
    Am. Rev. Respir. Dis., 123, 378-381, 1981.

8.  MACKLEM, P.T.
    Obstruction in small airways.  A challenge to medicine
    (Editorial).
    Am. J. Med., 52, 721-724, 1972.

9. KRUMHOLZ, R.A., Chevalier, R.B., Ross, J.C.
    Changes in cardiopulmonary functions related to
    abstinence from smoking.
    Ann. Int. Med., 62, 197–207, 1965.

10. WILHEMSEN, L., Orha, I., Tibblin, G.
    Decrease in ventilatory capacity between ages of 50 and
    54 in representative samples of swedish man.
    Brit. Med. J., 3, 553–556, 1969.

11. PETERSON, D.I., Lonergan, L.H., Hardinge, M.G.
    Smoking and pulmonary function.
    Arch. Environ. Health., 16, 215–218, 1968.

12. INGRAM, R.H.Jr., O´Cain, Ch.F.
    Frequency dependence of compliance in apparently
    healthy smokers versus non-smokers.
    Bull. Physiopath. Resp. (Nancy), 7, 195–208, 1971.

13. DIRKSEN, H., Janzon, L., Lindell, S.E.
    Influence of smoking and cessation of smoking on lung
    function.
    Scand. J. resp. Dis. (Suppl), 85, 266–274, 1974.

14. BODE, F.R., Dosman, J., Martin, R.R., Macklem, P.T.
    Reversibility of pulmonary function abnormalities in
    smokers.
    Am. J. Med., 59, 43–52, 1975.

15. BUIST, A.S., Seston, G.J., Nagy, J.M., Ross, B.B.
    The effect of smoking cessation and modification of
    lung function.
    Am. Rev. Respir. Dis., 114, 115–122, 1976.

16. McCARTHY, D.S., Craig, D.B., Cherniack, R.M.
    Effect of modification of the smoking habit on lung
    function.
    Am. Rev. Respir, Dis., 114, 103–113, 1976.

17.  BAKE, B., Oxhoj, M., Sixt, R., Wilhemsen, L.
         Ventilatory lung function following two years of
         abstinence.
         Scand. J. Resp. Dis., 58, 311–318, 1977.

18.  BUIST, A.S., Nagy, J.M., Sexton, G.J.
         The effect of smoking cessation on pulmonary function :
         a 30 month follow-up of two smoking cessation clinics.
         Am. Rev. Respir. Dis., 120, 953–957, 1979.

19.  JONES, N.L., Burrows, B., Fletcher, C.M.
         Serial studies of 100 patients with chronic airway
         obstruction in London and Chicago.
         Thorax, 22, 327–335, 1967.

20.  BURROWS, B., Earle, R.H.
         Course and prognosis of chronic obstructive lung
         disease.
         N. Engl. J. Med., 280, 397–404, 1969.

21.  JOHNSTON, R.N., mcNeill, R.S., Smith, D.H., Legge, J.S.,
         Fletcher, F.
         Chronic bronchitis – measurements and observations over
         10 years.
         Thorax, 31, 25–29, 1976.

22.  HUGHES, J.A., Hutchinson, D.C.S., Bellamy, D., Dowd, D.E.
         et al.
         The influence of cigarette smoking and its withdrawal
         on the annual change of lung function in pulmonary
         emphysema.
         Quart. J. Med., 702, 115–124, 1982.

23.  VANDERBERGH, E., Van de Woestijne, K.P., Gyselen, A.
         Weight changes in ther terminal stages of chronic
         obstructive pulmonary disease.  Relation to respiratory
         function and prognosis.
         Am. Rev. Respir. Dis., 95, 556–566., 1967.

24.  ROYAL College of Physicians.
         Smoking and Health.
         Tunbridge Wells, Pitman, 82, 1977.

25.  NEMERY, B., Moavero, N.E., Brasseur, L., Stanescu, D.C.
         Smoking, lung function and body weight.
         Br. Med. J., 286, 249–251, 1983.

26. NEMERY, B., Moavero, N.E., Brasseur, L., Stanescu, D.C.
        Significance of small airway tests in middle-aged
        smokers.
        Am. Rev. Respir. Dis., 124, 232-238, 1981.

27. KAUFFMANN, F.
        Smoking, lung function and body weight
        (correspondence).
        Br. Med. J., 286, 1280, 1983.

28. SCHOENBERG, J.B., Beck, G.J., Bouhuys, A.
        Growth and decay of pulmonary function in healthy
        blacks and whites.
        Respir. Physiol., 33, 367-393, 1978.

29. HUNTER, A.M.B., Carey, M.A., Larsh, H.W.
        The nutritional status of patients with chronic
        obstructive pulmonary disease.
        Am. Rev. Respir. Dis., 124, 376-381, 1981.

30. SMEPLE, P. d'A., Watson, W.S., Beastall, G.H.,
        Bethel, M.I.F., Grant, J.K., Hume, R.
        Diet, absorption, and hromone studies in relation to
        body weight in obstructive airway disease.
        Thorax, 34, 783-788, 1979.

31. GLAUSER, S.C., Glauser, A.M., Reidenterg, M.M., Rusy, B.F.,
        Tallanda, R.J.
        Metabolic changes associated with the cessation of
        cigarette smoking.
        Arch. Environ. Health, 200, 377-381. 1970.

32. D'SOUZA, S.W., Black, P., Richards, B.
        Smoking in pregnancy : associations with skinfold
        thickness, maternal weight gain, and fetal size at
        birth.
        Br. Med. J., 282, 1661-1663, 1981.

DISCUSSION

LECTURER: Stanescu                              CHAIRMAN: Cumming

BAKE:                One on my concerns with the study of cessation
                     of smoking is the fact that the investigators
                     are aware of which subject smokes and which has
                     stopped.  Of all the studies to which you have
                     referred how many have made a blind assessment
                     of the curves, this being particularly
                     important in the case of the single breath
                     nitrogen curve which are very subjectively
                     assessed.

STANESCU:            I cannot speak for other authors but the test
                     we applied was blind, but I can sympathise with
                     you regarding the assessment of the slope of
                     the nitrogen plateau but the $FEV_{1.0}$ curves were
                     analysed by technicians who had no idea of the
                     smoking status of the subjects.

RAWBONE:             You have described short term and long term
                     studies and in the short term studies there
                     appears to have been some improvement in
                     $FEV_{1.0}$, and the longer term studies shows a
                     deceleration in decline in $FEV_{1.0}$.  It could be
                     that in the longer term studies that the rate
                     of deceleration is minimised by the initial
                     improvement and the actual deceleration might
                     be greater.  Are there any longitudinal studies
                     which show both the initial improvement and
                     then the decline?

STANESCU:            The two types of study are on two different
                     populations, so I don't think that the results
                     can be directly compared between the groups.

HIGENBOTTAM:         How many of the abstension studies checked to
                     ascertain that smoking has stopped?

STANESCU:            We made no measurements of carbon monoxide, but
                     we cross-checked the history from colleagues.

GUYATT:              Do people stop smoking because they are having
                     symptoms, or are they relatively new smokers
                     who therefore find it easier to give up?

STANESCU:            There is no definite answer to this question, the data is apparently not available.   There is undoubtedly some bias in the self selection.

BAKE:

I have collaborated in the cessation of smoking clinic and it is surprising how many smokers who say they have stopped smoking in fact continue to do so, as demonstrated by COHb determiniation.   What morphological changes do you think are responsible for the changes in function seen both acutely and chronically or cessation of smoking.

STANESCU:

Auerbach in 1953 showed a lesser degree of pathological change in ex-smokers for 10 years than in smokers. In a recent study in 1980 Bertolan studied biopsy speciments and came to similar conclusions.  A study on goblet cells also showed no difference between ex-smokers and non-smokers after 10 years.

PRIDE:

A recent study from Hogg's group in Vancouver shows no difference in small airway pathology (in laboratory specimens) between ex-smokers and smokers.   In other words given a lobe from a group of patients they were unable to distinguish smokers from ex-smokers.  Regarding the misinformation which might be given by smokers, we have only on one occasion found a disparity between the statement made and the COHb.   We must distinguish between smoking cessation clinics where there is much pressure for the participants to do well and the vascular surgeons clinic.  If the surgeon says he will not operate unless smoking is stopped there is a great pressure to dissemble, so the context of the advice is important.   In our practice the advice is given as a matter of routine by a doctor with a full ashtray in front of him and the pressures to prevaricate are much less.

CUMMING:

I will draw the discussion to a close at that point.

# CHRONIC BRONCHITIS AND DECLINE IN PULMONARY FUNCTION

## WITH SOME SUGGESTIONS ON TERMINOLOGY

C. M. Fletcher
Postgraduate Medical School
University of London

## SUMMARY

The history of developments in terminology of obstructive pulmonary diseases is briefly reviewed. An account is given of two prospective studies which have shown that the rate of loss of ventilatory function is independent of indices of bronchial mucus hypersecretion and infections so that there is no casual connection between chronic bronchitis and development of generalised airflow limitation. Since it is also known that mucus hypersecretion arises in large bronchi and airflow limitation in small bronchi, it is inappropriate to use the term 'chronic obstructive bronchitis' to describe patients with expectoration and airflow limitation. The term 'chronic bronchiolitis would be more appropriate for subjects with chronic airflow limitation which is not due to asthma or emphysema.

Of all the harmful effects of smoking breathlessness due to irreversible air-flow limitation is perhaps the most distressing and disabling. In the United Kingdom, where death rates from this condition are the highest in the world, this condition is usually referred to as "Chronic Bronchitis" or "Emphysema". It was largely ignored by chest physicians until the post-war years when, with the decline in tuberculosis the importance of chronic bronchitis became apparent. Its relationship to smoking was completely ignored. At that time there was great confusion about terminology. The first step towards clarification came in 1959 when the report of a CIBA Guest Symposium was published (1). Its proposal that emphysema should be defined on an anatomical basis of dilatation and

397

destruction of alveoli has retained universal acceptance. It was also proposed that bronchitis should be defined as hypersecretion of bronchial mucus sufficient to cause persistent expectoration and that generalised limitation of airflow should be described simply by that term and should be separated into a reversible form or asthma – a definition which has been widely accepted but never precisely defined – and an irreversible type. It is in relation to this latter condition, when not manifestly due to emphysema, that confusion in terminology persists, the reasons for which I would like to discuss and about which I hope finally to make acceptable recommendations.

In 1965 a committee of the British Medical Council proposed a classification on bronchitis (2) (Figure 1). This was based on the epidemiological observation (3) of a marked association between volume of sputum, frequency of chest illnesses and reduction of FEV. So it was thought that these three abnormalities were part of a single disease process and it was widely believed in the U.K. (Figure 2) that hypersecretion encouraged infection which in turn damaged the lung causing airflow obstruction (AFO) which was thought to be due to emphysema. At that time it had already been shown that severe obstruction could occur in the absence of emphysema. The pathology of this sort of obstruction was not widely known; but severe bronchiolitis in one such case had been described by Harrison in 1951 (4) and simple methods for distinguishing 'emphysematous', 'mixed' and 'bronchial' types of case in 1966 (5). Pathologists had also described bronchiolar infection in association with emphysema (6, 7) and bronchiolar stenosis had also been described (8), but it was generally believed that excessive mucous in the bronchi caused much of the obstruction.

## A prospective study of chronic bronchitis

The scientific basis for this hypothesis was insecure being based only on cross-sectional surveys and retrospective speculation by clinicians and pathologists. A prospective survey of early cases of bronchitis seemed to be the best way of observing what actually happened and this is what I and my colleagues did by examining 792 working men in London twice yearly between 1961 and 1968 (9). Expectoration was assessed by the MRC questionnaire. (Regular sputum specimens proved less reliable but enables us to assess sputum purulence). At each survey we recorded chest illnesses and measured $FEV_1$ (standardised for age and for height by the index $FEV/H^3$ which is numerically similar to FEV/VC) so that we could calculate a linear rate of decline which we called FEV slope.

CLASSIFICATION OF CHRONIC BRONCHITIS (CB)
(MRC 1965)


SIMPLE CB
Increase in secretion of bronchial mucus
sufficient to cause expectoration.

MUCOPURULENT CB
CB with mucopurulent sputum.

OBSTRUCTIVE CB
CB with persistent expiratory bronchial narrowing
causing increased resistance to airflow.

FIGURE 1: Classification of Chronic Bronchitis
Medical Research Council 1965

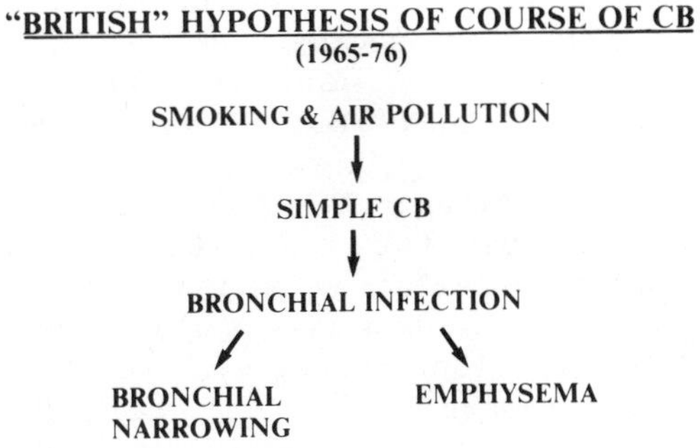

"BRITISH" HYPOTHESIS OF COURSE OF CB
(1965-76)

SMOKING & AIR POLLUTION
↓
SIMPLE CB
↓
BRONCHIAL INFECTION
↙     ↘
BRONCHIAL     EMPHYSEMA
NARROWING

FIGURE 2: British Hypothesis of Natural History of Chronic
Bronchitis

At our first survey we found that smokers had a lower mean FEV than non-smokers and that FEV was also reduced in relation to both sputum volume and history of chest illnesses (Figure 3). So our sample of men were suitable subjects among whom to study how these reductions of FEV were brought about during the subsequent 8 years of the survey.

The general pattern of decline of FEV which we found is shown diagramatically in Figure 4. Men with low levels of FEV must on average have had steep FEV slopes before our study and tended to continue with them during it, so that there was a highly significant correlation between FEV level and FEV slope. Because of this, smoking habits, sputum volume, and chest illnesses which correlate with FEV level must also be correlated with slope. To see what effect any changes in these variables during the survey had changes of slope we must use correlations with slope after allowing for FEV levels to detect those which have a casual relation to steepness of slope. This is a most important principle which needs to be observed in all prospective studies of variables with a steady rate of change such, for instance, as blood pressure, but which is often forgotten. Table 1 shows that after adjusting for FEV slope, smoking was the only factor which effected the rate of decline of FEV during our survey.

To check the surprising absence of any effect of expectoration and illnesses on FEV slope we looked at the immediate effect of changes in expectoration on contemporary measurements of FEV in individual men and found that on average FEV levels at surveys where expectoration was higher were no different from those when it was lower which confirmed the lack of any immediate effect of changes in mucous hypersecretion. To check whether illnesses had any permanent effect in lowering FEV levels we looked at these levels before and after individual illnesses (Figure 5). Permanent damage would have resulted in the pattern in Figure 5a. Temporary damage would have been as shown in Figure 5b. Examination of deviations of FEV from linear regression lines before and after chest illnesses showed that the latter pattern prevailed so we were confident in excluding the hypothesis that bronchial infections caused airflow obstruction. But we did confirm a significant correlation between hypersecretion and chest illnesses, showing that hypersecretion does increase liability to bronchial infections.

So this study showed that hypersecretion leading to infection and AFO leading to disability are two independent processes only associated by both being caused by smoking (Figure 6).

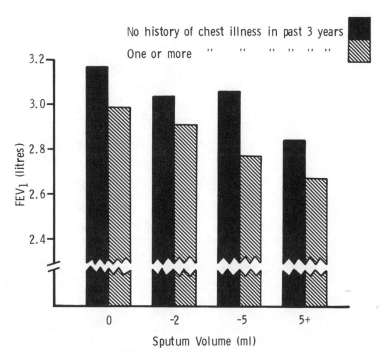

**FIGURE 3:**  Mean FEV (standardised for age and height) in 1961
according to sputum volume and history of chest
illnesses. (792 men aged 30–59 who were followed for
8 years.

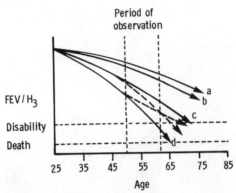

**FIGURE 4:** Effect of alterations during survey of factors which affect FEV slope.     FEV declines steadily during adult life at a rate which varies between individuals for a mixture of constitutional and environmental reasons.  By middle age, when our study was conducted, fast decliners will have lower levels of FEV and slow decliners will have higher levels of FEV, so that FEV slope and level are highly correlated.  Anything which is correlated with FEV level will thus correlate with slope. Factors which can alter slope in middle age will, if they change in the course of the survey, alter the slope as shown by the dotted lines in the figure, so that observed slopes of such individuals will be different from those expected from FEV levels.  They will therefore correlate with survey slope after adjustment for FEV level.   Factors which are correlated with level but do not cause changes in FEV slope during the surgey will not be correlated with this slope after adjustment for FEV level.

EFFECT ON FEV SLOPE OF PERMANENT AND TEMPORARY LOSS
OF FEV DUE TO AN ILLNESS IN 1965

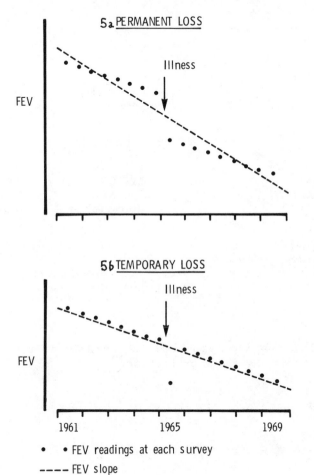

FIGURE 5

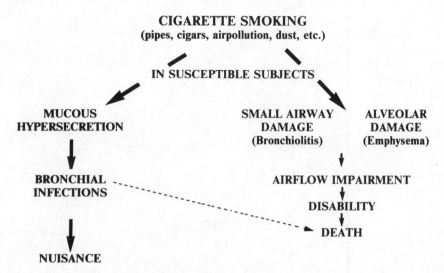

**FIGURE 6:** Two-disease hypothesis of chronic bronchitis and chronic airflow obstruction.

This conclusion is compatible with previous studies (for references see 10) and was confirmed by a subsequent large prospective study in France (11).

## Effects of smoking on FEV loss

The usual picture from prevalence studies are as we observed at our first survey in Figure 7. Smokers have only an 8.5% reduction of FEV (7% in light and 10% heavy smokers). But this conceals the truth. The greater reduction in the lightest smokers shown in Figure 7 was found to be due to smokers who had symptoms of AFO cutting down their smoking. But when we looked at all the smokers (including ex-smokers) we found only 15% had any significant reduction of FEV but the reduction among them averaged 45%. Presenting, as is usually done, the effects of smoking on FEV as the means of all smoking groups conceals, among the large majority whose lungs are not affected by smoking, the serious effect of smoking on the minority of smokers who are susceptible to it among the large majority whose lungs are not affected by smoking.

The effects of smoking on lifelong FEV loss which we derived from our study are summarised in Figure 8. Non-smokers and the majority of smokers have a slow, slightly curvilinear loss of FEV throughout their lives. Smokers who are susceptible to lung damage have a variable degree of accelerated loss (only one example is shown in the figure) which, as they get older, may prove fatal after a decade or more of disablement. Susceptible smokers who stop smoking do not regain their lost FEV, but their rate of loss after stopping reverts to that of non-smokers. If they stop smoking in middle age, while they still have a reasonably good FEV level, they will escape disability and death from AFO. If they do not stop until they are already disabled death from AFO may be slightly slightly postponed but is inevitable.

This picture of a steady lifelong increased rate of FEV loss is one which many clinicians find difficult to accept. They do see occasional patients with a rapid loss of FEV over a few years. The rarity of these sudden falls of FEV is shown by our having never seen, in a total of 5,000 man years of six monthly measurements, a single, sudden, permanent lowering of FEV. When clinicians ask their patients with severe AWO how long they have been breathless, the answer is often: ˊonly for a year or twoˊ. They conclude that the obstruction must have developed acutely, often after a chest illness. This is quite unacceptable evidence. A man aged 47 was admitted to my ward

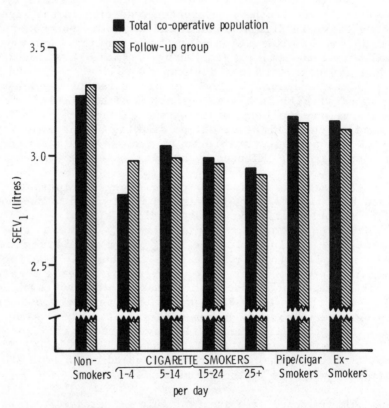

**FIGURE 7:** Mean FEV (standardised for age and height) in 1961 according to smoking habits.

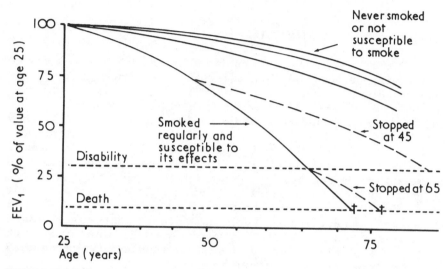

**FIGURE 8:** Effect of cigarette smoking on development of chronic airflow limitation.

Non-smokers, and smokers whose lungs are not susceptible to damage by smoke (top curve), have FEV levels which decline slowly as they get older. The FEV of smokers whose lungs are damaged by smoke decline much faster to varying degrees (only one example is shown) and if they continue to smoke they are likely to become disabled and may die from airflow obstruction. The FEV of susceptible smokers who stop smoking decline subsequently at the same rate as those of non-smokers: those who stop in middle age before the FEV is greatly reduced escape disability. Those who stop when already disabled only postpone a fatal outcome.

with FEV 0.51 in 1967.  He said he had only been breathless
since a pneumonia in 1964.  But he had been seen in our survey
in 1961 when he had FEV 1.01 and said he had breathlessness
Grade 3.  His FEV had fallen to 0.61 before his pneumonia after
which it declined slowly.  Accelerated falls over periods of
ten years or so are of course occasionally observed in clinical
practice.

## A twenty-year mortality follow-up

        Our study over 8 years did not last long enough to see
if men with reduced FEV in middle age really have an increased
risk of dying from AFO, but in 1981, twenty years after our
survey started, my colleagues (10) decided to examine this in
the 1136 men who had attended our first survey in 1961.  They
together with 1582 men aged 30 - 64 who had also had
measurements of FEV and had answered questions about
expectoration, recent chest illnesses and smoking habits in
surveys carried out by the Pneumoconiosis Research Unit of the
Medical Research Council between 1954 and 1958, making a total
of 2718 men all of whom were traced (except a few men who had
emigrated) and 1190 of whom had died.  All their death
certificates were obtained and were classified by the eminent
epidemiologist Sir Richard Doll without knowledge of the survey
findings into the causes shown in Table 2.

        The initial FEV values in all these men had been
obtained by different methods, so to make them comparable they
were classified according to their deviations from the mean at
each survey, adjusted for height and age and put into the four
groups shown in Table 3.  The increase in percentage of men
dying from severe airflow obstruction as initial FEV declines
is striking.  This can be a misleading index of relative resk
for men of different ages and with different periods at risk.
More precise is the ´logrank´ method of analysis of cause-
specific mortality (12), the results of which are also shown in
the table, and in Figure 9. In those with FEV less than 1 SD
below the risk is very small, but rises to twenty times and to
over fifty times the risk in those with FEV 1 - 2 and more than
2 SD below the mean.

        These data probably underestimate the true increased
risk of death from AFO because of inaccuracies in death
certificates.  AFO may have been important in some deaths
certified as due to other respiratory diseases.  Some of the
AFO deaths in men with previously normal FEV were found, on
obtaining clinical data, to have been wrongly certified as due
to AFO.

**TABLE 1**

Correlations of Major Factors with FEV Slope

| Factor | Simple Correlation | Given $FEV/H^3$ and Age |
|---|---|---|
| $FEV/H^3$ | < 0.001 | * |
| Smoking Habits | < 0.001 | < 0.001 |
| Phlegm Score | < 0.001 | NS |
| Chest Episode Frequency | < 0.001 | NS |
| H. I. Antibodies | < 0.05 | NS |
| Sputum Pus | NS | NS |
| Sputum Eosins | NS | NS |
| Personal or family allergy | NS | NS |

**TABLE 2**

Causes of Death and Smoking Habits in 2718 Men
Aged 25-64 Examined 20-25 Years Previously

|  | No | % |
|---|---|---|
| SEVERE AIRWAYS OBSTRUCTION | 99 | 8.3 |
| OTHER RESPIRATION DISEASES | 92 | 7.7 |
| LUNG CANCER | 103 | 8.7 |
| ALL VASCULAR DISEASES | 578 | 48.6 |
| OTHER CAUSES | 318 | 26.7 |
| ALL CAUSES | 1190 | 100 |
| NON-SMOKERS | 295 | 10.9 |
| EX-SMOKERS | 387 | 14.2 |
| CURRENT SMOKERS 1-14/DAY | 1181 | 43.5 |
| CURRENT SMOKERS 15 +/DAY | 855 | 31.5 |
|  | 2718 | 100 |

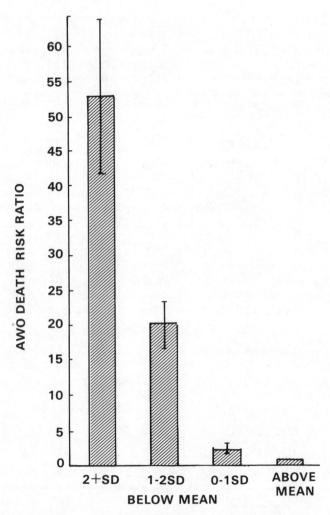

RELATIVE RISK OF AWO DEATH BY INITIAL FEV/H³

**FIGURE 9:** Means +SE of risk ratios of death from airflow
obstruction in 20 - 25 years according to standard
deviations of initial levels of FEV (standardised
for age and height), from means.

The effective risks of expectoration (placed in three groups) and in those with or without illnesses are shown in Table 4. Both of them have a small but significant prognostic value but these are solely due to their correlations with FEV level and they are virtually abolished after allowing for this.

We can safely conclude that the picture of a slightly accelerating fall of FEV during adult life is characteristic of the minority of smokers whose lungs are susceptible to being damaged by cigarette smoke and that this susceptibility is independent of that which leads many smokers also to develop a productive cough and a consequent increased liability to bronchial infections.

## Future research on epidemiology of AWO

These conclusions have an important bearing on future research in this field. They cast serious doubt on the relevance or value of continued research by standardised questions about mucous hypersecretion or chest illnesses. If the purpose of any survey is to detect a risk of AFO from exposure to dust, fumes, air pollution in association with the added risk of smoking, simple methods can now be used to assess early AFO and if subjects with consistent exposure are available, a simple prevalence study may detect a significant difference between those with and without exposure. But in such studies there is always the risk that the most susceptible individuals may, by stopping smoking or changing jobs, produce a biased reduction of mean FEV levels amongst those with a small exposure to the hazard. This bias will be particularly important in older subjects and could be lessened by concentrating on middle aged people. This bias, of course, could well distort any cross-sectional survey designed to discover the relative effects of less hazardous cigarettes. In many cases it may be necessary to carry out repeated measurements of AFO over as much as five to ten years to show the true relevance of continuing or modified exposure to development of this condition.

## Predictive value of FEV measurements

The second important implication of these studies is that the predictive power of FEV in middle age is so great that it might be used as a means for detecting men who are likely to develop airflow obstruction in later life. Unfortunately for this purpose (as is shown in Figure 10) the standard deviation

**TABLE 3**

EXPECTORATION AND AWO DEATH RISK

|                    | No OF MEN % | UNADJUSTED RELATIVE RISK | RR ADJUSTED FOR FEV/H$^3$ |
|--------------------|-------------|--------------------------|---------------------------|
| No USUAL SPUTUM    | 1424 (52)   | 1.0                      | 1.0                       |
| SPUTUM am or pm    | 644 (24)    | 2.5                      | 1.3                       |
| SPUTUM am ε pm     | 650 (24)    | 4.2                      | 1.3                       |
|                    |             | (p<0.001)                | (Not sig't)               |

**TABLE 4**

INITIAL FEV/H$^3$ GROUPS AND DEATHS FROM AWO

| SD OF FEV/H$^3$     | No OF MEN (%) | No DYING FROM AWO (%) | RELATIVE RISK |
|---------------------|---------------|-----------------------|---------------|
| ABOVE MEAN          | 1475 (54)     | 16 (1)                | 1.0           |
| 0-1SD BELOW MEAN    | 826 (30)      | 17 (1)                | 2.5           |
| 1-2SD BELOW MEAN    | 309 (11)      | 38 (12)               | 20.3          |
| 2+ SD BELOW MEAN    | 108 (4)       | 28 (26)               | 51.8          |

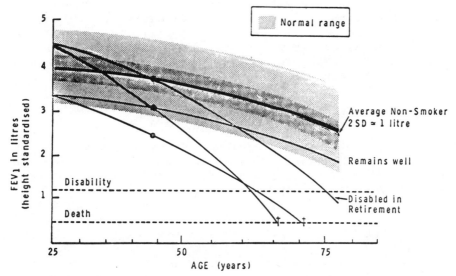

**FIGURE 10:**  Prognostic value of FEV measurement in middle age. The normal curve of slow decline of FEV in a non-smoker is shown with shading indicating the normal range of 2SD on each side. The prognostic significances of three FEV levels of smokers aged 40 are shown by solid circles.

An average level usually implies no risk of developing AFO, but if the level at 25 years had been above average disabling AFO may develop at age 75.

A level about 2SD below the mean may simply be due to a low normal level at age 25 and carries no risk, but if FEV at age 25 was high this level may indicate a risk of fatal AFO at age 60 if smoking is continued.

A level of more than 2SD below the mean indicates almost certain fatal AFO at age 70 if smoking is continued.

of a single measurement is so great that a man at the lower
level of the so-called normal range may be at high risk if he
previously had a high FEV but may be no risk at all if he has
always had a low FEV owing to having relatively small lungs for
his size and age.

For life insurance and health screening purposes, a
single measurement of FEV though informative, would not be
sufficient and it would be wise to confirm the presence of
early diseases in the peripheral airways in subjects with
reduced FEV by some more sensitive tests such as the nitrogen
slope or airflow at the end of a single expiration (13).

### Terminology of "obstructive lung diseases"

Although pathologists still debate among themselves
about the classification of emphysema the WHO definition of it
as 'dilatation of alveoli with destructive changes in their
walls' is clear, appropriate and universally accepted.

Terminology in relation to the other causes of
'irreversible' airflow obstruction remains confused:  the
irreversibility is seldom complete, the borderline between this
state and the reversible form called asthma has not yet been
precisely defined.  In my view early onset of asthma is easily
distinguished from the irreversible type if only on the basis
of age of onset and absence of emphysema.  Late onset asthma
can also be distinguished if it occurs in a non-smoker, is
associated with eosinophilia, has a relatively acute onset, or
is greatly reduced by steroid treatment.

The main type of non-emphysematous, late onset, largely
irreversible AFO is almost completely confined to cigarette
smokers, (past and present), and this is an important
definitive criterion.  The main confusion about terminology
arises in relation to the use of the term chronic bronchitis.
This should be used only in relation to the condition defined
in the CIBA proposals (1) as 'chronic or intermittent
hypersecretion of bronchial mucous sufficient to cause
expectoration'.  Different types can be described by adding
prefixes such as purulent, eosinophilic, pituitous (watery),
etc.  What is quite wrong is to use this word prefaced by
'severe' or 'obstructive' to refer to subjects with AFO.
'Advanced chronic bronchitis' should just mean that sputum is
profuse.  'Obstructive chronic bronchitis' should imply that

sputum is so profuse or expectoration so feeble that the secretions accumulate and actually obstruct the passage of air through the bronchi.  These terms cannot logically be used to describe the common form of irreversible AFO in the absence of emphysema for two main reasons:

1.  Because hypersecretion plays no part in its development: the two conditions are associated only because they are caused by smoking.

2.  Because mucous hypersecretion occurs chiefly in the larger bronchi (7) whereas expiratory airflow limitation is located chiefly in the smaller bronchi and bronchioles (14).  There is no justification for using a diagnostic term which is generally accepted for one condition to refer to another, quite distinct one.

What then should the non-emphysematous type of largely irreversible expiratory limitation be called?  Since the limitation is located in abnormal peripheral airways the term 'small airways disease' has become widely used.  But this term makes no reference to its obstructive effect and must include emphysema for this also impairs airflow in small airways. Those who introduced the term (14) rejected 'bronchiolitis' because the changes often involve small airways with cartilage in their walls which are strictly speaking small bronchi not bronchioles.  My own preference is to reject this anatomical purism, because the obstructive changes in these small airways (obliteration, stenosis or inflamatory exudate), are basically inflammatory to call the condition 'chronic obstructive bronchiolitis'.  The word bronchiolitis also agree with one of the earliest pathological descriptions of the condition by Harrison in 1951 (4) and I hope that this term may come into increasing use.  Meanwhile 'chronic airflow obstruction or limitation' is a perfectly acceptable term.  When the contribution to the obstruction by emphysema has been established by special investigations, the words 'largely due to emphysema', 'partly due to emphysema' or 'not due to emphysema' could be added.

The common American term 'chronic obstructive lung disease' is reasonable to include both emphysematous and non-emphysematous types but has the objection of a redundant word 'disease'.  'Chronic airflow obstruction' is shorter and 'chronic expiratory airflow limitation' (CEAL) more explicit.

## REFERENCES

1.    CIBA Guest Symposium.
      Terminology, definitions and classification of chronic
      pulmonary emphysema and related conditions.
      Thorax 1959, 14, 286-299.

2.    MEDICAL Research Council.
      Definition and classification of chronic bronchitis for
      clinical and epidemiological purposes.  A report to The
      Medical Research Council by their committee on the
      aetiology of chronic bronchitis.
      Lancet 1965, i, 775-780.

3.    FLETCHER, C.M. et al.
      The significance of respiratory symptoms and the
      diagnosis of chronic bronchitis in a working population.
      Br. Med. J., 1959, 2, 257-262.

4.    Harrison, C.V.
      In: Clinico-patholoical conference: Emphysema.
      Postgraduate Medical Journal, 1951, 27, 25-32.

5.    BURROWS, B. et al.
      The emplysematous and bronchial type of chronic airways
      obstruction.  A clinicopathological study of patients in
      London and Chicago.
      Lancet, 1966, i, 830-835.

6.    McLEAN, K.H.
      The pathogenesis of pulmonary emphysema.
      Amer. J. Med., 1958, 25, 62-66.

7.    REID, L.McA.,
      The pathology of chronic bronchitis.
      Lancet, 1954, i, 275-282.

8.    ESTERLEY, J.R., Heard, B.E.
      Multiple bronchiolar stenoses in a patient with
      generalised airways obstruction.
      Thorax, 1965, 20, 309-16.

9.    FLETCHER, .C.M. Peto, R.
      The natural history of chronic airflow obstruction.
      Brit. Med. J., 1977, i, 1645-1649.

10.   PETO, R. et al.
      The relevance in adults of airflow obstruction but not of
      mucous hypersecretion, to mortality from chronic lung
      disease: 20-year results from prospective surveys.
      Amer. Rev. Respir. Dis., 1983, 128, 491–500.

11.   KAUFFMANN, F. et al.
      Twelve years of spirometric changes among Paris workers.
      Int. J. of Epidemiology, 1979, 8, 201–212.

12.   PETO, R. et al.
      Design and analysis of randomised clinical trials
      requiring prolonged observation of each patient.
      Brit. J. Ca., 1977, 35, 1–39.

13.   TATTERSALL, A. et al.
      The use of tests of peripheral lung function for
      predicting future disability from airflow observation in
      middle-aged smokers.

14.   HODD, J.C., Macklem, P.T., Thurlbeck, W.M.
      Site and nature of airway obstruction in chronic
      obstructive lung disease.
      New Engl. J. Med., 1968, 278, 1355–1363.

DISCUSSION

LECTURER: Fletcher                                    CHAIRMAN: Cumming

HIGENBOTTAM:        Several of the patients you showed were
                    described as having bronchiolitis, I wonder if
                    you think this is a useful term to describe
                    small airways disease, particularly having
                    heard earlier from Dr. Crystal about macrophage
                    alveolitis.

FLETCHER:           We need some adjective, I would not mind it
                    being called airflow limitation. There are a
                    lot of other causes of bronchiolitis, Thurlbeck
                    in a recent paper listed six or seven and they
                    are not all obstructive.

PRIDE:              Several comments have implied that when small
                    airways are obstructed it is not expected that
                    the FEV would be reduced. The fact that the
                    $FEV_{1.0}$ is a test of large airways in normal
                    subjects is entirely a manifestation of the
                    distribution of the serial resistance. Once
                    that distribution of resistance is grossly
                    disturbed, as with pathological change then
                    $FEV_{1.0}$ cannot be used to infer the anatomical
                    site. Studies in which resistance has been
                    partitioned indicate that the major site of
                    increase is peripheral, and thus is in no way
                    incompatible with a marked reduction in $FEV_{1.0}$
                    and peak flow.

LEE:                You mentioned that bronchitis was a disease of
                    the working class, yet smoking habits were not
                    very different between the classes. How are
                    other factors concerned and is smoking
                    necessary to develop the disease?

FLETCHER:           Even in the most disadvantaged group they must
                    be smokers to develop the disease. You do, in
                    you first question raise a very important
                    problem about which I remain completely
                    mystified. Why the ratio of mortality rates
                    between the professional class and unskilled
                    workers in 1930 was 5:1, in 1971 it remained
                    5:1 and this is quite extraordinary and we have
                    no idea what the professional classes do to
                    obtain immunity, or what the unskilled workers
                    do to be susceptible.

CUMMING:          The medical geographers have addressed this problem and a paper by Girt, published in Medical Geography (Methuen) analysed the factors, environmental and ecological which contributed to chronic bronchitis. He showed that residence in the north east of an industrial town and birth in sub-standard housing each made a contribution.

FLETCHER:         Air pollution has changed dramatically in the past 20 years, and housing has also improved a great deal yet mortality ratios have not changed, although the rates have gone down.

STANESCU:         Is the same difference in mortality ratios also found in the United States of America?

FLETCHER:         It is, but not to the same extent. The poor in America do not appear to be so poor as in England.

SADOUL:           I am surprised that the ratio has remained the same in the United Kingdom during a period of such marked political change.

CUMMING:          There is built into that comment an important assumption, that the activity of government has an effect upon the life of the people. A very dubious assumption, and on that note I will bring this session to a close.

# DOES SMOKING HAVE ANY INFLUENCE ON THE COURSE OF

# RESTRICTIVE LUNG DISEASE

David Denison, Malcolm Law, Hadi Al-Hillawi
and Duncan Geddes

The Lung Function Unit, Brompton Hospital, London

The role of smoking in the initiation and aggravation of obstructive lung disease is well-established but its influence on restrictive lung conditions is by no means as certain. So, we have examined our records retrospectively to discover whether we should take account of smoking habits in the interpretation of routine lung function in such patients. To do this we have taken groups of 250 patients, with scoliosis, pulmonary sarcoidosis, or cryptogenic fibrosing alveolitis and compared their data with those of 165 healthy people, 100 patients with asbestosis and 50 patients with pleural tumours. The information available allowed us to put members of each group into two categories; those who presently or previously smoked, and those who had not.

Our approach to the interpretation of routine lung function tests is summarised by Figure 1, which illustrates a set of findings in a 40 year old lady with asthma and polychrondritis. Her maximal flow volume loop is on an absolute volume scale, in units of predicted TLC (TLC - Total Lung Capacity). The vertical flow scale is in units of 1 pred FVC/sec (FVC = Forced Vital Capacity). The maximal-flow volume loop of a healthy person of the same sex, age and height is shown by a dotted line. The patient's vital capacity is represented by the length of her flow volume loop. So, when she inhales a vital capacity of air marked with helium and carbon monoxide and breath-holds for 10 seconds, these gases diffuse into the residual volume to an extent shown by the additional length of the shaded horizontal bar marked VA (VA = accessible gas volume). The unshaded part to the right of the bar, represents the volume of lung, that could not be reached in 10 seconds,

D.S. ♀ aet 40     ASTHMA, POLYCHONDRITIS

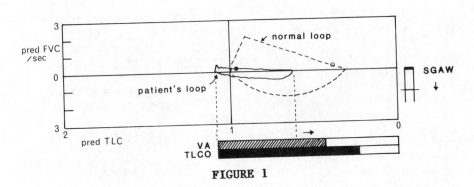

**FIGURE 1**

i.e. the inaccessible gas volume of the lung. The carbon monoxide uptake of the whole lung, over the same time period, is shown by the lower bar, marked $TL_{CO}$. Had it been appropriate to the accessible gas volume VA (i.e. had TLC/VA, which equals $K_{CO}$, been normal) it would have had an identical length. However, it is larger because $K_{CO}$ is higher than normal, a feature typical of asthma. The inspiratory airway conductance (SGaw) in this patient is shown by the vertical bar on the right. Had it been normal it would have reached down to the depth of the inspiratory limb of the normal flow volume loop. As it is, her conductance (shown by the shaded band at the top of the box) is markedly reduced, to a level roughly corresponding to the inspiratory limb of the actual loop, suggesting it is the major cause of inspiratory flow limitation. This technique of presenting routine flow volume loops pictorially is described in more detail by Denison et al, 1983).

Smoking initiates and aggravates obstructive lung disease and is normally associated with the functional changes illustrated in Figure 2, i.e. a displacement of the flow volume loop to the left, with flattening of the loop more marked in the expiratory than the inspiratory limb, and appearing first in late expiration. The inaccessible gas volume (TLC-VA) rises, airway conductance is reduced and $TL_{CO}$ is usually diminished at least in proportion to the fall in VA. As will be seen, these changes are quite distinct from those seen in restrictive conditions and if they are present should generally minimise or mask the latter, which could lead to underestimates of the presence or severity of restriction.

**Materials and Methods**

The characteristics of the people studied are given in Table 1. The 165 healthy volunteers were recruited from a group of 1500 people who responded to an advertisement for participants in a physical training programme. There were 83 men and 82 women. Their ages ranged from 31 to 59. None had a history of previous gross or recent trivial lung disease. They are proposed as a reasonable, but by no means perfect, yardstick against which to compare the groups that follow.

The 250 patients with sarcoidosis were those referred to the hospital for dyspnoea, in whom the chest radiograph had shown bilateral hilar lymphadenopathy alone (Stage 1), hilar lymphadenopathy with pulmonary shadowing (Stage II), or pulmonary shadowing alone (Stage III). In all cases the diagnoses were made histologically.

FUNCTIONAL CHANGES TYPICAL OF OBSTRUCTIVE DISEASE

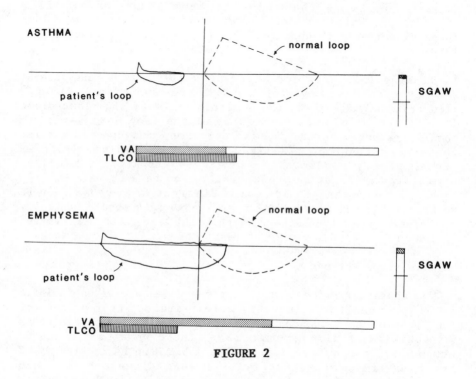

FIGURE 2

**TABLE 1:** Details of the People Studied

| | No | SMOKERS | NON SMOKERS | M | F | 17-20 | 21-30 | 31-40 | AGE 41-50 | 51-60 | 61-70 | 71-80 | MEAN AGE | % SMOKERS |
|---|---|---|---|---|---|---|---|---|---|---|---|---|---|---|
| Healthy | 165 | 84 | 81 | 83 | 82 | - | - | 86 | 64 | 15 | - | - | 40 | |
| Sarcoid | 250 | 80 | 170 | 143 | 107 | 4 | 71 | 75 | 63 | 23 | 11 | 3 | 38 | 32 |
| CFA | 250 | 131 | 119 | 168 | 82 | 1 | 8 | 9 | 40 | 75 | 94 | 25 | 58 | 52 |
| Asbestosis | 100 | 96 | 4 | 100 | 0 | - | - | 1 | 16 | 49 | 29 | 5 | 57 | 96 |
| Mesothelioma | 50 | 30 | 20 | 40 | 10 | - | - | 5 | 12 | 20 | 10 | 3 | 54 | 60 |
| Scoliosis | 250 | 71 | 179 | 109 | 141 | 37 | 94 | 61 | 23 | 12 | 9 | 1 | 31 | 28 |

The 250 patients with cryptogenic fibrosing alveolitis were diagnosed clinically, on the basis of end-inspiratory basal lung crackles together with persistent diffuse interstitial shadowing on the chest radiograph, in the absence of a known cause of lung fibrosis. The diagnosis was confirmed histologically in 130 of these 250 patients. Asbestosis was diagnosed in the group of 100 patients on the same clinical criteria, in association with a history of prolonged asbestosis exposure. In the group of 50 patients with mesotheliomas or adenocarcinoma of pleural distribution, there was no radiographic or histological evidence of interstitial lung disease, despite asbestos exposure in some. The 250 patients with scoliosis presented with that deformity as their primary or sole defect.

Everyone came to the laboratory for routine assessments on several occasions during their physical training programme or clinical management. All of the measurements reported below refer to their first attendance. Each patient completed the following test sequence. A series of maximal flow volumes loops were recorded from an Ohio rolling-seal spirometer linked to a Prime computer. The three most reproducible loops were stored and analysed. The forced expired volume in one second (FEV1) and the forced vital capacity (FVC) were derived automatically and scaled as a fraction of those predicted from the age, height and sex of the person, using the regression equations given by Cotes (1978).

Total lung capacity (TLC), residual volume (RV), and specific airway conductance (SGaw) were measured using a Fenyves & Gut whole-body plethysmograph. Single-breath whole lung transfer of carbon monoxide ($TL_{CO}$), the accessible gas volume in 10 seconds (VA) and the carbon monoxide transfer per litre of accessible gas volume ($K_{CO}$) were measured with a PK Morgan Model C resparameter. All of these measurements, (but for SGaw), were also expressed as fractions of these predicted, from age, sex and height, by Cotes (1978).

The results of the individual measurements were then compiled in two forms:- (i) as histograms, in order that all data could be viewed at one time; and (ii) in the pictorial form described earlier.

**RESULTS**

**(a) The healthy subjects**

Figure 3 shows frequency histograms of the FEV1, FVC, VC,

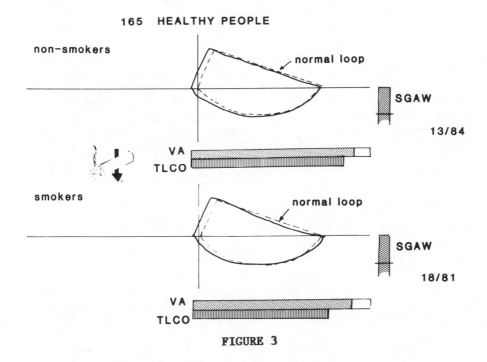

FIGURE 3

TLC, VA, RV, $TL_{CO}$ and $K_{CO}$, observed in this group of 165
people. In each case the non-smokers are represented by the
unshaded histogram above the abscissa and the current and ex-
smokers are depicted by the shaded histogram below the
abscissa. The abscissa is scaled as the observed/predicted %
for each variable. The vertical line (at 100%) represents
predicted means for the group (weighted for age, height and
sex). The small horizontal bar is the length equivalent to 20%
displacement on an abscissa

FEV1, FVC and VC are, on average, some 10% above predicted
mean, confirming previous findings from this laboratory that
the population local to London in 1980/1983 is 10% 'better' in
these regards than those referred to by Cotes (1978). TLC is
slightly larger, VA and RV are close to predicted and $TL_{CO}$is
slightly low. The little arrows below each abscissa indicate
the direction in which habitual smoking would be expected to
move the dark histograms. No such displacement is apparent,
although more detailed analysis, in this group, where precise
smoking habits were known, showed that people who smoked more
than 10 cigarettes a day had slightly reduced end-expiratory
flows and CO transfer. This finding suggests that it is not
necessary to take smoking habits into account in the
description of group characteristics, but it is necessary to
consider it in the interpretation of an individual's results.

The lung function data for this group is summarised
pictorially in Figure 4, which shows that typically flow-volume
loops were very close to normal, and airway conductance was
usually greater than the lower limit of normal (but in 13 or 84
non-smokers and 18 of 81 smokers it was not). The inaccessible
gas volume (TLC-VA) was, on average, 13% of predicted TLC, and
$TL_{CO}$slightly lower than predicted.

## Patients with Scoliosis

The results from the patients with scoliosis are presented
in Figures 5 and 6 which indicate that almost all showed
massive reductions in FEV1, FVC, TLC, RV and $TL_{CO}$, but a
substantial increase in $K_{CO}$. These findings are typical of
lungs compressed from without, expelling proportionately more
air than blood, and are similar to those seen in normal people
on progressive expiration. There are no distinctions of
importance between smokers and non-smokers.

## Patients with Fibrosing Alveolitis

The findings of these people are summarised in Figures 7 and
8. There are substantial falls in FEV1 ( FVC), TLC, RV, CO and

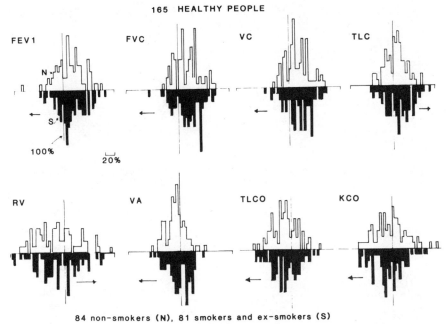

165 HEALTHY PEOPLE

84 non-smokers (N), 81 smokers and ex-smokers (S)

**FIGURE 4**

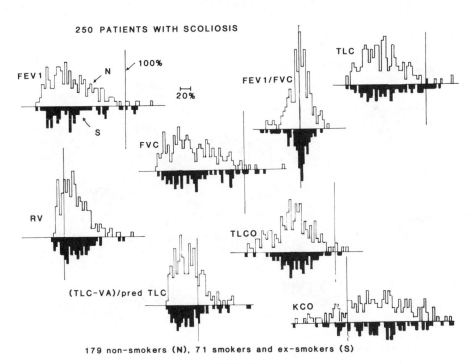

250 PATIENTS WITH SCOLIOSIS

179 non-smokers (N), 71 smokers and ex-smokers (S)

**FIGURE 5**

250 PATIENTS WITH SCOLIOSIS

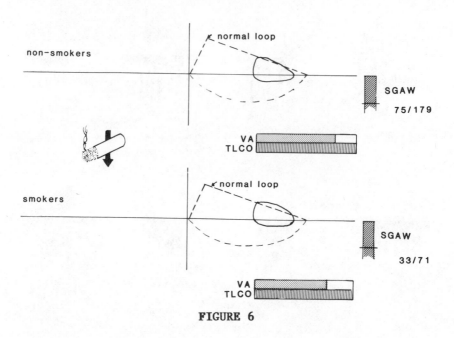

FIGURE 6

250 PATIENTS WITH FIBROSING ALVEOLITIS

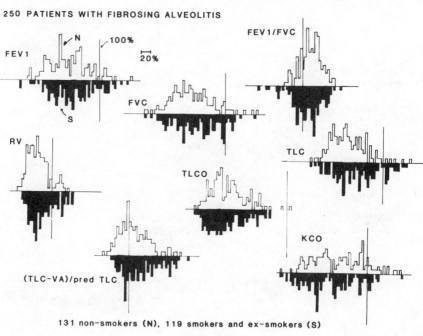

131 non-smokers (N), 119 smokers and ex-smokers (S)

FIGURE 7

PATIENTS WITH FIBROSING ALVEOLITIS

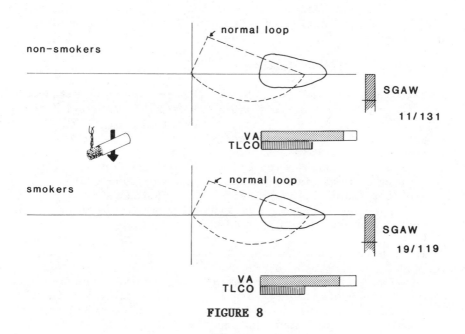

FIGURE 8

$TL_{CO}$ but in contrast to scoliotics, they also show a considerable fall in $K_{CO}$, which is typical of lungs that are contracted from within. The proportion of patients with lower than normal airway conductance was less than in the healthy group (30/250 of 31/165). However, in this group also, there are no important functional differences between those who smoke and those who do not.

**Patients with Pulmonary Sarcoidosis**

These results are given in Figures 9 and 10 and are essentially the same as in fibrosing alveolitis although the falls are less marked and gas inaccessibility (TLC-VA) is more evident. The latter probably reflects more small airway involvement. Rather more showed reduced conductance (42/250 of 30/250) which may reflect large airway distortion. Again there are no important differences between those who have smoked and those who have not.

**Patients with Pleural Tumours**

These patients exhibited the findings shown in Figures 11 and 12, namely considerable falls in dynamic and absolute lung volumes and in carbon monoxide transfer. Eight of the 50 showed reduced airway conductance (which is the same proportion as the healthy people and those with pulmonary sarcoidosis). However, 7 of these 8 were in the 30 that smoked. This group had larger lungs and a lower $K_{CO}$. The non-smoker, like scoliotics, appear to have their lungs compressed from without. Relative to them, those who smoked, were different, in the direction expected for obstructive lung disease.

**Patients with Asbestosis**

These patients consisted of 96 smokers and 4 non-smokers. However, the 4 non-smokers are functionally quite distinct, as indicated in Figures 13 and 14, i.e. they have substantially smaller lungs, less trappings and lower $TL_{CO}$.

**CONCLUSIONS**

These findings support the view that functional descriptions of the lungs of patients with pleural tumours or pulmonary asbestosis are distorted by some additional effects of smoking whereas the functional descriptions of scoliosis, pulmonary sarcoidosis and cryptogenic fibrosing alveolitis are not. However, this conclusion applies to the description of group characteristics and does not exclude a distortion of the

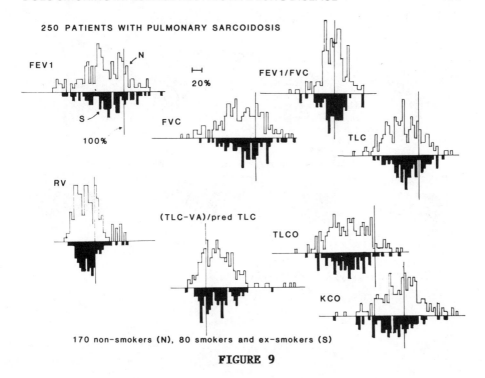

250 PATIENTS WITH PULMONARY SARCOIDOSIS

170 non-smokers (N), 80 smokers and ex-smokers (S)

**FIGURE 9**

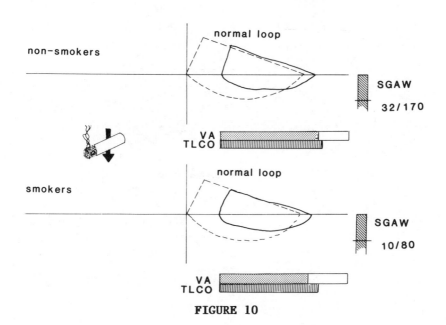

250 PATIENTS WITH PULMONARY SARCOIDOSIS

**FIGURE 10**

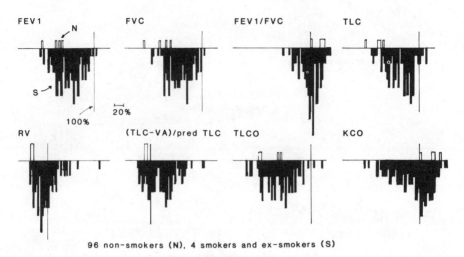

FIGURE 11

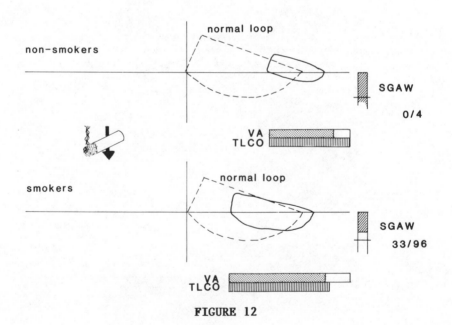

FIGURE 12

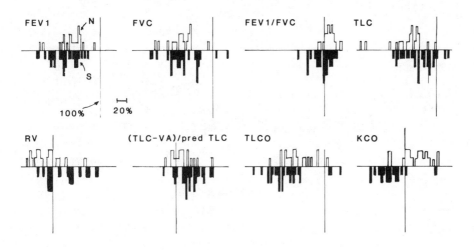

FIGURE 13

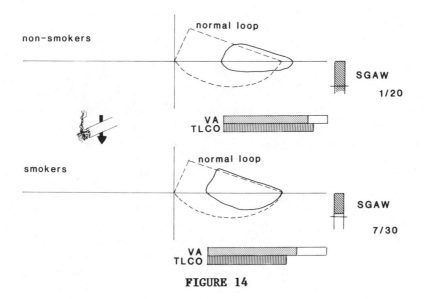

FIGURE 14

functional picture in <u>individual</u> patients with the latter
conditions.


**REFERENCE**

DENISON, D.M., Du Bois, R., Sawicka, E. (1983),
    Does the Lung Work?  6. Pictures in the mind.
    Br. J. Dis. Chest, 77, 35-50.

## DISCUSSION

LECTURER: Denison                    CHAIRMAN: Cumming

LEE:                How is smoking related to the onset of these
                    conditions.  I have not heard that it might be
                    negatively related for instance to sarcoidosis
                    and scoliosis, but you had far fewer smokers in
                    these groups than in the normals.

DENISON:            I don't know how significant these figures are,
                    51% of the normal population smoke, and in the
                    sarcoid group 32% and I 'don't know if this is
                    significant.

LEE:                I have just done the calculation and the p
                    value is less than 0.0001.

DENISON:            Thank you for pointing out this observation.
                    All the results are at the time of first
                    attendance and the age distribution in the
                    groups may difer.  How would you comment on the
                    28% with scoliosis who smoked?  Could it be
                    that they contracted their disease whilst young
                    and were then under pressure not to smoke.

LEE:                I know little about scoliosis, but that could
                    be an explanation.

PRIDE:              The model of scoliosis must include a component
                    of under development of the lung.

DENISON:            The increase in airways resistance and in $K_{CO}$
                    as an indicator of the lung being overstuffed
                    with blood, follows the path of normal subjects
                    as they breathe out.  Whether this implies that
                    the lung might be normal but enclosed in a
                    tight box is unknown.

PRIDE:              The most provocative finding which you have
                    shown is the presence of the normal $K_{CO}$ in
                    asbestosis in non-smokers.  This is surprising,
                    since the change is thought to resemble
                    cryptogenic fibrosing alveolitis, and you
                    showed that smoking of itself has only a small
                    effect on the $K_{CO}$.  What is your
                    pathophysiological explanation?

DENISON:          A surprising number of patients with sarcoidosis and fibrosing alveolitis have a normal $K_{CO}$. Total carbon monoxide transfer is greatly reduced, and this figure reflects the small lungs and the $K_{CO}$ may be illusory, representing only the diversion of blood to adequately ventilated areas. It may be incorrect to suppose that a normal $K_{CO}$ implies healthy lungs.

FLETCHER:         How did you manage to recruit this large group of smokers who were just as good as the non-smokers?

DENISON:          This touches on the question of social class which has been raised previously. We advertised in the Sunday Times (which is an upmarket newspaper) for those people who would like to take part in a study. In no way could they be considered normal, though it is surprising how closely they fitted into the groups. They were people in a particular social group who lived close enough to London to come to the Brompton Hospital regularly, to accept this for one year and to undergo physical training two nights a week, with regular exercise tests and pulmonary function tests whilst in addition paying a fee of one hundred pounds. The group is thus not representative but in comparison with a much larger group of normal people they show an identical functional picture. Three separate normal studies have give the same results although in the others we did not distinguish between smokers and non-smokers. It could be that these smokers are not showing any effect because they come from the upper social classes. That was something I did not appreciate until I heard you talk this morning.

FLETCHER:         It does appear that the non-smokers with asbestos is empty their lungs almost completely at the end of a normal expiration. How much do they have left in?

DENISON:          They have a residual volume which is 40% of normal, so that if it is normally 1.5 litres, that would make it 600 ml.

FLETCHER:          They would have sufficient surfactant to open their lungs again after expiring?

DENISON:           The surprising thing about these restrictive diseases is that the lungs are remarkably open. The flow-volume loop is not one of progressive airway collapse; the limit to expiration (as it is in young people) is the elastic limit of the thoracic cage.

STANESCU:          That was a beautiful general picture, but there are some details which surprise me. It is difficult to compare a group of four non-smokers with asbestosis with a group of 96 smokers with asbestosis.

DENISON:           I was hesitant to do this, but when I looked at all the variables they all moved in the expected direction. It could have been predicted early in the study and was there everytime, which I did find significant. Some things are more obvious to the eye than to the mind of a statistician and if the only result of the study is the mounting of a larger one in which it becomes evident to a statistician, then that would be a good outcome.

STANESCU:          I suspect that there is not a great difference in the specific airways conductance in smokers and non-smokers in fibrosis. You mentioned the effect of jogging on total lung capacity but did not go into details, would you like to elaborate on this?

DENISON:           It is a result with which I am completely happy, I am convinced that it is a real observation, in both groups. On the first visit to the laboratory they had three measurements of T.L.C. which agreed very closely, four months later after training the values again agreed very closely. The standard error was about 1.5%, whilst the increase in T.L.C. was 1%. If one performs a paired t-test for each person this was significant at the one in a thousand level. The increase was more marked in the people who did anaerobic exercise than the aerobic, and I suspect it to be due to the increased strength of the chest wall. The curious thing is that the residual volume decreased and we believe this to be due to progressive airway closure; the change was more marked in the smokers than the non-smokers, which was a surprise.

LAURENT:    What was the mean age of your normal population, and of the mesothelioma and sarcoid populations.

DENISON:    The mean age of the normals is 47, of the sarcoid 38, fibrosing alveolitis 58, scoliosis 31, asbestosis 57, mesothelioma 54. However this functional picture is normalised for age, sex and height, in respect only of lung function and it does not normalise history or other events so it is correct only for the known changes in lung function, or in smoking history.

BAKE:       I wonder about your interpretation. How do you know that a group undergoing sham training would not produce a 1% increase in T.L.C.

DENISON:    We wondered if this could be a learning phenomenon, but the four month interval led us to dismiss this possibility. It is difficult to obtain normal values in exercise studies since many people give up exercise during the study and others take it up so one ends up with two identical groups. A very large Finnish study ended in this way, and we also felt inhibited from asking people to pay one one hundred pounds for the privilege of not exercising.

SPADOFORA:  Sarcoidosis and fibrosing alveolitis are two conditions which are preceded by an alveolitis and I wonder whether it is correct to consider patients with one particular pathology as either having fibrosis or not, since the progress of the disease may be different in different cases.

DENISON:    I set myself a very limited question, namely what should be the policy of the laboratory towards the tests which we do for clinicians in respect of smoking. This is a first attempt to address this question and has raised many problems like the one you have suggested upon which further work needs to be done.

We have a lot a clinical information on these patients and they have all had biopsies so we may be in a position to study them in detail, this is only the first stage in the analysis of the results.

CUMMING:          The interest in the discussion has persuaded me to let it run over time, for which I hope you will forgive me.   I would like to make a brief comment on nomenclature.  Mesothelioma is an anatomic/histologic descriptor, asbestosis is an aetiological descriptor, sarcoid is shorthand for pulmonary sarcoidosis which is an anatomical/pathological descriptor, scoliosis is purely anatomical, C.F.A. − cryptogenic is aetiologic, fibrosing is pathalogic and alveolitis is anatomical/pathological.  This brings us to the descriptor normal, which may be defined as a subject who has been inadequately investigated as this paper has clearly shown.

# THE SINGLE BREATH NITROGEN TEST

Bjorn Bake

Department of Clinical Physiology

University of Goteborg, Sweden

In the search for sensitive tests for the early detection of airways obstruction, the single breath $N_2$-test is one of the most promising. It is furthermore relatively inexpensive and easy to perform. The subject expires maximally to residual volume (RV) and then inspires a vital capacity (VC) of oxygen. From total lung capacity (TLC) he expires slowly (expired flow rate about 0.3 l/s) back to RV. The $N_2$-concentration versus inspired and expired volume are recorded on an X-Y-recorder (1, 2). Four phases are usually identified on these curves: phase I represents pure dead space, phase II represents a combination of dead space and alveolar gas, phase III represents alveolar gas and finally phase IV indicates the existence of airway closure and this phase is also called closing volume (CV). It is particularly the slope of phase III which appears to be promising in terms of sensitivity to presumed early changes in the airways or in the lung parenchyma.

The mechanisms for the development of a slope of phase III are not yet fully understood but two requirements are usually thought of. Firstly, a concentration difference within the lungs between lung units must result from the dilution of nitrogen by the inspired vital capacity of oxygen and secondly, these units with unequal $N_2$-concentration must empty asynchronously during expiration to cause a sloping phase III. Various pathological changes in the airways or lung parenchyma can conceivably affect the slope of phase III which therefore may be regarded as an unspecific test of asynchronous or uneven ventilation.

The slope of phase III changes with age in normal subjects who have never smoked. We have through the years accumulated data covering normal male subjects in the age range of 7 to 70 years. The slope is relatively steep in the youngest children, about 2% $N_2$/1 in a 7-year-old child and decreases with increasing age or height of the subject until 18 years when it is about 1% $N_2$/1 (3). With increasing age to about 50 years the slope remains relatively constant but above the age of 50 it increases in an exponential fashion at least to 70 years (4).

The regression equations for children against height are for girls; slope of phase III (%$N_2$/1) = 5.093 - 2.474 x height (m). For boys: slope of phase III (%$N_2$/1) - 3.288 - 1.324 x height (m) (3).

For adult males (age 30 to 70 years) the corresponding regression equation was found to be: slope of phase III (%$N_2$/1) = 0.85 + $e^A$, where A = 0.0939 x age (years) - 6.302 (4).

The upper limit of normality roughly corresponds to 180 % of the predicted normal value.

In clinical experience almost all patients with airflow limitation have abnormally increased slope of phase III with the exception of those with very centrally located obstruction, for example in the larynx. It is also typical that the slope increases with increasing obstruction, the cardiogenic oscillations diminish and the definition of phase IV, i.e. the onset of airway closure, becomes obscure. Similar changes are observed in patients with emphysema and patients with airways disease. In lung fibrosis the changes appear less consistent in our experience, sometimes the $N_2$-curve is normal and sometimes quite pathological with steep slope of phase III. Smokers often have an abnormally increased slope of phase III (see below), although other lung function tests may be normal.

The factors causing the normal change of the slope of phase III with age or the abnormally steep slope observed in smokers are not fully known. It is not clear to what extent the airways are responsible or to what extent the lung parenchyma may contribute. We have in a simple lung model explored the possibility that uneven elastic properties can account for the slope of phase III (5). The major assumption in this model is that the elastic properties of small units of lung tissue are unequal. We believe that there is a distribution of elastic properties within the lungs. A simple way to model such a system is to consider the system as composed by two compartments of unequal elastic properties in accordance with measured data as assessed by the P-V curve. Essentially the shape or curvilinearity of the compartmental P-V curves were varied but always in such a way the resulting overall shape corresponded to the shape of a given overall P-V curve. Furthermore, the RV/TLC of each compartment was varied but always in such a way that the overall RV/TLC was in accordance with a given predetermined value. Finally, the size of the compartments was defined so that their VC's were always equal, i.e. each compartmental VC was half of the over-all VC. This model appeared powerful in the sense that any slope of phase III could be simulated for any given set of overall characteristics, i.e. TLC, RV/TLC and P-V curve. Thus, the extremely steep slope of phase III as seen for example in emphysema as well as normal flat slopes could be simulated. We also found that the normal change of slope of phase III with age from 7 to about 60 years could be simulated with a constant magnitude of uneveness between the compartments, i.e. the age effect could be explained solely by the change in overall TLC, RV/TLC and P-V curves. In older ages, however, an increased degree of unevenness had to be assumed to account for the steep increase in slope of phase III observed above the age of 60. We would conclude from our model calculations that uneven elastic properties within the lungs is a powerful potential mechanism behind the slope of phase III and it seems conceivable that the effects of tobacco smoking on the slope of phase III rather depends on alveolar disease than "small airways disease".

As chronic airflow limitation in chronic bronchitis or emphysema is usually caused by tobacco smoking, investigators have used smokers as a test model assuming the existance of a certain proportion of susceptible smokers with early abnormal changes. For example, one has argued that if smokers have a normal airway resistance and pressure volume curve (P-V curve) then the larger airways and lung parenchyma are normal and an abnormal test result of any particular test indicates increased resistance in peripheral airways (6). Similarly a normal forced expired volume in one second ($FEV_1$) has been taken to indicate normal larger airways and therefore an abnormal CV or slope of phase III in smokers would indicate abnormalities in peripheral airways (7). However, a normal $FEV_1$ cannot exclude subtle changes in larger airways nor for that matter in the lung parenchyma. Therefore, a pathological $N_2$-test in the presence of a normal $FEV_1$ should not be interpreted in terms of localisation of the abnormality but rather in terms of a higher sensitivity.

At least in a statistical context <u>sensitivity</u> is defined as the proportion of abnormal test results of the total

number of abnormals.  At first glance this definition may seem
straight forward but there is a problem in that an abnormal
test result is not defined.  Thus the limits of normality must
be defined because the number of abnormals, i.e. the
sensitivity is dependent on this definition.  One way to
express the effect of a given borderline test value is in terms
of specificity which is defined as the proportion of normal
test results of the total number of normals.  The normal range
of many tests in medicine is defined by plus or minus two
standard deviations ($\pm$ 2 SD) of a normal population.  If only a
lower limit of normality is considered as for example for $FEV_1$
and if the test results indeed are normally distributed, the
specificity would be 97.5%.  Because the sensitivity of a test
is dependent on the specificity any comparison between
sensitivities of different test requires equal specificity – a
condition of crucial importance but often hard to meet.

Oxhoj and co-workers (8) compared at equal
specificities the sensitivity of various tests including the
single breath nitrogen test in tobacco-induced abnormalities.
They investigated a random sample of 631 men who were born
either 1913 or 1923, i.e. they were either 60 or 50 years at
the time of the study.  They found that the slope of phase III
was clearly the most sensitive test, being abnormal in about
40% of the current smokers (i.e. sensitivity = 40%) when the
specificity was chosen to 97.5% as defined by the healthy
never-smokers.

Whatever the mechanism for an abnormally steep slope of
phase III in smokers, it is its prognostic value which decides
if the test qualifies as a test for early detection of chronic
airflow limitation.  Is a pathological slope in a smoker really
of importance in predicting future deterioration of lung
function or is it only an indicator of tobacco consumption?
Doctor Olofsson and co-workers in Goteborg (9) have conducted a
longitudinal study over 7 to 8 years with the basic aim to
determine if subjects with a pathological slope of phase III
have an increased annual decline of $FEV_1$ compared to subjects
with a normal slope of phase III.  We found that smokers with
equal age, sex, smoking habits and $FEV_1$ at the onset of the
study but with different slope of phase III indeed differed
significantly with respect to decline of $FEV_1$:  those who had
an abnormally steep slope had an annual decline of $FEV_1$ of
about 70 ml/year, whereas those with a normal slope had an
annual decline of $FEV_1$ of about 53 ml/year.  We conclude from
these results that smokers with an abnormal slope of phase III
run an increased risk of developing chronic airflow limitation.
The single breath $N_2$-test, therefore, appears to qualify as a
test for early detection of airways obstruction.

**REFERENCES**

1.   ANTHONISEN, N.R., Danson, J., Robertson, P.C. &
       Ross, W.R.D.
       Airway closure as a function of age.
       Resp. Physiol., 8 : 58, 1969-70.

2    OXHOJ, H. & Bake, B.
       Measurement of closing volume with the single breath
       nitrogen method.
       Scand. J. Resp. Dis., 55, 320, 1974.

3.   SOLYMAR, L., Aronsson, P.R., Bake, B & Bjure, J.
       Nitrogen single breath test, flow-volume curves and
       spirometry in healthy children, 7 - 18 years of age.
       Eur. J. Resp. Dis., 61, 275-186, 1980.

4.   SIXT, R., Bake, B. & Oxhoj, H.
       The single-breath $N_2$-test and spirometry in healthy
       non-smoking males.
       Eur. J. Resp. Dis.  In Press.

5.   NIU, S.F., Bake, B & Sixt, R.
       Non-uniform lung elastic properties and the slope of
       the alveolar plateau.
       In manuscript.

6.   WOOLCOCK, A.J., Vincent, N.J. & Macklem, P.T.
       Frequency dependence of compliance as a test for
       obstruction in the small airways.
       J. Clin. Invest, 48, 1097, 1969.

7.   McCARTHY, D.S., Spencer, R., Green, R. & Milic-Emili, J.
       Measurement of "closing volume" as a simple and
       sensitive test for early detection of small airway
       disease.
       Am. J. Med., 52, 747, 1972.

8.   OXHOJ, H., Bake, B. & Wilhelmsen, L.
       Ability of spirometry, flow-volume curves and the
       nitrogen closing volume test to detect smokers.  A
       population study.
       Scand. J. Resp. Dis., 58, 80, 1977.

9.   OLOFSSON, J., Bake, B. & Skoogh, B.E.
       The prognostic value of an abnormal single breath $N_2$-
       test in a longitudinal study.
       In preparation.

DISCUSSION

LECTURER: Bake                                   CHAIRMAN: Cumming

DENISON:         I would like to apply your curve to two model
                 patients, one with a high and one with a low
                 RV/TLC ratio with the same degree of
                 inhomogeneity. The former will mark their
                 alveoli with only small quantities of oxygen,
                 when they breath out the slope will be shallow
                 because the percentage of marker will be low.
                 The same manoeuvre with a low ratio marks the
                 alveoli strongly and the slope would be
                 greater. I was not clear whether you took
                 account of this by expressing the slope as a
                 fraction of the mean.

BAKE:            In the model we assume that the vital
                 capacities are always the same so that in that
                 respect the compartments are identical. So far
                 as the ratio is concerned, a low ratio gives a
                 small slope and a high ratio a high slope as
                 you suggested. To correct for this is
                 difficult since the curve is not linear.

DENISON:         I have the suspicion that it would still be
                 wise to normalise in the way I suggest.

GUINTINI:        You presented two groups one with a normal and
                 the other with an abnormal Phase 3 slope, and
                 showing a difference in the rate of decline of
                 $FEV_{1.0}$ with time. Did you compare the two
                 groups statistically and was the difference
                 significant?

BAKE:            One group had a value of 72 ml and the other 53
                 ml per year and this was statistically
                 significant.

GUYATT:          You seem to be using the slope of Phase 3 to
                 predict the rate of decline of FEV. Is it not
                 more important to use the test in its own
                 right?

BAKE:            I am not sure of your point, we have used the
                 test about which we have most information – the
                 FEV and we know that its rate of decline has
                 prognostic significance.

STANESCU:      Phase 3 of the expired nitrogen curve is a very
               sensitive test and in a crossover study we have
               found the same correlation with $FEV_{1.0}$. In a
               more general way, the sensitivity of a test is
               dependant on the prevalance of the disease, the
               greater the prevalance the greater the
               sensitivity.

LEE:           I wonder whether sensitivity is a good measure
               of the value of a test. You have identified
               40% of heavy smokers as suffering lung damage,
               yet we know that a much smaller percentage of
               smokers die from the disease, suggesting that
               the test is over sensitive. It would be quite
               easy to define a test which would pick up 98%
               of smokers.

               In your matched pairs analysis of normals and
               abnormals was there any difference in the
               smoking habits of the matched pairs?

BAKE:          Unfortunately we do not have that information.

PRIDE:         We have some evidence of ever increasing slopes
               but do not know how to quantify it, how do you
               resolve this technical problem?

BAKE:          We define a point at 750 ml and try to identify
               the point of airway closure and calculate the
               slope in between.

ASSANATO:      Sensitivity and specificity are unrelated to
               the prevalance of the disease whilst predicted
               value is dependant and I would ask Dr. Stanescu
               if I am in error about my basic concepts of
               epidemiology. Sensitivity and specificity are
               generally applied to risk factors in the
               screening of disease. If we wish to
               discriminate a proportion of smokers of 50%
               what would be the predicted value for such a
               case?

BAKE:          I am afraid I cannot calculate that in my head,
               though I have the information at home. I did
               not understand Dr. Stanescu's comment about
               specificity, but as I understand it I thought
               he was wrong.

CUMMING:       I was interested in your interpretation and you

model.   A good test of models is to apply a
mass balance for nitrogen, measuring the volume
of nitrogen contained under your predicted
curve and the volume actually recovered.

BAKE:               I have a slide to answer that question.   We
                    were initially rather concerned and this shows
                    what happens in the model, and these results
                    are interesting.   These are three simulated
                    curves wih different degrees of inhomogeneity.
                    The curves came out at different heights
                    suggesting different RV´s which they were not.
                    Then we made the model empty completely with
                    both compartments emptying at the same rate,
                    the smaller compartment finished emptying
                    whilst the larger one continued.   What this
                    means is that the amount of nitrogen predicted
                    as remaining in the lung is unreliable.   It is
                    only when the RV/TLC ratio is identical that
                    one gets a correct impression of what remains
                    in the lung.

CUMMING:            Thank you Bjorn, with that I will bring the
                    discussion to a close.

# THE RELATION BETWEEN INCREASED BRONCHIAL REACTIVITY

# AND ANNUAL DECLINE IN AIRWAY FUNCTION IN SMOKERS

N. B. Pride

Department of Medicine
Royal Postgraduate Medical School
Hammersmith Hospital, London W12 0HS

A relationship between cigarette smoking and the development of chronic airflow obstruction is already established but there remains a very wide range of individual susceptibility to progressive airflow obstruction among smokers, the cause of which remains unknown. More than 20 years ago, Orie (1) and van der Lende (2) in the Netherlands proposed that smokers with chronic and largely irreversible airflow obstruction shared with asthmatic patients a common allergic constitution and increased bronchial reactivity. This was termed the 'Dutch hypothesis' by Fletcher, Peto, Tinker and Speizer (3) in contrast to the then prevailing 'British hypothesis' (which they subsequently disproved) that progressive airflow obstruction was a consequence of repeated broncho-pulmonary infections. In their original studies, Fletcher and co-workers failed to find evidence of a relation between increased bronchial reactivity to inhaled histamine and accelerated annual decline of $FEV_1$, but more recently several groups, including our own, have re-investigated this problem, and usually have found a positive relationship. In this paper we review these more recent data and assess the overall importance of increased reactivity as a risk factor for progressive airflow obstruction in smokers.

The revival of interest in this topic started with a study by Barter and Campbell in Melbourne which was published in 1975(4). They found that bronchial reactivity to a standard dose of inhaled methacholine correlated well (r = 0.76) with the annual decline in $FEV_1$ over the previous five years in a group of 34 men with chronic expectoration and mild airflow obstruction. This stimulated us to study bronchial reactivity to inhaled histamine in 50 non-asthmatic smokers from the youngest cohort of men originally recruited to Fletcher's study at London Transport in 1961(5). We used doubling doses of histamine and defined reactors as those men who showed a fall in $FEV_1$ of 20% or more at a provocation concentration ($PC_{20}$) of histamine $\ll$ 8 mg/ml. We confirmed Barter and Campbell's results (Fig. 1) but noted that baseline $FEV_1$ at the time of testing bronchial reactivity was lower in the histamine reactors. If the Dutch hypothesis were correct, and if, as seems to be the case, accelerated decline of $FEV_1$ in susceptible smokers takes place over many years (3), a positive relation between enhanced reactivity and low baseline $FEV_1$ will inevitably be found in late middle age. We, therefore, decided to study bronchial reactivity in a younger and larger group of non-asthmatic men we had recruited in West London in 1974. Confirming our earlier study we found that annual decline in height-corrected $FEV_1$ in $ml/yr/m^3$ was 14.1, SEM 1.4 in 34 smokers with $PC_{20}$ $\ll$ 16 mg/ml (6) (Fig. 2). However, again there was a close relationship with the baseline $FEV_1$ (Fig. 3).

These studies indicate that an accelerated rate of decline in $FEV_1$ is consistently associated with increased bronchial reactivity (as assessed by change in $FEV_1$) to inhaled histamine and metacholine. Such an association is a necessary consequence of the Dutch hypothesis. However, all the studies have been conducted when baseline $FEV_1$ is already reduced, and it is possible that this abnormal reactivity is a consequence rather than a cause of the accelerated decline.

Clearly, however, the simplest hypothesis is that smokers who subsequently develop progressive airflow obstruction have a predisposing increased bronchial reactivity dating from childhood, well before the onset of smoking. For the present purpose it does not matter whether this increased reactivity is genetic or acquired in the first few years of life. In either case the increased reactivity would be expected to be accompanied by evidence of allergy, such as a positive personal or family history of allergic disease, increased blood or sputum eosinophilia and annual decline in $FEV_1$ in their smokers with chronic bronchitis. Similarly we observed in the cohort of men from London Transport that if we divided the 60 regular cigarette smokers into those with and those without some marker of allergy, the 7 men with the most rapid decline in $FEV_1$ all showed some evidence of allergy (5) (Fig. 4). Indeed smokers without any evidence of allergy showed rates of decline in $FEV_1$ which were very similar to that of non-smokers. However, in this study we used a very broad definition of allergy and the evidence of allergy was extremely weak in most of the allergy positive smokers. In our

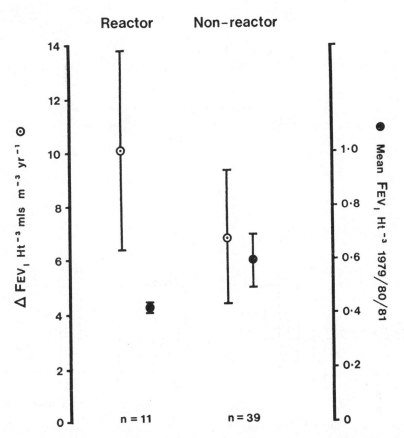

**FIGURE 1:** Annual decline in $FEV_1/Ht^3$ measured over preceding 20 years for smokers (open circles) whose $PC_{20}$ for inhaled histamine was $\leqslant$ 8 mg/ml (reactor) or $>$ 8 mg/ml (non-reactor). Also shown are the baseline values of $FEV_1/Ht^3$ at time of reactivity measurements (closed circles). Data of Connellan et al (5) obtained in the cohort of London Transport men born 1916-1931 originally studied by Fletcher et al (3). Mean $\pm$ SD.

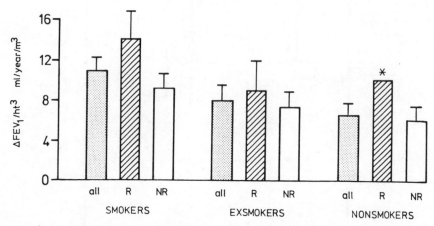

**FIGURE 2:** Annual decline in $FEV_1/Ht^3$ measured over 7.5 years for 117 smokers, 72 ex-smokers and 39 non-smokers studied by Taylor et al (6). No men with a clinical diagnosis of asthma were studied; R = reactor to inhaled histamine with $PC_{20} >$ 16 mg/ml, NR = $PC_{20}$ for histamine $\leqslant$ 16 mg/ml.

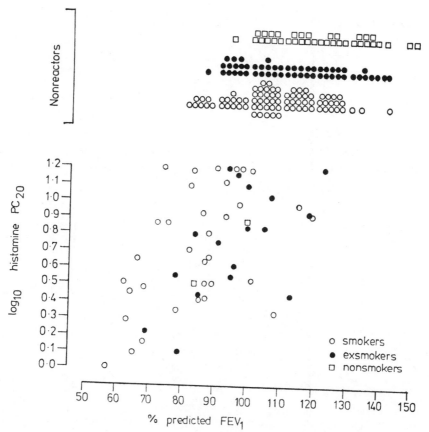

**FIGURE 3:** Relation of baseline $FEV_1$ (expressed as % predicted value) to log $PC_{20}$ for inhaled histamine. Nonreactors = $PC_{20} > 16$ mg mg/ml. Data of Taylor et al (6).

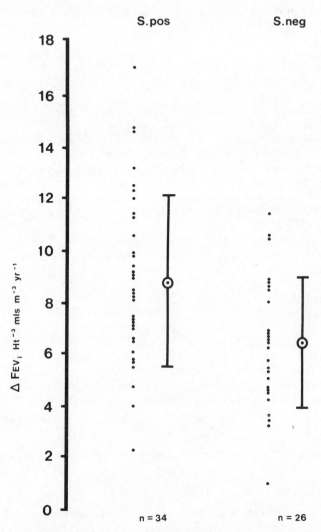

**FIGURE 4:**     Individual (and mean ± SD) values of annual decline
in $FEV_1 Ht^3$ measured over 20 years for smokers born
1916-1931 divided into those with positive allergic
features in personal or family history (S.pos) and
those without any such allergic features (S.neg).
Data of Connellan et al (5).

subsequent study (6) we again found a weak relation between rapid decline in $FEV_1$ and allergy but, of the individual allergic factors examined, only a personal history of allergic rhinitis (without overt diagnosis of asthma) was positively related to accelerated annual decline in $FEV_1$. Several markers of allergy have been found to be related to smoking history without being consistently established as related to particularly accelerated rate of decline in $FEV_1$. Thus it has been suggested by Burrows and co-workers in Tucson that smokers show lower rates of positive prick skin tests to inhalant allergens because atopic subjects are less likely to start smoking and more often to stop smoking if they ever begin (7). In our own studies (Fig. 5), we have found a significant increase in positive skin tests in ex-smokers although we found no difference between smokers and lifelong non-smokers (8). In the same subjects we have also found an increase in blood eosinophil counts in smokers (Fig. 6) which appears to be greater than can be accounted for by the well-known increase in total white blood cell count found in smokers. Perhaps the most provocative recent finding has been that smokers with negative prick skin tests show a higher total serum IgE than non-smokers. This has been shown in recent years in 4 studies, including our own (Fig. 7). The cause of this rise in total IgE has not been identified; an interesting speculation has been that it might be an IgE specific for some component of tobacco smoke (9). But, as can be seen in Fig. 6 and 7, the absolute values of blood eosinophils and the differences in total IgE in smokers are small compared with the values commonly seen in young asthmatic subjects. Similarly the skin test scores recorded in Fig. 3 are very low compared with those often found in subjects with a clinical diagnosis of asthma.

The hypothesis that progressive airflow obstruction in smokers particularly occurs in individuals with an allergic predisposition is also reflected in current interest in the role of childhood respiratory illness as a risk factor. Burrows and Taussig (10) have pointed out that the childhood illnesses thought to be a risk factor are those associated with wheezing and asthma rather than with recurrent acute infection. This distinction between obstructive and infective illness in childhood, resembles that already studied in adults by Fletcher and colleagues (3). By analogy with Fletcher's results, clearly it cannot be assumed that repeated infections in childhood necessarily result in airflow obstruction. In the absence of the very lengthy prospective follow-up study required to provide a definitive answer to this problem, current studies are based on recall of childhood illness by adults, which introduces the hazard of selective recall (10, 11). Nor surprisingly, therefore, there is no agreement on the

|  | % Negative | % Positive |  |
|---|---|---|---|
| SMOKERS (n ≃ 120) | 67 | 33 | p < 0.001 |
| EX-SMOKERS (n ≃ 73) | 41 | 59 | p < 0.02 |
| NON-SMOKERS (n = 44) | 66 | 34 | |

**FIGURE 5:**  Results of skin prick tests to 9 common inhalant allergens.  Any individual with one or more positive tests was regarded as positive.  Data of Taylor et al (8).

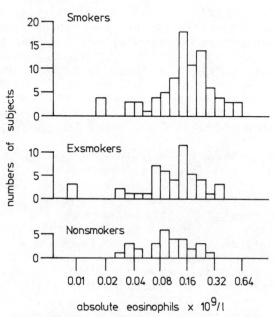

absolute eosinophils × 10⁹/l

**FIGURE 6:**  Histograms of absolute blood eosinophil counts (plotted on logarithmic scale) in men studied by Taylor et al (8).

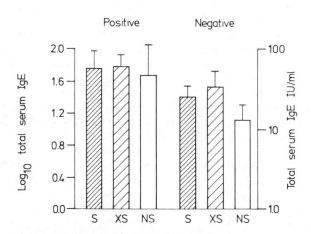

**FIGURE 7:** Total serum IgE (mean ± SEM) in smokers (S), ex-smokers (XS) and non-smokers (NS) with positive and negative prick skin tests to inhalant allergens. Data of Taylor et al (8).

absolute importance of childhood respiratory illness in predisposing to airflow obstruction in adult smokers (3, 10, 11). This remains an attractive, but unestablished hypothesis.

Hence although changes in some individual allergic markers are found in groups of smokers, these changes are small and have not been consistently shown to be significant risk factors for airflow obstruction in an individual smoker. Because the evidence for a predisposing allergic "host factor" is rather weak, we need to consider the possibility that the increase in bronchial reactivity is acquired following the onset of smoking. Several mechanisms have been proposed which could result in such a change (Table 1).

Perhaps the strongest possibility is that increased reactivity arises from altered geometry of the airways, i.e. that the size of the response depends on the initial size of the airways before challenge (12). Using $PC_{20}$ to assess response normalizes for the initial baseline $FEV_1$ but it can be anticipated that the amplifying factors of altered geometry could considerably exceed this normalization. Our finding that reactivity is increased in all men with $FEV_1$ less than 80% of predicted values indicates this could be an important mechanism; indeed we know of no published reports of subjects with reduced baseline airway function but without enhanced reactivity. This is not to say that altered geometry is the only factor involved; our full analysis is not complete but it certainly appears that for a given reduction in baseline $FEV_1$ the increase in reactivity is less in non-asthmatic smokers than in subjects with asthma. Another reported difference between the reactivity in non-asthmatic smokers and subjects with asthma is that smokers appear less responsibe to hyper-ventilation with cold air (13, 14). The obvious need is to study smokers sequentially, ideally first before starting smoking, shortly after starting smoking and at intervals thereafter. More practically, young smokers with normal baseline function can be compared with non-smokers; at least 8 such studies have been reported in recent years and most show little difference between smokers and non-smokers, although studies have been reported showing that smokers are both more reactive and that they are less reactive. Such studies, particularly if they are conducted some years after the smoking habit is established, are themselves subject to the bias that strongly reactive smokers may stop smoking while other continuing smokers with increased reactivity develop reductions in baseline airway function. Hence a comparison at identical and normal baseline lung function tends to compare the best smokers with the less good non-smokers (see baseline values of $FEV_1$ for smokers and non-smokers in Fig. 3). Interestingly, in the only study where these types of errors were avoided by allocating smoking and sham-smoking regimes for 3 years randomly (to baboons) the smoking animals were subsequently less reactive to inhaled metacholine (15). The importance of altered geometry is also supported by our finding that enhanced bronchial reactivity persists in ex-smokers who have reduced baseline $FEV_1$ (Fig. 3) even although their annual rate of decline in $FEV_1$ is similar to that of non-smokers.

Another physiological mechanism increasing reactivity may be a change in the pattern of deposition of inhaled aerosols. In subjects with established airways obstruction deposition is more central, so resulting in a greater proportion of an inhaled drug depositing on central airways. This in turn may amplify the effect of histamine and methacholine on airway function, in part because cholinergic and irritant receptors are more numerous in central airways but also because the methods used to assess airway function ($FEV_1$, airway conductance) are often more sensitive to changes in central than in peripheral airways. We think this mechanism is probably less important than the geometric factor already considered because most studies suggest that a change to more central deposition only develops when $FEV_1$ is reduced to 60% or less of predicted values (16) while enhanced reactivity is found when baseline function is less reduced. Some attempts have been made to examine this question by altering deposition pattern by varying aerosol size and the pattern of breathing.

These two mechanisms for amplifying reactivity are not specifically related to smoking and would also apply when airway narrowing was due to asthma or cystic fibrosis. A more specific effect might be related to the increased capillary permeability found in smokers (17, 18). Assuming this change in permeability affects airway as well as alveolar epithelium, this leads to the attractive hypothesis that cigarette smoking loosens epithelial tight junctions, enhancing access of drugs or allergens to bronchial muscle or receptors and so increases direct and reflex responses to bronchoconstrictors (19). This change would be analogous to that observed after acute viral infections in asthma with the difference that the irritant is applied daily over many years. The difficulty with this hypothesis is that the increased permeability appears to be seen in virtually all smokers and reverses within a few weeks on stopping smoking (18, 20) - in contrast to the persistent enhanced bronchial reactivity found in ex-smokers many years after stopping smoking.

Finally reactivity might be acquired during the smoking years by an immunological mechanism, analogous to the type I allergy found to pollen and the house dust mite. As discussed above Burrows (9) has speculated that the small rise in blood IgE might be related to some component of inhaled tobacco smoke. We have already discussed the small rises in blood eosinophils and in total serum IgE which appear to be acquired in smokers and revert (sometimes very slowly indeed (7)) on stopping smoking. A variation of this immunological hypothesis is that the effect of smoking is less specific and amplifies a pre-existing allergy to common inhalant antigens which previously had not given important clinical symptoms. The chief argument against this hypothesis is the failure to find an increased incidence of positive prick skin tests to inhaled allergens in smokers.

It is clear that we have a great deal to learn about the 'Dutch hypothesis'. Increased bronchial reactivity is undoubtedly common in smokers with accelerated annual decline of $FEV_1$ and reduced baseline $FEV_1$ but the increased reactivity may be a consequence rather than a cause of airway narrowing. We think the available evidence strongly supports the importance of reduced baseline dimensions of the airways but this still leaves room for further important amplifying factors. The most interesting allergic factor is the increased total IgE in smokers with negative prick tests to common inhalant allergens, but its origins are quite unknown. Some other allergic features - e.g. blood eosinophils, incidence of positive prick skin tests - appear to be slightly altered in smokers or ex-smokers but the change is usually small and of unproven significance as a risk factor for accelerated decline in lung function. But even if it turns out that only the increased reactivity part of the Dutch hypothesis is true, the hypothesis will have served a useful function in stimulating further studies. There are many features of chronic airways obstruction it cannot explain, such as the predominance of males and the social gradient. We agree with Burrows that it is probably only relevant for a subgroup of such patients, broadly corresponding to those with predominant intrinsic airway disease - it can hardly directly account for the development of emphysema (Fig. 8). But enhanced reactivity could well be important for some of the clinical features. Possibly bronchial infections enhance non-specific reactivity in smokers, just as they do in asthma. Perhaps the decline in baseline lung function that lasts for several weeks after infection indicates that the changes are due to reactive airway changes rather than the conventional idea of retained sputum. Furthermore, reactive changes in smokers, which are likely to involve a wide range of airway size, including the largest airways, may account for some of the discrepant physiological results in the literature - such as the reductions in $FEV_1$

TABLE 1

POSSIBLE MECHANISMS FOR AN ACQUIRED INCREASE
IN BRONCHIAL REACTIVITY IN SMOKERS

Reduced baseline airway dimensions (geometric effect)

More central deposition of inhaled bronchoconstrictors

Increased permeability of bronchial mucosa

Immunological response to some component of tobacco smoke
(tobacco "allergy")

Amplification of pre-existing sub-clinical allergy to other
inhalants

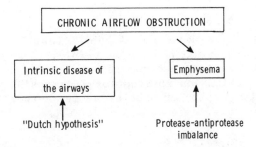

**FIGURE 8:**   Possible roles of the 'Dutch hypothesis' and
protease-antiprotease imbalance in determining the
major site of lung damage in smokers.

found in young smokers or the tendency of smokers with chronic bronchitis to respond to atropine better than to beta-agonists. Loss of reactive airway changes may be responsible for the rapid (but small) improvement in airway function that follows stopping smoking. None of these suggestions are any more proven than the original 'Dutch hypothesis' but all indicate that the interface between asthma and the airway changes in smokers still requires further investigation and clarification.

In summary, therefore, we found increased bronchial reactivity to inhaled histamine in smokers, particularly those with reduced baseline $FEV_1$. Although this is comparible with the 'Dutch hypothesis' the evidence (increased blood eosinophil count, increased positive allergic skin tests in ex-smokers and raised blood IgE in skin test negative smokers and ex-smokers) for an associated allergic 'host factor' was relatively weak. The observed changes in eosinophil count and IgE may be acquired after the onset of smoking. We suspect that in part the increase in airway reactivity is a consequence of altered airway geometry and altered site of aerosol deposition. Regardless of its origins, increased bronchial reactivity may be an important determinant of changes in airway function on giving up smoking, and of the deterioration of lung function occurring during the after bronchial infection.

We thank H. Joyce, R. Carson, E. Gross and F. Holland for excellent administrative and technical help with these studies.

**REFERENCES**

1. ORIE, N. G. M., Sluiter, H. J., de Vries, K.,
       Tammeling, G.J., Witkop, J. (1961),
       The host factor in bronchitis.
       In: Bronchitis an international symposium, 27-29 April
       1960, University of Groningen, The Netherlands.
       Assn: Royal Van Gorcum, 43-59.

2. VAN DER LENDE, R., de Kroon, J. P. M.,
       van der Muelen, G. G., Tammeling, G. J., Visser, B.F.,
       de Vries, K., Orie, N.G.M. (1970),
       Possible indicators of endogenous factors in the
       development of CNSLD.
       In: Orie, N. G. M., van der Lende, R, eds. Bronchitis
       III proceedings of the third international symposium on

bronchitis, 23-26 September 1969, Groningen, The
Netherlands. Assen: Royal Van Gorcum, 52-70.

3.   FLETCHER, C. M., Peto, R., Tinker, C.M., Speizer, F.(1976)
     The natural history of chronic bronchitis and
     emphysema.
     Oxford, Oxford University Press.

4.   BARTER, C. E. and Campbell, A. H. (1975),
     Relationship of constitutional factors and cigarette
     smoking to decrease in 1-second forced expiratory
     volume.
     Am. Rev. Resp. Dis., 113 : 305-314.

5.   CONNELLAN, S. J., Joyce, H., Holland, F., Carson, R. and
     Pride, N.B. (1982),
     Factors determining susceptibility to chronic airway
     narrowing in smokers.
     Thorax 37: 232.

6.   TAYLOR, R. G., Gross, E., Joyce, H., Holland, F. and
     Pride, N. B.
     Bronchial reactivity to inhaled histamine and rate of
     decline in $FEV_1$ in male smokers and ex-smokers.
     In preparation.

7.  BURROWS, B., Halonen, M., Barbee, R. A., Lebowitz, M.D.
     (1981).
     The relationship of serum immunoglobulin E to cigarette
     smoking.
     Am. Rev. Resp. Dis., 124: 523-5.

8.   TAYLOR, R. G., Gross, E., Richards, J. D. M., Martin, P.
     and Pride, N.B.
     Smoking, allergy and the differential white blood cell
     count.
     In preparation.

9.   HALONEN, M., Barbee, R. A. Lebowitz, M. D. and
     Burrows, B. (1982).
     An epidemiologic study of the interrelationships of
     total serum immunoglobulin E, allergy skin-test
     reactivity and eosinophilia.
     J. All. Clin. Immunol., 69, 221-228.

10.   BURROWS, B. and Taussig, L. M. (1980).
     "As the twig is bent, the tree inclines" (perhaps).
     Am. Rev. Resp. Dis., 122: 813-816.

11.  SAMET, J. M., Tager, I. B. and Speizer, F. E. (1983).
     The relationship between respiratory illness in
     childhood and chronic air-flow obstruction in
     adulthood.
     Am. Rev. Resp. Dis., 127 : 508-523.

12.  BENSON, M. K. (1975),
     Bronchial hyper-reactivity.
     Br. J. Dis. Chest, 69 : 227-239.

13.  ARNUP, M. E., Mendella, L. A. and Anthonisen, N.R. (1983).
     Effects of cold air hyperpnea in patients with chronic
     obstructive lung disease.
     Am. Rev. Resp. Dis., 128 : 239.

14.  RANSDALE, E., Morris, M., Hargreave, F. (1984),
     The relation between bronchial responsiveness to
     metacholine and respiratory tract heat loss in patients
     with chronic bronchitis.
     Chest, in press.

15.  ROEHRS, J. D., Rogers, W. R. and Johanson, W. G. Jr. (1981)
     Bronchial reactivity to inhaled methacholine in
     cigarette-smoking baboons.
     J. Appl. Physiol., 50 : 754-760.

16.  AGNEW, J. E., Pavia, D. and Clarke, S. W. (1981).
     Airways penetration of inhaled radioaerosol: an index
     to small airways function.
     Eur. J. Respir. Dis., 62 : 239-255.

17.  JONES, J. G., Minty, B. D., Lawler, P., Hulands, G.,
     Crawley, J. C. W., Veall, N. (1980).
     Increased alveolar epithelial permeability in cigarette
     smokers.
     Lancet 1 : 66-68.

18.  MASON, G. R., Uszler, J. M., Effros, R. M. and Reid, E.
     (1983).
     Rapidly reversible alterations of pulmonary epithelial
     permeability induced by smoking.
     Chest 83 : 6-11.

19.  HOGG, J. C. (1981).
     Bronchial mucosal permeability and its relationship to
     airways hyper-reactivity.
     J. All. Clin. Immunol., 67 : 421-425.

20. MINTY, B. D., Jordan, C., Jones, J. G. (1981).
    Rapid improvement in abnormal pulmonary epithelial
    permeability after stopping cigarettes.
    Br. Med. J., 282 : 1183-1186.

DISCUSSION

LECTURER: Pride                                        CHAIRMAN: Cumming

CUMMING:            One of the great enigmas of cigarette smoking
                    is that it seems to be so selective, the bulk
                    of smokers escape the pulmonary consequences
                    but some do not, and what constitutes the
                    difference.   As this paper was presented I
                    first saw what I thought to be a good
                    explanation  of  the  difference  in
                    susceptibility, but as he entered the realms of
                    immunology my faith in the explanation was
                    somewhat shaken.

BAKE:               Considering the relationship between dimensions
                    and resistance in a tube, it is only linear if
                    the flow is laminar.   The bronchial tree is
                    very complex and the relationship between
                    dimensions and resistance is very difficult to
                    establish.   What is the significance of a 20%
                    change in $FEV_{1.0}$ in two subjects starting from
                    different lung volumes and what does this mean
                    in terms of bronchial sensitivity in the two
                    subjects?

PRIDE:              The difference between laminar flow and other
                    regimes would not perturb me unduly, the fourth
                    pwer of the radius is relevant in laminar flow,
                    and it is a higher power in other regimes.
                    Ther serial distribution down the airways is a
                    much more difficult concept and it is difficult
                    to relate a 35% change in $FEV_{1.0}$ and, for
                    instance, a 20% change in airways resistance.

GUYATT:             We have had a similar problem about baselines.
                    In a study of 200 non-smokers we attempted to
                    transform the values which enabled us to
                    compare the values in the absence of a
                    baseline.   Can you do something similar?

PRIDE:              One reason for using the $FEV_{1.0}$ was because the
                    test was carried out in the field, and a second
                    was that all studies that have attempted to
                    distinguish asthmatics from normals have found
                    that the separation using $FEV_{1.0}$ is greater
                    than when using specific conductance.

CUMMING:       I should say that a few years ago I offered a prize of a magnum of champagne to the first worked who devised a dimensionless expression for airflow limitation, and the offer still holds.

ANTHONY:       I would like to comment on the immunology, since this is an attractive hypothesis to explain airflow limitation. Several mechanisms might operate such as the products of activated mast cells, the products of lymphocytes and activated complement, and there are a number of ways in which the polymorphs might be attracted to the lungs to participate in the protease/anti-protease systems and active oxygen radicals of which we have heard so much this week. Mast cells increased by a factor of four have been reported in smokers as have mediators in the sputum. Sensitivity of glycoproteins to tobacco smoke constituents have been reported, and this report seems genuine, so that the immunology of smoking deserves more attention. The increase in white cell count reported by Dr. Pride has been found consistently in smokers and deserves further study.

# A COMPARISON OF THE ABILITY OF DIFFERENT LUNG FUNCTION

# TESTS TO DISCRIMINATE ASYMPTOMATIC SMOKERS AND NON-SMOKERS

R. Saracci, C. Giuntini, P. Paoletti, E. Fornai
F. Di Pede, P. Fazzi, R. Da Porto, M. Cipriani
G. Pistelli, G. Giuliano, and A. Dalle Luche

C.N.R. Clinical Physiology Institute
University of Pisa, Italy

## Summary

We studied 120 volunteers, 20 to 49 year old, equally divided into sex and smoking habit groups to assess the ability of lung function tests to discriminate smokers from non smokers. The subjects performed routine spirometry including helium dilution, body plethysmography, single breath nitrogen test and forced expiratory manoeuvre. Admission criteria were: 1) normal chest x-ray, 2) absence of cardiorespiratory diseases, 3) absence of repiratory symptoms and exposures (by a questionnaire), 4) normal routine spirometry, and 5) $FEV_1/VC\% > 70$. The F-value was calculated for the difference between means in smokers and non-smokers, adjusted for age and height by covariance analysis on transformed variables. Results showed that CV, CV/VC%, $N_2\%1$, FEF75-85 and Vmax75 had the highest ability to discriminate smokers and non smokers both in males and females. Other tests, such as plethysmography and helium dilution, were not sensitive. A ranking approach, with the determination of percentage of misclassification for each test using a statistical method, confirmed this trend for both sexes. Assumptions of Gaussian distribution, use of equations based on other population, and use of arbitrary criteria of abnormality were thus avoided.

## Introduction

Chronic obstructive lung disease (COLD) is recognized as an increasing health problem in many industrialized countries (35, 39, 40, 45). It reflects pathological processes

471

which develop over several decades before producing symptoms
and disability (2, 11, 21, 29).

In recent years much effort has been spent in
developing more sensitive tests to identify obstructive disease
in its earliest stages, when the conventional tests of
pulmonary function are normal and the pathological processes
may still be reversible. Some of these tests are simple enough
for use as screening tests and in epidemiological studies (5,
7, 8, 9, 25, 26, 28, 30, 31, 32). However, in the evaluation
of the relative merit of pulmonary function tests, several
difficulties still persist. These are related, at least in
part, to the lack of clear knowledge about the natural history
of COLD. Recent works tend to favour the concept of a
continuous rather than abrupt deterioration of pulmonary
function (11, 17, 23).

The choice of tests, especially when intended for long-
term epidemiological research on the natural history and
determinants of COLD, ought to be based on explicit and
unequivocal criteria. This is preferable to any vague hope
that a more sophisticated test than those in use will turn out
to be in some sense "better" in the end.

Only longitudinal validation, relating the values of
the test to the later outcome of disease is fully informative.
However, since this is not applicable to the problem of initial
tests selection, two types of cross-sectional validation may be
adopted:

1)      by relating the values of each test under investigation
        to some indicator of overt disease (e.g. chronic
        productive cough), to a physicians diagnosis, or to the
        values of some other test, typically $FEV_1$, as a
        "reference" for disease (14, 15, 21, 26, 43). However,
        this provides in itself little guidance on which test,
        out of a battery, is likely to prove the most sensitive
        when studied longitudinally. Of course, without well
        defined pre-existing disease, one cannot measure
        sensitivity and specificity.

2)      by studying the test values in relation to conditions
        which alter function such as smoking and selecting the
        test which provides the best discrimination between the
        normal and the altered conditions. Thus two groups
        could be distinguished in relation to smoking: either
        healthy non-smokers before and after smoking, or
        asymptomatic smokers and non-smokers; this second
        option was adopted in the present study.

The assessment of the relative merit of tests may be distorted by an inappropriate treatment of the data or by incorrect statistical methods of analysis. Using regression equations calculated on a different population may be inappropriate, since regression coefficients need not be the same in the two populations (1, 3, 6, 41). A more appropriate and valid solution is to employ a regression equation entirely derived from the data in the study population itself. One well known accepted statistical method is covariance analysis (4). It has the advantage of providing comparisons of the adjusted mean values of a test in smokers and non-smokers.

Within the framework of a national program established by the Italian National Research Council we have started an investigation of the tests most suitable for early detection of airway obstruction. The purpose of the present study, which represents the first phase of the investigation, is to assess the relative merit of a number of tests. Thus, selection of tests for subsequent usege in a large cross-sectional and longitudinal study is optimized. The present assessment was done by ranking the tests according to their ability to discriminate volunteer asymptomatic smokers from non-smokers, after adjusting the test values for height and age by covariance analysis. This analysis is a new approach in that it does not assume that one test is better than another simply based on different percentages of smokers and non smokers who fall below arbitrary criteria of abnormality (based on assumptions of Gaussian distributions) as do most previous studies.

**Materials and Methods**

a)      Subjects and tests
        We studied 120 volunteer subjects, 60 males and 60 females, in the age range between 19 and 49 years. They all came from the Pisa and Lucca areas of Tuscany (near sea level), where measured levels of air pollutants are below the legal standards. "Non-smokers" had never smoked, while "smokers" have been regularly smoking; they had to have smoked at least 20.000 cigarettes; only 7 subjects (19 to 25 year old) had smoked less than 30.000 cigarettes.

        The subjects were recruited among physicians, nurses, medical students and their acquaintances. They were admitted to the study only if they had:

        a) negative history for cardiopulmonary diseases;

b)  no respiratory symptoms: chronic cough and sputum, wheezing and breathlesness (assessed by a questionnaire) (13, 19);

c)  a normal chest x-ray in two projections;

d)  a normal routine evaluation of the following indexes: vital capacity (VC), functional residual capacity (FRC), residual volume (RV), total lung capacity (TLC), forced expiratory volume in one second ($FEV_1$);

e)  $FEV_1/VC\%$ ratio equal to or over 70%.

Out of all the subjects meeting criteria (a) and (b), 5 were discarded because of criterion (c) was not satisfied, and 3 because criterion (e) was not met. Since this was a physiological study, population representation is not necessary.

Spirography for the VC, FRC, TLC and RV by Helium dilution technique were obtained using a water spirometer (Repo-test 602, Dargatz).

Thoracic gas volume (TGV), airway resistance (Raw) and specific conductance (SGaw) were determined using a body plethysmograph (Fenyves & Gut) by the methods of Dubois et al. (16) and Jager and Otis (24).

Closing volume was assessed using a modified single breath nitrogen test (air bolus $SBN_2$ method in the attempt to increase the base to apex nitrogen gradient) (22); nitrogen concentration was determined by a mass spectrometer (Varian-Mat SM3). Volume and flow were measured by a pneumotachograph (Fleisch n. 3).

The two manoeuvres of inspired and expired vital capacity (IVC, EVC) were performed at a subject and operator controlled flow between 0.3 and 0.5 1/sec. The manoeuvres IVC and EVC were accepted if the EVC was within 5% from the previously spirometer measured VC and equal or 5% greater than IVC in the same manoeuvres. Each tracing was examined by two readers and the closing point was selected fitting a transparent straight edge to the alveolar plateau and choosing "the first convincing permanent departure" (33) from this straight line.

Closing volume was selected from the greatest EVC of three acceptable manoeuvres and the ratio CV/VC% was computed. Furthermore, CC and the ratio CC/TLC% were calculated using RV and TLC measured by the helium dilution technique. The slope of the alveolar plateau was determined on the final 2/3 of the III phase, starting from the closing point (28). Finally, FVC manoeuvers were performed, using a pneumotachograph (Fleisch n. 3).

At least 2 manoeuvres were accepted which corresponded to the specifications later defined by the "Epidemiological Standardization Project" of the American Thoracic Society (20).

b)      Statistical analysis

For statistical analysis, we retained for each test and subject: "best" values, and means of the replicate values. Three replicate tracings for the $SBN_2$ were accepted and recorded and two for forced vital capacity and derived indexes. "Best" values, for the $SBN_2$ curve, are defined as those based on the curve with highest (expiratory) vital capacity and, for forced expiratory flows and volumes, as the highest value among the replicates.

Frequency distributions of all variables (age, height, mean and "best" value of each test) were plotted and inspected separately for each of the four subgroups. Most of the distributions of mean and "best" values appeared skewed towards the right, sometimes markedly. These were made approximately symmetrical and with equal scatter for a given test, in each of the four sex-smoking subgroups, by the usual logarithmic transformation of the original measurements. This ensured Gaussian distributions for purposes of statistical analysis. The transformed measurements of mean and "best" values were used to assess the ability of each test to discriminate between smokers and non-smokers, in males and females separately. We first tested that the linear multiple regression of the lung function values (dependent variable) on age and height (independent variables) was statistically significant and not different in smokers and non-smokers.

Discriminating ability was then determined by

calculating the F-value to test the significance of the
difference between means in smokers and non-smokers,
adjusted for age and height, by the covariance
analysis, a classical approach (4). For statistical
significance, the t values and F values had to have a
$p < 0.05$.

For purpose of completeness and comparison with
existing literature, "% of predicted" values were also
calculated for some tests, using original
(untransformed) measurements and various published
equations (8, 9, 25, 28, 36, 37). Differences between
the means of these "% of predicted" values in smokers
and non-smokers were tested by a Student's t-test.
Finally, to get a better insight into the comparative
discriminating ability of the various tests, a
"threshold" value was calculated at the mid-point of
the distance between the means (based on transformed
values) for smokers and non-smokers, based on the
method of Cochrane for examining screening tests (12).
This method avoids arbitrary criteria of abnormality.
This "threshold" value and tables of the normal
distribution integral (42) were used to compute the
percentage of the population of smokers which would be
misclassified as non-smokers and, symmetrically, the
percentage of non-smokers which would be misclassified
as smokers, by a given lung function test. The tests
were then ranked based on these percentages.

## Results

Table I presents the distribution of the 120 subjects
in the study by age, sex, physical characteristics and smoking
habits. The mean age was similar for all 4 groups. Mean
height and weight were similar for smokers and non smokers in
each sex group.

Tables II (males) and III (females) summarize the main
results of the analysis of covariance for the "best" values
(log transformed) of each lung function test. The mean values
are not presented because they were almost always less
significant than the "best" values. The regression
coefficients of the lung function test values on age ($b_1$) and
height ($b_2$) are reported, together with their statistical
significance by t test. The coefficient for height was only
significant for volumes (FVC and $FEV_1$ in both sexes, RV and TLC
in females) and for CC before connection for volume (i.e.,
CC/TLC%) in both sexes. The F-values ($F_1$) testing the
significance of the multiple regression ("within groups") are

Table 1.  Distribution of the 120 Subjects by Sex, Smoking
          Habits, Age and Physical Characteristics.

|  | MALES | | FEMALES | |
| --- | --- | --- | --- | --- |
|  | Smokers | Non-smokers | Smokers | Non-smokers |
| Number | 30 | 30 | 30 | 30 |
| Age (years)* | | | | |
| mean | 34.4 | 31.8 | 34.1 | 33.8 |
| S.D. | 9.6 | 7.5 | 8.2 | 8.4 |
| Height (cm) | | | | |
| mean | 172.3 | 173.6 | 161.5 | 159.4 |
| S.D. | 7.09 | 7.17 | 6.40 | 5.64 |
| Weight (Kg) | | | | |
| mean | 75.9 | 75.5 | 61.3 | 58.2 |
| S.D. | 10.45 | 10.63 | 9.05 | 8.34 |

(*) Age at last birthday

Table 2.    Covariance Analysis of Logarithm of Lung
Function Values in Asymptomatic Smoking
and Non-Smoking Males.

| Test | $b_1$ | $b_2$ | $F_1$ | $F_2$ | Adjusted mean | $F_3$ |
|------|-------|-------|-------|-------|---------------|-------|
| RV |  |  |  |  |  |  |
| S | .012* | .022** | 6.35** |  | 1.49 |  |
| NS | .011 | .006 | 1.46 | 1.36 | 1.50 | .02 |
| TLC(He) |  |  |  |  |  |  |
| S | -.000 | .012** | 8.13** |  | 6.61 |  |
| NS | -.001 | .010** | 5.19* | .08 | 6.62 | .003 |
| CV |  |  |  |  |  |  |
| S | -.023** | .006 | 5.70** |  | .70 |  |
| NS | .022 | .006 | .93 | .00 | .50 | 6.78* |
| CV/VC% |  |  |  |  |  |  |
| S | .026** | -.003 | 11.75** |  | 13.8 |  |
| NS | .027 | -.006 | 1.56 | .02 | 9.8 | 7.26** |
| CC |  |  |  |  |  |  |
| S | .014** | .018** | 7.79** |  | 2.23 |  |
| NS | .012* | .006 | 3.56* | 1.20 | 2.08 | 1.67 |
| CC/TLC% |  |  |  |  |  |  |
| S | .015** | .006 | 13.85** |  | 33.7 |  |
| NS | .015** | .001 | 7.50** | .50 | 30.8 | 5.52* |
| N2%/1t |  |  |  |  |  |  |
| S | .021 | .030 | 1.13 |  | 1.19 |  |
| NS | .001 | .010 | .67 | 1.58 | .96 | 2.78 |
| FVC |  |  |  |  |  |  |
| S | -.006** | .008* | 11.11** |  | 4.94 |  |
| NS | -.007* | .010** | 9.50* | .31 | 5.03 | .34 |
| FEV1 |  |  |  |  |  |  |
| S | -.011** | .006 | 16.65** |  | 3.84 |  |
| NS | -.010** | .009* | 9.15** | .23 | 4.00 | 1.65 |
| FEV1/FVC% |  |  |  |  |  |  |
| S | -.005** | -.001 | 5.27* |  | 78.4 |  |
| NS | -.003 | -.002 | 1.37 | .41 | 80.5 | 2.13 |
| FEF25-75 |  |  |  |  |  |  |
| S | -.024** | .005 | 8.90** |  | 3.58 |  |
| NS | -.022** | .003 | 5.73** | .26 | 4.06 | 2.50 |
| FEF75-85 |  |  |  |  |  |  |
| S | -.045** | .000 | 21.05** |  | .98 |  |
| NS | -.040** | .006 | 12.21** | .18 | 1.26 | 8.06** |
| Vmax50 |  |  |  |  |  |  |
| S | -.016* | .000 | 3.47* |  | 5.05 |  |
| NS | -.012 | .000 | 1.40 | .07 | 5.05 | .000 |
| Vmax75 |  |  |  |  |  |  |
| S | -.034** | -.004 | 10.34** |  | 1.68 |  |
| NS | -.026** | -.001 | 4.46* | .09 | 2.00 | 3.46 |

S = smokers; NS = non-smokers; $b_1$, $b_2$ = regression
coefficients for age and height respectively; $F_1$ = F value
for regression "Within Group" (2 and 14 df - degree of
freedom - for smokers, 2 and 21 df for non-smokers for
$N_2$%/1t; 2 and 27 df for all other tests); $F_2$ = F value
for difference of "Within Group" regressions (2 and 35 df
for $N_2$%/1t; 2 and 54 df for all other tests); $F_3$ = F value
for difference between adjusted means (1 and 37 df for
$N_2$%/1t; 1 and 56 df for all other tests); * = p 0.05; **
= p 0.01.

Table 3.  Covariance Analysis of Logarithm of Lung
Function Values in Asymptomatic Smoking
and Non-Smoking Females.

| Test | $b_1$ | $b_2$ | $F_1$ | $F_2$ | Adjusted mean | $F_3$ |
|------|------|------|------|------|------|------|
| **RV** | | | | | | |
| S | .014** | .024** | 10.22** | | 1.33 | |
| NS | .070 | .004 | .63 | 1.76 | 1.32 | 1.26 |
| **TLC(He)** | | | | | | |
| S | -.000 | .017** | 25.91** | | 4.92 | |
| NS | -.001 | .007 | 2.13 | 2.73 | 5.06 | 1.10 |
| **CV** | | | | | | |
| S | .029** | .014 | 10.60** | | .45 | |
| NS | .72** | .013 | 6.36** | 2.11 | .38 | 4.45* |
| **CV/VC%** | | | | | | |
| S | .035** | -.002 | 21.95** | | 12.3 | |
| NS | .075** | .006 | 6.73** | 1.72 | 8.3 | 4.82* |
| **CC** | | | | | | |
| S | .019** | .021** | 13.56** | | 1.71 | |
| NS | .013* | .003 | 3.77** | 2.14 | 1.74 | .11 |
| **CC/TLC%** | | | | | | |
| S | .019** | .003 | 19.73** | | 34.8 | |
| NS | .014** | -.005 | 10.41** | 1.15 | 34.8 | .14 |
| **N2%/1t** | | | | | | |
| S | .008 | -.017 | .78 | | 1.30 | |
| NS | .005 | -.007 | .20 | .11 | .98 | 4.63* |
| **FVC** | | | | | | |
| S | -.005 | .014** | 13.20** | | 3.60 | |
| NS | -.002 | .008* | 3.76* | 1.29 | 3.60 | .00 |
| **FEV$_1$** | | | | | | |
| S | -.10** | .010* | 12.09** | | 2.86 | |
| NS | -.008** | .007* | 15.03** | .41 | 2.91 | .35 |
| **FEV$_1$/FVC%** | | | | | | |
| S | -.004** | -.002 | 6.35** | | 80.5 | |
| NS | -.006** | -.001 | 5.54** | .42 | 82.1 | 1.45 |
| **FEF25-75** | | | | | | |
| S | -.020** | .001 | 5.67** | | 2.78 | |
| NS | -.017** | .011 | 8.64** | .44 | 2.98 | 1.32 |
| **FEF75-85** | | | | | | |
| S | -.042** | .010 | 17.56** | | .78 | |
| NS | -.038** | .015 | 10.60** | .08 | .94 | 4.02* |
| **Vmax50** | | | | | | |
| S | -.009 | .001 | 1.54 | | 3.94 | |
| NS | -.012* | .012 | 6.09** | .80 | 4.28 | 2.02 |
| **Vmax75** | | | | | | |
| S | -.028** | .004 | 11.99** | | 1.47 | |
| NS | -.025** | .017 | 9.05** | .54 | 1.75 | 5.47* |

S = smokers; NS = non-smokers; $b_1$, $b_2$ = regression
coefficients for age and height respectively; $F_1$ = F value
for regression "Within Group" (2 and 19 df - degree of
freedom - for smokers, 2 and 20 df for non-smokers for
N$_2$%/1t, 2 and 27 df for all other tests); $F_2$ = F value
for difference of "Within Group" regressions (2 and 39 df
for N$_2$%/1t, 1 and 56 df for all other tests); $F_3$ = F
value for difference between adjusted means (1 and 41 df
for N$_2$%/1t; 1 and 56 df for all other tests); * = p
0.05; ** = p 0.01.

presented.  Some of these are not significant, most frequently
in non-smokers.  The t and F values for Raw and SGaw for males
were completely non significant.  For female, the t test was
significant for age for Raw and SGaw, so the F test was
significant also.  However, even in females, this method does
not add to the discrimination.  Therefore the results of these
tests are not presented.  The F-values ($F_2$), testing the
significance of the difference between the regressions within
smokers and non-smokers, are presented.  Non of the $F_2$ values
were statistically significant.  Thus, the adjusted means were
calculated, and are shown in the tables.  The F-values ($F_3$)
testing the statistical significance of the difference between
the adjusted means in smokers and non-smokers are also shown.
For males and females, the differences between adjusted means
for RV, TLC, Raw and SGaw are never significant; they are not
analyzed further.

For males (Table II), of the indexes derived from the
single breath nitrogen test, CV, CV/VC%, and CC/TLC% show a
statistically significant difference, while no such difference
appears to be present for CC and $\Delta N_2\%/1$.  However, it should
be noted that determinations of the slope of the alveolar
plateau were performed in only 41 (instead of 60) subjects, and
this smaller number of determinations may account for he lack
of statistical significance; and increase in the sample size
would yield a significant difference.

Of the indexes derived from the flow volume curve for
males, the only one showing a statistically significant
difference between the adjusted means for smokers and non-
smokers for "best" (and mean) values is FEF75-85.  However, the
$\mathring{V}$max 75 also has a high $F_3$ value (higher than that of
the $\Delta N_2\%/1$ and might also be significant with larger sample
sizes.

Table III refers to female smokers and non-smokers.
Statistically significant differences are present between the
adjusted means for smokers and non-smokers for CV,
CV/VC%, $\Delta N_2\%/1$, as well as for FEF75-85 and $\mathring{V}$max75 "best" (and
mean) values.

For the purpose of comparison with existing literature
we calculated the percent predicted means, standard deviations,
and t-values for the difference of the means between smokers
and non-smokers.  The "% of predicted" values are derived from

the original (untransformed) measurements ("best" and mean values) using published prediction equations for the indexes derived from flow-volume curves, forced expirograms and $SBN_2$ (8, 9, 25, 28, 36, 37). The results show that the subjects in this study were very close to normal as fas as FVC and $FEV_1$, with the use of equations of Morris (36) and of Knudson (25). The FEF25-75 and the V̇max50 were also "normal" in these subjects, and always higher in non-smokers. The "best" value was usually better than the mean value, and they are described and used further. Differences between smokers and non-smokers, even in tests whose t value was not significant, were always in the direction expected. Differences of results in this population compared to the other populations, especially for the $SBN_2$ test, may reflect different techniques.

Table IV shows the significant results for smokers and non-smokers; the difference between means is significant for: 1) FEF75-85, "best" (and mean) values from Morris et al. (37); 2) V̇max75 "best" values from Knudson et al. (25); 3) CV/VC% "best" (and mean) values from Buist et al. (8) and from Knudson et al. (28); 4) CC/TLC%, "best" (and mean) values from Buist et al. (8); 5) Δ $N_2$%/1t "best" values from Buist et al. (9) and Knudson et al. (28). For smoking and non-smoking females; the difference between means is significant only for: 1) V̇max75, "best" values from Knudson et al. (25); 2) Δ$N_2$%/1t, "best" values from Knudson et al. (28).

The results presented in the previous tables allows a ranking of the tests according to their capacity to discriminate between smokers and non-smokers. This is done, separately for males and females, in Table V, where the tests are ranked in order of decreasing values of $F_3$, a statistic whose magnitude is directly proportional (for a constant number of observations) to the degree of separation between the means of a given lung function test in smokers and non-smokers. Also in the same table the percentages of smokers which would be misclassified as non-smokers (and, symmetrically, the percentages of non-smokers misclassified as smokers) by a given lung function test are reported. Among values obtained from flow volume and $SBN_2$ tests, only FVC lacks any discriminatory ability.

It is notable that all these percentages are included within a rather narrow range, indicating that whatever differences exist between the different tests, they are small when assessed in terms of percentage of misclassification.

Table 4.   Significant Differences Between Smokers and
           Non-Smokers in Males and Females Using Published
           Percent of Predicted for "Best" Values From
           Flow Volume Curves and $SBN_2$ Test.

| Author | Test | Mean | S.D. | t | p |
|--------|------|------|------|---|---|
| **MALES** | | | | | |
| Morris | FEF75-85 | | | | |
| | S | 76.3 | 30.2 | 3.05 | .005 |
| | NS | 102.4 | 35.9 | | |
| Knudson | Vmax75 | | | | |
| | S | 59.3 | 22.5 | 2.05 | .05 |
| | NS | 72.1 | 25.9 | | |
| Buist | CV/VC% | | | | |
| | S | 120.7 | 32.3 | 3.56 | .001 |
| | NS | 92.5 | 28.9 | | |
| Knudson | CV/VC% | | | | |
| | S | 181.7 | 50.2 | 3.48 | .001 |
| | NS | 137.4 | 39.4 | | |
| Buist | CC/TLC% | | | | |
| | S | 109.7 | 16.4 | 2.21 | .05 |
| | NS | 101.0 | 14.0 | | |
| Buist | $N_2\%/1t$ | | | | |
| | S | 128.9 | 64.0 | 2.18 | .05 |
| | NS | 96.6 | 29.4 | | |
| Knudson | $N_2\%/1t$ | | | | |
| | S | 281.3 | 159.9 | 2.23 | .05 |
| | NS | 203.0 | 55.4 | | |
| **FEMALES** | | | | | |
| Knudson | Vmax75 | | | | |
| | S | 59.6 | 19.5 | 2.03 | .05 |
| | NS | 72.0 | 27.0 | | |
| Knudson | $N_2\%/1t$ | | | | |
| | S | 248.3 | 146.0 | 2.05 | .05 |
| | NS | 179.7 | 65.3 | | |

Table 5.  Rank of Lung Function Tests According to Ability to Discriminate Between Smokers and Non-Smokers.

| | MALES | | | | FEMALES | | |
|---|---|---|---|---|---|---|---|
| RANK | TESTS | $F_3$ "best value" | % MISCLASSIFICATION | RANK | TESTS | $F_3$ "best value" | % MISCLASSIFICATION |
| 1 | FEF75-85 | 8.06** | 35.6 | 1 | Vmax75 | 5.47* | 38.2 |
| 2 | CV/VC% | 7.26** | 36.3 | 2 | CV/VC% | 4.82* | 38.8 |
| 3 | CV | 6.78* | 36.7 | 3 | $N_2$%/1t | 4.63* | 37.4 |
| 4 | CC/TLC% | 5.52* | 37.8 | 4 | CV | 4.45* | 39.4 |
| 5 | Vmax75 | 3.46 | 40.5 | 5 | FEF75-85 | 4.02* | 39.7 |
| 6 | $N_2$%/1t | 2.78 | 39.5 | 6 | Vmax50 | 2.02 | 42.7 |
| 7 | FEF25-75 | 2.50 | 41.7 | 7 | $FEV_1$/FVC% | 1.45 | 44.3 |
| 8 | $FEV_1$/FVC% | 2.13 | 42.5 | 8 | FEF25-75 | 1.32 | 44.0 |
| 9 | CC | 1.67 | 43.2 | | | | |
| 10 | $FEV_1$ | 1.65 | 43.6 | | | | |

Vmax 50 and FVC of less significance ($F<1.0$) and value

$FVC, FEV_1$, CC/TLC% of less significance ($F<1.0$) and value

** $p<0.01$
* $p<0.05$
based on 41 determinations instead of 60

## Discussion and Conclusions

The treatment of the data in terms of "% of predicted" values using prediction equations borrowed from the literature is, as already remarked, less appropriate than the one here adopted based on a covariance analysis, because of differences in populations and techniques. However, results obtained from the "% of predicted" approach (Table IV) broadly concur with the covariance approach in identifying the most discriminating tests. These, in our case, turn out to be FEF75-85, CV/VC%, $\dot{V}$max75, $\Delta N_2\%/1$ in males and females: this consistency increases the likelihood that the most discriminating tests have been correctly identified in the present work. Results not consistent in both sexes were reported in a recent investigation (3, 18) which, however, differed from our study in design and method of analysis. However our results are basically consistant with those of other authors examining more restricted sets of tests in their populations (7, 9, 10, 26, 27, 28, 37, 38, 44). In a comparative study, $\Delta N_2\%/1$ and $\dot{V}$max75 have been shown to have similar sensitivity (27).

By restricting our sample to those with normal TLC, RV, FVC and $FEV_1$, we could not evaluate fully the parameters. Furthermore, larger cross-sectional studies, with wider age ranges, are needed to evaluate age, as well as other risk factors, in relation to the discriminating ability of these tests.

Only the results of longitudinal studies can firmly establish the merit of the tests emerging as the most discriminating from the present initial validation study. Longitudinal validation would provide the tests which most clearly indicate the outcome and which do so the earliest. The outcome could be the presence of obstruction based on previously validated and accepted criteria (e.g. $FEV_1$, and/or $FEV_1$/FVC below a certain level, or clinical appearance of the disease). The use of the $FEV_1$ as a reference, because of its demonstrated ability to mark disease progression (14, 15, 21, 26, 43) has its difficulties. The absence of a correlation with $FEV_1$ might cast serious doubts on the value of an alternative test, even if the alternative test measures something different (e.g. inhomogeneity). A perfect correlation with $FEV_1$, however, would indicate the alternative test is simply redundant. Thus, other criteria are necessary as well.

This investigation is a worthwhile preparation for longitudinal validation because of its statistical treatment of "normal" data, using a covariance analysis (linear or non-linear). These methods avoid assumptions and also arbitrary criteria used in other studies. Thus, they could be usefully incorporated into other studies aimed at cross-sectional validation of functional tests.

Furthermore, the methods utilized have allowed us to choose the most likely tests (i.e. potentially most sensitive) for our longitudinal studies. They have shown, through comparisons, that several parameters from the flow-volume curve and from the $SBN_2$ test may be important, rather than only one or two parameters from only one of the tests. This study has shown as well the tests which might be ignored as being less useful. Thus, the Raw and SGaw, and the RV and TLC, were not significant discriminators, as well as not being significant in the other respects. They are less practical to perform in epidemiological studies as well. Thus, the choices for such studies are improved.

From a substantive view point, the principal importance of our results lie in showing differences in groups of asymptomatic never smokers and smokers of either sex. Moreover, they were selected by being symptomless on medical exam and, more important, normal in their chest x-ray film and on routine spirography (including VC, FRC, RV, TLC, $FEV_1$ and $FEV_1/VC\%$). The method used has shown these statistically significant differences in lung function detectable in respect to different tests of function. These smoking related alterations in lung function are best detected in groups using FEF75-85, CV/VC%, Vmax75, $N_2\%/1$ pointing to a potential of these tests in longitudinal studies for early detection of COLD.

**REFERENCES**

1.  ANDERSON, T.W., Brown, J.R., Hall, J.W., Stephard, R.J. (1968).
    The limitations of linear regressions for the prediction of vital capacity and forced expiratory volumes.
    Respiration, 25, 140-158.

2.  BATES, D.V. (1973).
    The fate of the chronic bronchitic: A report of the ten year follow-up in the Canadian Department of Veteran's Affairs Coordinated study of chronic bronchitis.
    Am. Rev. Resp. Dis., 108, 1043-1065.

3.    BECKLAKE, M.R., Permutt, S. (1979).
      Evaluation of tests of lung function for "screening"
      for early detection of chronic obstructive lung
      disease.
      In: The Lung in Trasition between Health and Disease.
      Edited by P. T. Macklem and S. Permutt.  Marcel Dekker
      Inc., New York, 1979, 365 p., 380 p.

4.    BLISS, C.I. (1970).
      Statistics in Biology.
      McGraw-Hill, New York, 509 p.

5.    BODE,  F.R.,  Dosman,  J.,  Martin,  R.R.,  Macklem,  P.T.
      (1975).
      Reversibility of pulmonary function abnormalities in
      smokers.  A prospective study of early diagnostic tests
      of small airways disease.
      Am. J. Med., 59, 43-52.

6.    BOUHUYS, A., Beck, G.J., Schoenberg, J.B. (1979).
      Lung function: normal values and risk factors.
      In: Small Airways in Health and Disease.  Edited by P.
      Sadoul, J. Milic-Emili, B.G. Simonsson, T.J.H. Clark.
      Proceeding of a Symposium, Copenhagen, 29th-30th March.
      Excerpta Medica, 196 p.

7.    BUIST, A.S., Van Fleet, D.L., Ross, B.B. (1973).
      A comparison of conventional spirometric tests and the
      test of closing volume in an emphysema screening
      center.
      Am. Rev. Resp. Dis., 107, 735-743.

8.    BUIST, A.S., Ross, B.B. (1973).
      Predicted values for closing volumes using a modified
      single breath nitrogen test.
      Am. Rev. Resp. Dis., 107, 744-752.

9.    BUIST, A.S., Ross, B.B. (1973).
      Quantitative analysis of the alveolar plateau in the
      diagnosis of early obstruction.
      Am. Rev. Resp. Dis., 108, 1078-1087.

10.   BUIST, A.S. et al. (1979).
      Relationship between the single breath test and age,
      sex and smoking habit in three North American cities.
      Am. Rev. Resp. Dis., 120, 305-318.

11.  BURROWS, B., Earle, R.H. (1969).
     Course and prognosis of chronic obstructive lung
     disease.
     New Engl. J. Med., 280, 397-304.

12.  COCHRANE, A.L. (1972).
     Effectiveness and Efficiency.  Random Reflections on
     the Health Service.
     Nuffield Provincial Hospitals Trust.  London.

13.  CONSIGLIO NAZIONALE DELLE RICERCHE - Italy - Questionario
     per Adulti.  Progetto Finalizzato Medicina Preventiva -
     Subprogetto Broncopneumopatie Croniche - Dr. P. Fazzi -
     Istituto Fisiologia Clinica del C.N.R., Via Savi, 8,
     Pisa, Italy.

14.  DAVIS, A.L., McClement, J.H. (1968).
     The course and prognosis of chronic obstructive
     pulmonary disease.  Current Research in Chronic Airways
     Obstruction.
     U.S. Department of Health, Education and Welfare,
     Publications No. 1717, 235-246.

15.  DIENER, C.F., Burrows, B. (1975).
     Further observations on the course and prognosis of
     chronic obstructive lung disease.
     Am. Rev. Resp. Dis., 111, 719-724.

16.  DU BOIS, A.B., Botelho, S.Y., Bedell, G.N., Marshall, R.,
     Comrol, J.H. Jr. (1956).
     A rapid plethysmographic method for measuring thoracic
     gas volume: a comparison with a nitrogen washout method
     for measuring functional residual capacity in normal
     subjects.
     J. Clin. Invest., 35, 322-326.

17.  EMIRGIL, C., Sobol, B.J., Varble, A., Waldie, J.,
     Weinheimer, B. (1971).
     Long term course of chronic obstructive pulmonary
     disease.
     Am. J. Med., 51, 504-512.

18.  ENJETI, S., Hazelwood, B., Permutt, S., Menkes, H.,
     Terry, P. (1978).
     Pulmonary function in young smokers: male-female
     differences.
     Am. Rev. Resp. Dis., 118, 667-676.

19.   FAZZI, P., Viegi, G., Paoletti, P., Giuliano, G.,
      Begliomini, E., Fornai, E.,  Giuntini, C. (in press).
      Comparison between two standardized questionnaires in a
      group of workers.
      Eur. J. of Resp. Dis.

20.   FERRIZ, B.G. (editor) (1978).
      Epidemiology Standardization Project.
      Am. Rev. Resp. Dis., 118 part 2, 57-62.

21.   FLETCHER, C., Peto, R., Tinker, C., Speizer, F. (1976).
      The natural history of chronic bronchitis and
      emphysema.
      London and New York, Oxford Univ. Press.

22.   HAMASH, P., Taveira Da Silva, A.M. (1974).
      Air bolus method compared to single breath method for
      determination of closing volume.
      Am. Rev. Resp. Dis., 110, 518-520.

23.   HUHTI, E., Ikkala, J. (1980).
      A 10 years follow-up study of respiratory symptoms and
      ventilatory function in a middle-aged normal
      population.
      Eur. J. Resp. Dis., 61, 33-45.

24.   JAEGER, M., Otis, A.B. (1964).
      Measurement of airways resistance with a volume
      displacement body plethysmograph.
      J. Appl. Physiol., 19, 813-820.

25.   KNUDSON, R.J., Burrows, B., Lebowitz, M.D. (1976).
      The maximal espiratory flow-volume curve: its use in a
      population study.
      Am. Rev. Resp. Dis., 114, 871-879.

26.   KNUDSON, R.J., Slatin, R.C., Lebowitz, M.D., Burrows, B.
      (1976).
      The maximal expiratory flow-volume curve.  Normal
      standards, variability and effects of age.
      Am. Rev. Resp. Dis., 113, 587-600.

27.   KNUDSON, R.J., Armet, D.B., Lebowitz, M.D. (1980).
      Reevaluation of tests of small airways function.
      Chest, 77, 284-286.

28.    KNUDSON, R.J., Lebowitz, M.D., Burton, A.P., Knudson,
       D.E. (1977).
       The closing volume test: evaluation of nitrogen bolus
       methods in a random population.
       Am. Rev. Resp. Dis., 115, 423-433.

29.    MACKLEM, P.T., Thurlbeck, W.M., Fraser, G.R. (1971).
       Chronic obstructive lung disease of the small airways.
       Ann. Intern. Med., 74, 167-177.

30.    MACKLEM, P.T. and the staff of the Division of Lung
       Disease (N.H.L.I.) (1974).
       Workshop on screening programs for early diagnosis of
       airway obstruction.
       Am. Rev. Resp. Dis., 109, 567-571.

31.    McCARTHY, D.S., Spencer, R., Greene, R., Milic-Emili, J.
       (1972).
       Measurement of "closing volume" as a simple and
       sensitive test for early detection of small airways
       disease.
       Am. J. Med., 52, 747-753.

32.    McFADDEN, E.R. Jr., Linden, D.A. (1972).
       A reduction in maximum mid- expiratory flow rate: A
       spirographic manifestation of small airways disease.
       Am. J. Med., 52, 725-737.

33.    McFADDEN, E.R. Jr., Holmes, B., Kiker, R. (1975).
       Variability of closing volume measurements in normal
       man.
       Am. Rev. Resp. Dis., 111, 135-140.

34.    MEAD, J. (1972).
       The lung's quiet zone.
       New Engl. J. Med., 23, 1318-1319.

35.    MORK, J. (1962).
       A comparative study of respiratory disease in England
       and Wales and Norway.
       Acta. Med. Scand., 172 (suppl.), 384.

36.    MORRIS, J.F., Koski, A., Johnson, L.C. (1971).
       Spirometric standards for healthy non-smoking adults.
       Am. Rev. Resp. Dis., 103, 57-67.

37.   MORRIS, J.F., Koski, A., Breese, J.D. (1975).
      Normal values and evaluation of forced end-expiratory
      flow.
      Am. Rev. Resp. Dis., 111, 755-767.

38.   OXHOJ, et al. (1977).
      Ability of spirometry, flow-volume and the nitrogen
      closing volume test to detect smokers. A population
      study.
      Scand. J. Resp. Dis., 58, 80-96.

39.   REID, D.D., Anderson, D.O., Ferris, B.G. Jr., Fletcher,
      C.M. (1964).
      An Anglo-American comparison of the prevalence of
      chronic bronchitis.
      Br. Med. J., 2, 1487-1491.

40.   Respiratory Diseases, Task Force Report on Problems,
      Research Approaches Needs.
      Lung Program, NHLI, DHEW Publ. No. (NIH), 1972, 73-432.

41.   SOBOL, B.J. (1976).
      The early detection of airways obstruction: Another
      perspective.
      Am. J. Med., 60, 619-624.

42.   TABLE SCIENTIFIQUES (1963).
      Documenta Geigy, sixieme Edition, Ed. J.R. Geigy,
      Department Pharmaceutique, 28 p.

43.   TATTERSALL, S.F., Benson, M.K., Mansell, A., Pride, N.B.,
      Fletcher, C.M., Peto, R., Gray, R., Humphreys, P.R.R.)
      (1978).
      Tests of peripheral lung function: their potential
      value for predicting future disability in middle aged
      smokers.
      Am. Rev. Resp. Dis., 118, 1035-1050.

44.   TOCKMAN et. al. (1976).
      A comparison of pulmonary function in male smokers and
      non-smokers.
      Am. Rev. Resp. Dis., 114, 711-722.

45.   VAN DER LENDE, R. (1969).
      Epidemiology of chronic non specific lung disease.
      vol. I and II, Assen Neth., Van Gorcum e Co., vol. I,
      8-20 p., Vol. II, 3-7 p.

DISCUSSION

LECTURER: Guintini                    CHAIRMAN: Cumming

CUMMING:        The molecule you have used bears some similarity to noradrenaline and is injected intravenously so I suppose it to be fixed in some way within the lungs, perhaps to the endothelial cells where metabolism proceeds. Noradrenaline is not metabolised in the lungs yet this very similar molecule is, would you comment on this?

GUINTINI:       The compound has a greater similarity to serotonin than to noradrenaline, so the metabolism is also similar.

BAHKLE:         I do not recognise your formula for serotonin, it is my recollection that there should be a free ethylamino group, your molecule appears to be completely closed.

GUINTINI:       Only the thick arms in the formulae are real, the thin ones are imaginary.

BAHKLE:         Do you know how this amine is actually metabolised?  Amines which are metabolised in the lung like adrenaline and serotonin are very rapidly lost from the lung so that there is no accumulation.  When accumulation occurs it is usually associated with no metabolism, as is the case with propanalol, so this compound is very interesting but does raise the question of how it is metabolised, and by which enzyme.

CUMMING:        Would you like to say what you mean by rapidly?

BAHKLE:         In isolated lung, clearance is less than 30 minutes, and in human lung it is probably faster.

GUINTINI:       The residence time of the amine in the lung is about ten hours, the data points on a straight line on a semi-logarithmic plot for many hours and this gives me confidence that we are looking at a unique process. The automatic analysis gives a correlation above 0.99, it is a specific uptake and a saturable phenomenon.

ASSENATO:       You showed an analysis of variance showing that two groups were different and this has little to do with discrimination. The discriminant power of the parameters you mentioned was very low ranging from 32% to 43%. You have equal sized groups and if all were labelled as smokers the efficiency would be 50%.

GUINTINI:       All these were asymptomatic smokers, as I emphasized in my talk.

ASSENATO:       In terms of cost effectiveness my method would be better. What I learned from your presentation was that it is impossible to discriminate smokers from non-smokers at present.

GUINTINI:       It is if we take into account the amine metabolism. The aim of the presentation was ranking the various tests and this can be done by the co-variant analysis and also by using the Cochrane approach which is not committed to any arbitrarily chosen threshold.

BAKE:           I regret that I cannot discuss with you the statistics, but I say that the expired flow rate on your nitrogen wash-out curves appeared to be two or three times greater than tidal volume.

GUINTINI:       The expired flow rate was always kept to 0.5 litres per second as well as the inspired rate.

DENISON:        I found the second part very exciting but I am only competent to discuss the first part. In your list of best buys all miscalculated by about 30% as you pointed out. Was it the same 30% in all the tests, or if you had combined them would you have improved your ability to discriminate? I recognise the physiological link between the tests which might make them the same.

GUINTINI:       Our purpose was to establish a ranking order, and I pointed out that the tests has a poor discriminating ability and could be improved considerably by combining them.

LEE:            From the number of variables which you studied, and the number of subjects in each group even random results would produce a combination of tests which would discriminate between smokers and non-smokers.

CUMMING:        Are you saying Peter that if you took two groups of 20 people, by an appropriate selection of tests producing random results you could discriminate the two groups?

LEE:            Yes.

GUINTINI:       This is well known, but some of the variables seemed to be unrelated. We tried to stay out of the tricks of discriminant analysis, it seems to us to be impossible and we applied standard co-variant analysis and the outcome was the one shown.

CUMMING:        Thank you Carlo, you put the position clearly and I noted the statisticians nodding their heads and on that note of agreement let us return to endothelial metabolism and Mick Bahkle.

BAHKLE:         What was the dose which you gave?

GUINTINI:       0.05 millimoles.

BAHKLE:         This small dose probably accounts for the absence of spillover. What was the radioactive tracer which you used?

GUINTINI:       Iodine 123 in position 5 on the benzene ring wih a half life of a few hours and a peak at 150 Kev, so it is suitable for gamma camera detection.

BAHKLE:         I am very glad to see this work since you have shown the advantages of a biochemist method as compared with the more physiological measurments.

CHAIRMAN:       A very narrow definition of physiology and brings to an end this session and also the course.

# INDEX